Proceedings of the Intern.
on Metallurgical Engineering and Centenary
Celebration

Sudipta Patra · Subhasis Sinha · G. S. Mahobia ·
Deepak Kamble

Editors

Proceedings of the International Conference on Metallurgical Engineering and Centenary Celebration

METCENT-2023, 26–28 October,
Varanasi, India

Springer

Editors
Sudipta Patra
Department of Metallurgical Engineering
IIT(BHU)
Varanasi, India

Subhasis Sinha
Department of Metallurgical Engineering
IIT(BHU)
Varanasi, India

G. S. Mahobia
Department of Metallurgical Engineering
IIT(BHU)
Varanasi, India

Deepak Kamble
Department of Metallurgical Engineering
IIT(BHU)
Varanasi, India

ISBN 978-981-99-6862-6 ISBN 978-981-99-6863-3 (eBook)
https://doi.org/10.1007/978-981-99-6863-3

This Springer imprint is published by the registered company Springer Nature Singapore Pte Ltd.
The registered company address is: 152 Beach Road, #21-01/04 Gateway East, Singapore 189721, Singapore

Paper in this product is recyclable.

Preface

The Department of Metallurgical Engineering, Indian Institute of Technology (BHU), Varanasi, India, was established in the year 1923 and has completed 100 glorious years in 2023. The first ever undergraduate program in Metallurgy in India was started here in 1923, and the department was also among the oldest in the country to award PhD and M.Tech degrees in 1955 and 1957, respectively. Till date, 2794 B.Tech, 692 postgraduates (including M.Tech and dual degree) and 197 Ph.D. scholars have graduated from the department.

An international conference named METCENT 2023 was held from 26–28th October 2023 to celebrate this auspicious occasion. The conference was attended by around 300 delegates consisting of academicians and industry professionals including several alumni of the department. More than 150 oral and poster presentations were made at the conference spread over five technical sessions.

The conference program comprised of five technical sessions covering the areas of Extractive metallurgy, Mechanical Behaviour, Physical Metallurgy and Characterization, Modelling and Simulations, Functional Materials, Welding, etc. Five Plenary speakers delivered lectures at the conference. The Plenary talks discussed the "use of biogas in iron and steel making", "Nano-Engineered Materials", "Computationally designed alloys", "novel nitrides and Chalcogenide materials" and "Iron making without using fossil fuel".

This proceeding contains 33 original research papers and five review papers covering various areas of metallurgical and materials research. The full research papers are categorized into Ferrous Process Metallurgy (eight papers), Computational Materials Science (seven papers), Advanced Materials (five papers), Development and Characterization of Steels (five papers), Processing and Structure-Property Correlation (six papers) and Review (five papers). Fifty-five manuscripts were received, out of which 38 were accepted for publication in the current proceeding.

We would like to thank all the members of the conference organizing committee, sponsors, authors, presenters, session chairs, student volunteers, delegates and reviewers for their support and contribution in making this event successful

October 2023

Sudipta Patra
Subhasis Sinha
G. S. Mahobia
Deepak Kamble

Organization

Members

Rajiv Kumar Mandal
Nilay Krishna Mukhopdhyay
N. C. Santhi Srinivas
B. Nageswara Sarma
Kamalesh Kumar Singh
Chhail Kumar Behera
Rampada Manna
Kausik Chattopadhyay
Joysurya Basu
Vikas Jindal
Jayant Kumar Singh
Nand Kishore Prasad
Ashok Kumar Mondal
Bratindranath Mukherjee
Randhir Singh
Surya Deo Yadav
Subhasis Sinha
Sree Harsha Nandam
Deepak Kamble
Lakhindra Marandi
Praveen Sathiyamoorthi
Ameya K. Kadrolkar
L. Sankara Rao
Arun P. Singh
Rana P. Yadav
Pramod Kumar Jain (Patron)
Sunil Mohan (Chairman)
Girija Shankar Mahobia (Convener)
Sudipta Patra (Treasurer)

Advisory Committee

Alberto Conejo   Morelia Technological Institute, Mexico
S. Seetharaman   Royal Institute of Technology, Sweden
Paulo Santos Assis   Universidade Federal de Ouro Preto, Brazil
Amit Chakravarty   National Water Company, Saudi Arabia

Antonello Astarita	University of Naples "Federico II", Italy
Amit Misra	University of Michigan, USA
K. Linga (KL)	Murty NC State University, USA
Fathi Habashi	Laval University, Canada
Kedar Nath Gupta	.
J. Paulo Davim	The University of Aveiro (UA), Portugal
Rajeshwar P. Wahi	Technical University Berlin
E. N. Suleimenov	Kazakh-British Technical University, Republic of Kazakhstan
Pallav Chattopadhyay	Lincoln Electric, Deutschland GmbH, Germany
Lokesh Kumar Singhal	California, USA
Govind S. Gupta	IISc Bangalore
Tata Narasinga Rao	ARCI, Hyderabad
K. Narasimhan	IIT Bombay
S. K. Jha	MIDHANI, Hyderabad
Sanjay Sharma	L&T Special Steels & Heavy Forgings Pvt. Limited, Surat
A. Nagesha	IGCAR, Kalpakkam
T. A. Abinandanan	IISc Bangalore
B. P. Kashyap	IIT Jodhpur
R. C. Gupta	IIT (BHU) Varanasi
T. R. Mankhand	IIT (BHU) Varanasi
S. Lele	IIT (BHU) Varanasi
Vakil Singh	IIT (BHU) Varanasi
Paresh Haribhakti	TCR Advanced Pvt Ltd., Vadodara
Manish Raj	Jindal Steel and Power, Raigarh
Narayana Murty	SVS, Liquid Propulsion Systems Centre, Trivandrum
Amit Rastogi	Institute of Medical Sciences (BHU), Varanasi
Dhiren Kumar Panda	JSW Steel & Coated Products Vidyanagar

Contents

Advanced Material

Development and Characterization of Advanced Steels

Processing and Structure-Property Correlation

Review Papers

Ferrous Process Metallurgy

Development of a Low Silica Calcium Aluminate Based Mould Flux for Casting High Al/Mn Steels

Aman Nigam[(✉)] and Rahul Sarkar[(✉)]

Indian Institute of Technology, Kanpur, India
{amann21,rsarkar}@iitk.ac.in

Abstract. The third generation of AHHS is a new type of advanced high-strength steel that has improved ductility and can be used in a variety of applications in automobiles. Because of its high Al (0.5–2 wt.%) % and Mn (5–7 wt.%) contents, Al and Mn easily react with the oxides such as SiO_2 present in the conventional calcium silicate mold flux which affects the lubrication and crystallization ability provided by the slag film resulting in casting defects such as Break Out Prediction (BOP) alarms, Transverse and Longitudinal depressions. In this study, a calcium aluminate-based alternative mold flux containing different fluxing agents (Na_2O, B_2O_3, SiO_2, and CaF_2) has been developed. Using a differential scanning calorimeter (DSC/DTA) at three different heating rates, 10 K/min, 15 K/min, and 20 K/min, the onset and peak temperatures of crystallization for the composition having $w(CaO)/w(Al_2O_3) = 1$ were determined. The melt structure of the vitreous calcium aluminate-based mold flux was also examined using Raman spectroscopy, and it was shown to have several $[AlO_4]$ structural units with $[BO_3]$ pyroborate units. The deconvolution results of Raman spectra reveal the degree of polymerization (DOP) of the aluminosilicate network present in calcium aluminate-based mold flux. Utilizing structural data such as the area fraction of various structural units found in the melt structure allowed us to examine the tendency of these fluxes to crystallize.

Keywords: AHSS · Mould Flux · Melt Structure · Degree of Polymerization · Raman Spectroscopy

1 Introduction

The continuous casting of steel requires the use of mold flux or casting powders. These mould fluxes usually contain CaO and SiO_2 as their major components with further addition of fluxing agents to engineer the properties required for the steel to be cast. During the continuous casting of steel, the mold flux spreads out on the steel meniscus at the top of the mold, forming a powder bed. These powders warm up and finally melt, creating a pool of slag as shown in Fig. 1. This slag pool acts as a reservoir that infiltrates between the solidified steel shell and the water-cooled copper mold as the mold descends. The slag partially solidifies to create a slag film that is made up of a liquid layer (0.1–0.3 mm thick) and a solid layer (1–2 mm thick) (Fig. 1). Due to the

S. Patra et al. (Eds.): METCENT 2023, *Proceedings of the International Conference on Metallurgical Engineering and Centenary Celebration*, pp. 3–12, 2024.
https://doi.org/10.1007/978-981-99-6863-3_1

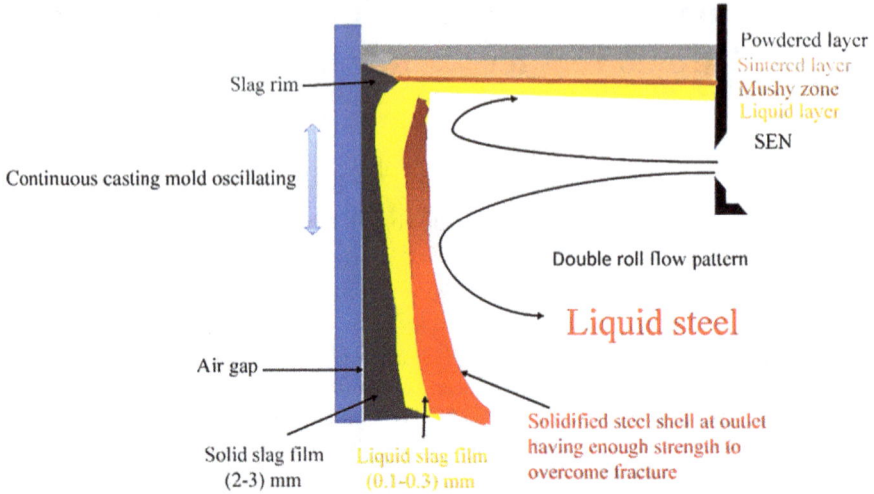

Fig. 1. Schematic diagram illustrating the function of mold flux in continuous casting.

rapid cooling, the slag layer will initially be glassy, but as time passes, the proportion of crystalline phases steadily rises until it achieves a stable state. The amount of lubrication provided to the shell is determined by the thickness of the liquid slag film. The amount of heat extracted from the shell is controlled by how thick the solid layer is and how much of it is crystalline. In addition to lubrication and heat extraction, the mold flux protects the steel from oxidation and prevents freezing by acting as thermal insulation on the top of the steel meniscus.

The third generation of AHHS is a new type of advanced high-strength steel that has improved ductility and can be used in a variety of applications in automobiles. Because of its high Al (0.5–2 wt.%) % and Mn (5–7 wt.%)[1], Al and Mn easily react with the oxides such as SiO_2 present in the conventional $CaO\text{-}SiO_2$-based mold flux [Eq. (1)-(2)] resulting in an increase in the viscosity and melting temperature of the mold flux due to a sudden rise in the basicity and Al_2O_3/SiO_2 ratio[2], which in turn will deteriorate the crystallization and lubrication provided by the slag film causing a variety of issues and flaws, including BOP (Break Out Prediction) alarms, transverse and longitudinal depressions.

$$3(SiO_2) + 4[Al] = 3[Si] + 4(Al_2O_3) \tag{1}$$

$$3(MnO) + 2[Al] = 3[Mn] + 2(Al_2O_3) \tag{2}$$

Numerous research has concentrated on the crystallization and lubricating tendencies of mold fluxes since they offer a proper understanding of the flux, which will help remove casting flaws. As a consequence, there is an increasing demand for utilizing calcium aluminate-based mold powders for casting high Al/Mn steels replacing conventional calcium silicate-based mold powders. The calcium aluminate mold fluxes contain little or no Silica present in them called "Non-Reactive" mold flux. The only problem with using this kind of mold flux is that it melts at a high temperature, which contradicts the purpose of using it. Several fluxing agents, such as Li_2O, Na_2O, B_2O_3, and MgO, are added to these mold fluxes to adjust their melting points to a desirable range and may also increase their ability to crystallize and lubricate. In the same context Gao et al.[3] developed a calcium aluminate mold flux with a low amount of silica having a certain addition of Na_2O which will eventually help in the enhancement of crystallization tendency and also will lead to the depolymerization of the melt structure. In their study of the non-isothermal crystallization kinetics of the non-reactive F-free mold flux, Shu et al.[4] found that the tendency for crystallization increased with an increase in the ratio of $w(CaO)/w(Al_2O_3)$ for constant B_2O_3 and Na_2O. Ryu et al.[5] investigated the impact of basicity and the alumina content on crystallization temperature and incubation period. They discovered that as the Al_2O_3 and C/S ratio increased, the temperature of crystallization increased while the incubation period decreased. According to Zhou et al.[6], the capacity to crystallize the mold powder utilized in the casting of high-Al steels was found to be initially increased and then inhibited with increasing Al_2O_3 concentration.

The tendency of the mold flux to crystallize is widely dependent on the melt structure of the mold fluxes which can be observed through various spectroscopic techniques like FTIR, Raman & NMR spectroscopy. Several researchers tried to explain the effect of crystallization with the help of the degree of polymerization (DOP) of the melt structure[6]. The amount of bridging and non-bridging oxygen in the melt structure corresponds to the degree of polymerization. The tendency of the mold flux for crystallization increases as the DOP of the structure decreases, and vice versa. Shu et al. [7] looked at how the structure of the melt affected the crystallization of F-free mold flux. The results showed that as the amount of Na_2O in the mold flux increases, the DOP of the melt structure decreases, making it easier to crystallize.

To prevent interfacial reactions and to achieve a proper balance between mold flux lubrication and heat extraction ability, the objective of this research is to create a specific type of mold flux that is suitable for third-generation AHSS steel with a high Al/Mn content. A non-reactive CaO-Al_2O_3-based mold flux with little to "No Silica" or by altering the $w(CaO)/w(Al_2O_3)$ & basicity ratio in conventional mold flux can be used to achieve this goal.[8].

2 Experimental Methodology

2.1 Preparation of Mold Fluxes

Reagent-grade chemicals of CaO, Al_2O_3, Na_2CO_3, SiO_2, CaF_2, and B_2O_3 with a purity of 99.99% were used as raw materials to prepare mold flux samples used in this investigation. They have compositions according to Table 1. Na_2CO_3 was calcinated to Na_2O at 873 K for 2 h in the air to remove moisture. The raw materials were thoroughly mixed in an agate mortar before being heated and melted in a graphite crucible inside the muffle furnace at 1673 K for one hour. The pre-melted samples were then pulverized into powder for observation in X-Ray Diffraction (XRD: PANanalytical Empyrean), Induced Coupled Plasma-Mass Spectroscopy (ICP-MS), and X-Ray fluorescence (XRF) after being quenched in a bucket of water to produce an altogether amorphous phase. XRD aims to study the phases in the phases formed in the mold fluxes and thus to determine the crystallinity whereas ICP-MS & XRF were used to determine the chemical composition of pre-melted samples so that there should not be any significant evaporation loss.

Table 1. Chemical constitution of the mold flux

Composition in weight %						
Sample	CaO	Al_2O_3	Na_2O	B_2O_3	SiO_2	CaF_2
w(CaO)/w(Al_2O_3)=1	35	35	15	5	5	5

2.2 Measurements Using DSC

Glassy samples were ground up and analyzed through a DSC analysis. With argon serving as the purge gas under dynamic conditions, the measurements were carried out using a thermal analyzer, model STA 2500 Regulus from the manufacturer Netzsch. To reduce sample volatile loss, platinum crucibles with platinum covers were used. The baseline was created using empty platinum crucibles exposed to variable heating rates (20 K/min, 15 K/min, and 10 K/min).

2.3 Raman Spectroscopy

The melt structure of the slag was determined by Raman spectra analyses on the ground-up quenched samples at room temperature. The Raman spectrometer (Princeton Instruments Acton Spectra Pro 2500i) was utilized to capture the Raman spectra in the frequency range of (100–2500 cm^{-1}) using a 532 nm (DPSS Laser (Laser Quantum gem 50 mw)) wavelength laser as the excitation source. However, the Raman shifts were mainly concentrated between (300–1700) cm^{-1} which is important for defining the structure of the slag samples. The baseline subtraction was done in the origin and the spectra were deconvoluted by the Gaussian functions with a minimum correlation coefficient of $R^2 \geq 0.969$. The area fraction of the observed peaks was used to compute the abundance or population of the structural units.

3 Results and Discussion

3.1 Sequence of Crystal Precipitation

Figure 2 displays the DSC data for the mold flux system having a w(CaO)/w(Al_2O_3) ratio of 1, for various heating rates between 20 K/min and 10 K/min. The thermographs shown represent the exothermic crystallization peak (Exothermic occurs in a downward direction). Table 2 shows the onset and peak temperatures for the first and second crystallization events (Peak 1 and Peak 2) for the w(CaO)/w(Al2O3) = 1 slag sample. The peak analysis in the Origin 2021b software was used to determine the various temperatures listed in Table 2. The temperature at which crystallization first starts during a non-isothermal crystallization process correlates to the crystallization temperature for the crystalline phase, which can be identified as the temperature at which exothermic peaks first appear during heating. The onset and the peak temperature of a particular crystallization event depend upon the heating rate provided, as the transition from glass to a crystal phase is dependent on the nucleation and growth rate. The glass phase provides ample time for nucleation and crystal growth when the heating rate is low, resulting in a comparatively low crystallization temperature. Higher heating rates result in shorter times for crystal nucleation and growth. The crystallization temperature will be higher when the rate of transition from the glass phase to the crystal phase is at its highest[9]. When the sample is heated to the various target temperatures listed in Table 2, XRD and SEM/EDS will figure out the order in which different crystals form.

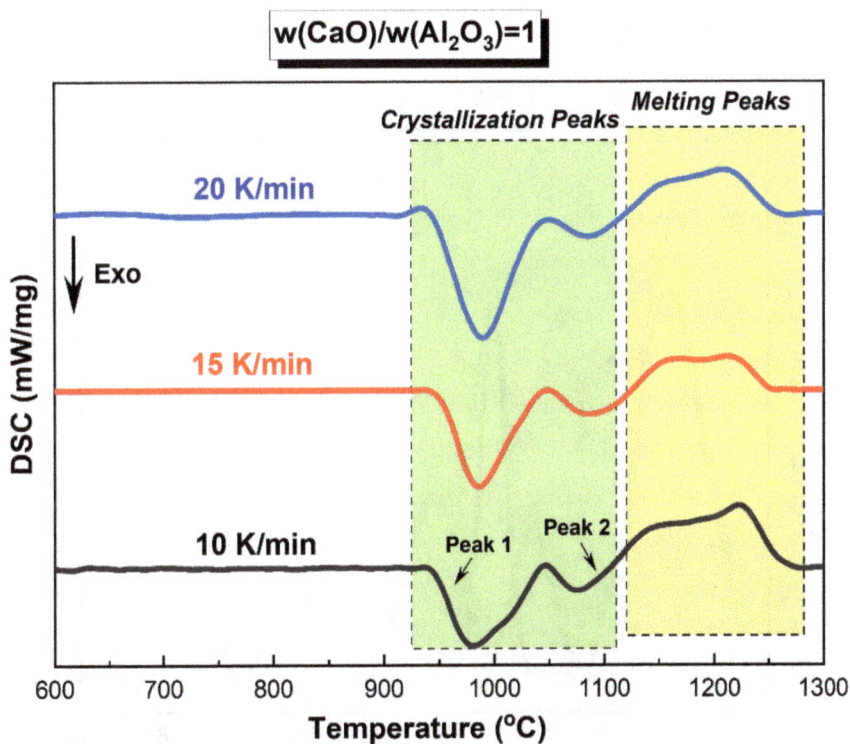

Fig. 2. DSC curve of $w(CaO)/w(Al_2O_3) = 1$ sample at different heating rates

Table 2. Onset and Peak temperature at different heating rates

Crystallization Events		Heating Rates		
Peak Information (ºC)		10 K/min	15 K/min	20 K/min
Peak 1	Onset Temperature	934	936	938
	Peak Temperature	981	986	990
Peak 2	Onset Temperature	1045	1049	1046
	Peak Temperature	1076	1086	1089

3.2 Raman Spectroscopy

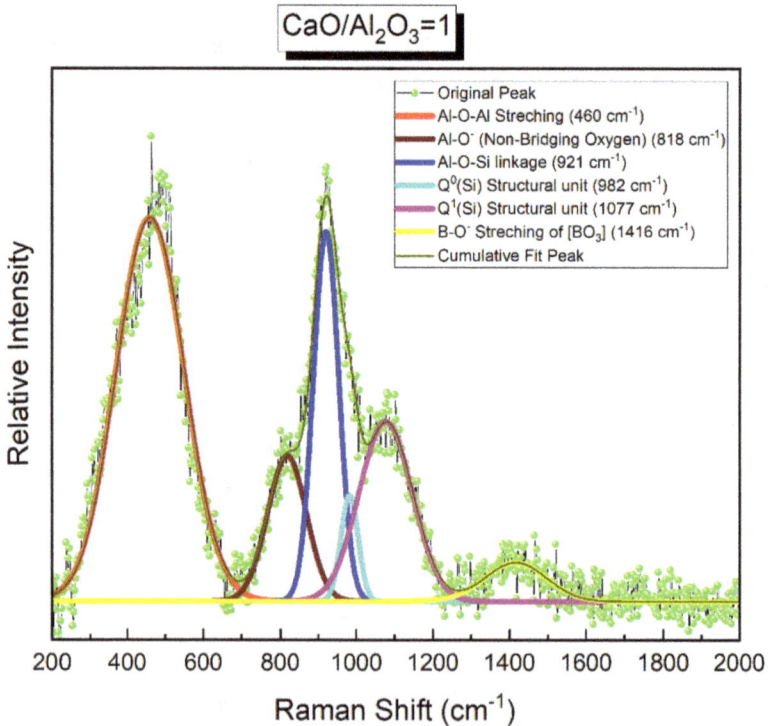

Fig. 3. The deconvoluted Raman spectra of $w(CaO)/w(Al_2O_3) = 1$

The melt structure of the calcium aluminate-based mold fluxes can be related to its viscosity and the crystallization tendency which in turn can influence how well the mold fluxes lubricate and transfer heat to the steel shell. The structural functionality performed by the Al_2O_3 in the melt structure is complicated as it is amphoteric. In other words, the number of constituents in the slag system correlates to its structural role. These components can be network formers or network breakers depending on the type of steel to be cast. To comprehend the melt structure of the calcium aluminate slag system and how the structure changes as SiO_2 is replaced by Al_2O_3, many studies have been conducted.

As mentioned above transition from the calcium silicate slag system to the calcium aluminate slag system will typically lead to a more intricate melt structure. It is indicated that Al_2O_3 mainly exists in the form of $[AlO_4]^{5-}$,.$[AlO_5]^{7-}$, and.$[AlO_6]^{9-}$ structural units in the melt structure. These units can act as network former or network breakers depending on the composition of Al_2O_3 present in the slag system. It is to be said that when the Al_2O_3 is present in the low composition in the calcium silicate mold flux it may behave as a network breaker in the form of $[AlO_5]^{7-}$ and.$[AlO_6]^{9-}$ structural units but as the composition of Al_2O_3 increases to a certain extent these structural units will be converted to $[AlO_4]^{5-}$ units and the Al_2O_3 will act as the network former. During this whole transition as the silicate structure changes to an aluminate structure, the structural units present in the silicate structure in the form of the $[SiO4]^{4-}$ will be transformed to the linkages like Al-O^0 (O^0 denotes the bridging oxygen) and Al-O-Si, with the aid of certain basic oxides like CaO, Na_2O, Li_2O, MgO & BaO. This will result in the depolymerization of the silicate structure which will eventually be transformed into the aluminate structure[9]. Thus the complexity of the melt structure will be enhanced and this will introduce variable crystalline phases.

Several researchers reported that in the calcium aluminate slag system with the increase in w(CaO)/w(Al_2O_3) ratio, the bridging.oxygen present in the form of Al-O^0 and Al-O-Si linkages will be destroyed and will be transformed to Al-O^-(.O^- is the bridging oxygen) also the silicate units present having higher polymerization degree in the form of Q^1(Si) and Q^2(Si) will be destroyed to the lower polymerization degree i.e. to the Q^0(Si) structural unit (the number in the superscript denotes the quantity of bridging oxygen in the $[SiO_4]^{4-}$ unit). Furthermore, with the aid of O^{2-} (free oxygen) generated by added CaO or Na_2O, the network former unit $[AlO_4]^{5-}$ will be transformed into the network breaker units (Eqs. 3–5) $[AlO_5]^{7-}$, $[AlO_6]^{9-}$ [10].

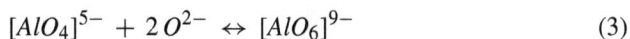

$$[AlO_4]^{5-} + 2O^{2-} \leftrightarrow [AlO_6]^{9-} \tag{3}$$

Yang et al. reported that adding B_2O_3 to the calcium aluminate mold flux system has the exact opposite effect to that of adding Na_2O.[10] The Raman spectra of CaO-Al_2O_3–7 wt.% SiO_2 illustrate the $[BO_4]$ related 3D pentaborate structure, where $[BO_4]$ units are connected to the borate structures and.$[AlO_4]^{5-}$ tetrahedral unit. When B_2O_3 is added, there will be more of these units and links as $[BO_4]$ units promote the transition of higher polymerization structural units to lower polymerization structural units. It indicates that the B_2O_3 will behave as the network former in the calcium aluminate slag system. So, it will make the structure of the melt more intricate as well as raise the degree of polymerization. But, in the case of adding Na_2O, the O^{2-} that is given off will cause the aluminosilicate structure to change from bridging oxygen (O^0) to non-bridging oxygen (O^-) (Eq. 4), which will make the structure of the melt less complicated. Similar findings were made by Wang et al. for the CaO-Al_2O_3–10 wt% SiO_2-based slag system with varying Na_2O composition, where it was found that the aluminosilicate structure depolymerization would be further enhanced with an increase in Na_2O content.

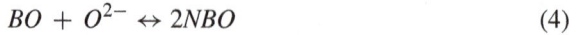

$$BO + O^{2-} \leftrightarrow 2NBO \qquad (4)$$

To analyze the change in the degree of polymerization of the aluminate or aluminosilicate network Raman spectroscopy was collected for the present mold fluxes. The original Raman spectrums of all samples are shown in Fig. 3.

The spectra pattern around 460 cm^{-1} (lies in the low-frequency region of the Raman spectra) is corresponding to Al-O-Al stretching characteristics peak representing the bridging oxygen in the aluminate network having $[AlO_4]^{5-}$ unit. The mid-frequency region ($700-1300 \text{ cm}^{-1}$) of the spectra shows the depolymerization of the aluminosilicate structure denoting Al-O$^-$ or Si-O$^-$ (O$^-$ indicates the depolymerization of the network) telescopic vibrations as shown in the figure. The characteristics peak at 818 cm^{-1} and 921 cm^{-1} stand for the symmetric.Al-O$^-$ and Al-O-Si bonds, respectively. The degree of polymerization or depolymerization is represented by Q^n (n represents that number of bridging oxygen), where Q^0 refers to $[AlO_4]^{5-}$ or $[SiO_4]^{4-}$ unit. The structural units $Q^0(Si)$, and $Q^1(Si)$ correspond to the peaks at 982 cm^{-1} and 1077 cm^{-1} respectively.

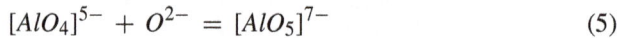

$$[AlO_4]^{5-} + O^{2-} = [AlO_5]^{7-} \qquad (5)$$

The band at 1416 cm^{-1} is attributed to the symmetric stretching vibrations of terminal oxygen atoms in orthoborate units $[BO_3]$, which means that B^{3+} is mainly forming $[BO_3]$ groups in the mold flux.

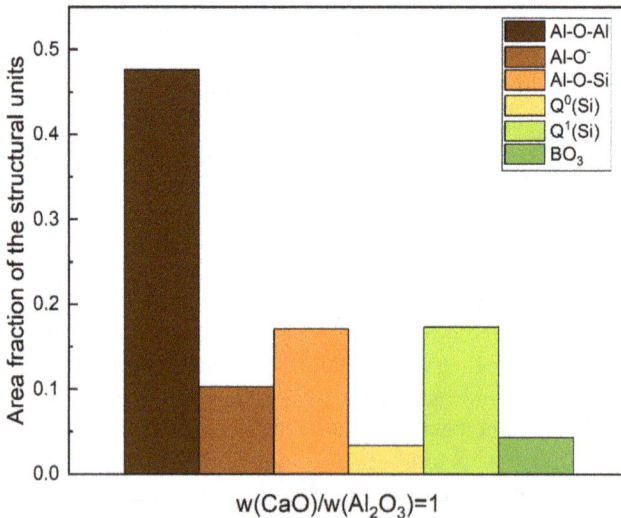

Fig. 4. Area fraction of the various structural units

The area fraction of the structural units determines the approximate population density of that unit, and this has been found out with the help of the area integration results of the deconvoluted Raman spectra of the various units. Figure 4 shows that these results suggest that the depolymerization of aluminate and silicate structures happens when specific network breakers are added. This means that more complicated aluminate and silicate structures are turned into simpler ones when more CaO and Na_2O are added to the melt structure as they will release more O^{2-} ions. This is to say that by adding more Na_2O or increasing the C/A, the depolymerization of the aluminosilicate structure increases, which aids in crystallization because the crystallization temperature rises.

4 Conclusion

The crystallization tendency and the subsequent melt structure were studied for the low Silica calcium aluminate mold fluxes and the following conclusions can be drawn:

1) The onset crystallization temperature is rising with the heating rate, which means that at higher heating rates, the crystallization tendency will also rise. This will aid in the extraction of heat from the solidified steel shell, but it will also result in the formation of a slag rim, which will affect the infiltration of slag into the space between the solidified steel shell and slag film.
2) The depolymerization of the aluminate and silicate structure by the addition of Na_2O is revealed by Raman spectroscopy of the mold slag. A solid-state ^{27}Al NMR study will be used in the later part of this study to determine the type of structural units, whether they are $[AlO_4]$, $[AlO_5]$, or $[AlO_6]$ in the aluminate structure. NMR will also be used to look at the impact of fluorine on the structure.
3) To find out if adding Al_2O_3 to the mold flux will act as a network-forming agent or a network-breaking agent, the link between the structure of the melt and the ability of the mold flux to crystallize must be determined in the later stages of this study.

Acknowledgment. The author would like to express gratitude to the several labs at the Department of Materials Science and Engineering at IIT Kanpur for providing the various facilities needed for the current research.

References

1. Aydin, H., Essadiqi, E., Jung, I.H., Yue, S.: Development of 3rd generation AHSS with medium Mn content alloying compositions. Mater. Sci. Eng. A **564**, 501–508 (2013). https://doi.org/10.1016/j.msea.2012.11.113
2. Kim, M.S., Lee, S.W., Cho, J.W., Park, M.S., Lee, H.G., Kang, Y.B.: A reaction between high Mn-High Al Steel and CaO-SiO2-type molten mold flux: part I. composition evolution in molten mold flux. Metall. Mater. Trans. B **44**(2), 299–308 (2013). https://doi.org/10.1007/s11663-012-9770-z
3. Gao, J., Wen, G., Sun, Q., Tang, P., Liu, Q.: The influence of Na2O on the solidification and crystallization behavior of CaO-SiO2-Al2O3-based mold flux. Metall. Mater. Trans. B **46**(4), 1850–1859 (2015). https://doi.org/10.1007/s11663-015-0366-2
4. Shu, Q., Li, Q., Medeiros, S.L., Klug, J.L.: Development of non sreactive f-free mold fluxes for high aluminum steels: non-isothermal crystallization kinetics for devitrification. Metall. Mater. Trans. B **51**(3), 1169–1180 (2020). https://doi.org/10.1007/s11663-020-01838-4

5. Ryu, H.G., Zhang, Z.T., Cho, J.W., Wen, G.H., Sridhar, S.: Crystallization behaviors of slags through a heat flux simulator. ISIJ Int. **50**(8), 1142–1150 (2010). https://doi.org/10.2355/isijinternational.50.1142

6. Zhou, L., Wu, H., Wang, W., Luo, H., Li, H.: Crystallization behavior and melt structure of typical CaO–SiO2 and CaO–Al2O3-based mold fluxes. Ceram. Int. **47**(8), 10940–10949 (2021). https://doi.org/10.1016/j.ceramint.2020.12.213

7. Shu, Q., Klug, J.L., Medeiros, S.L.S., Heck, N.C., Liu, Y.: Crystallization control for fluorine-free mold fluxes: effect of Na2O content on non-isothermal melt crystallization kinetics. ISIJ Int. **60**(11), 2425–2435 (2020). https://doi.org/10.2355/isijinternational.ISIJINT-2020-132

8. Zhang, L., Wang, W.L., Shao, H.Q.: Review of non-reactive CaO–Al2O3-based mold fluxes for casting high aluminum steel. J. Iron Steel Res. Int. **26**(4), 336−344 (2019). https://doi.org/10.1007/s42243-018-00226-2

9. Wang, W., Xu, H., Zhai, B., Zhang, L.: A Review of the melt structure and crystallization behavior of non reactive mold flux for the casting of advanced high strength steels. Steel Res. Int. **93**(3), 1 (2022). https://doi.org/10.1002/srin.202100073

10. Yang, J., et al.: Effect of B2O3 on crystallization behavior, structure and heat transfer of CaO-SiO2-B2O3-Na2O-TiO2-Al2O3-MgO-Li2O mold fluxes. Metall. Mater. Trans. B **48**(4), 2077–2091 (2017). https://doi.org/10.1007/s11663-017-0997-6

Some Aspects of the Chemistry of Slags Containing Cr and V

Seshadri Seetharaman[(✉)]

Emeritus Royal Institute of Technology, Stockholm, Sweden
raman@kth.se

Abstract. The present paper provides some pioneering experimental information regarding the chemistry of steelmaking slags containing Cr and V. Initially, the thermodynamic activities of CrO and $VO_{1.5}$ in slags at very low oxygen potentials were determined by gas-equilibration technique involving CO-CO_2-Ar gas mixtures at 1600 °C. While the activity of CrO as a function of X_{CrO} showed a positive deviation from ideality, the trend in the case of $VO_{1.5}$ was found to be the opposite. In the case of Cr-slags, it is important to determine the valence states of Cr in the slag as a function of basicity, oxygen partial pressure prevailing and temperature. In view of the contradictory results in the literature, experiments were conducted by XANES (X-ray Absorption Near-Edge Spectroscopy) as well as the high temperature Knudsen cell mass spectroscopy (pioneering experiment). The ratio of Cr^{2+}/Cr^{3+} as functions of basicity, p_{O2} and temperature were determined. Cr emission from slags were determined by a thin film method, using SHTT (Single Hot Thermocouple Technique) set-up. It was found that at CrO_3 emissions are significant in air at steelmaking temperatures.

In the case of V-containing slags, the ratio of the valence states of vanadium was determined as functions of temperature and basicity by the same methods as in the case of Cr. The kinetics of evaporation of pure V_2O_5 was examined by thermogravimetry. Thin film experiments showed that vanadium losses as V_2O_5 are significant in oxidizing atmospheres. Further experiments were conducted by simultaneously applying vacuum and oxygen purging in order to force the micro bubbles of V_2O_5 formed and entrapped in the viscous slag. These experiments demonstrated the potential of this method to extract vanadium from slags, low grade ores and even pet-coke ash.

Keywords: Chromium · Slags · Steelmaking · Valences · Vanadium

1 Introduction

With the increasing need for light weight, high strength steel, the need for alloying elements has increased significantly world-wide. Correspondingly, the amounts of these valuable alloying elements are also lost as emissions in slags and dust. A study by the Swedish Steel Producers Association from 2012 showed (Table 1) that the amount alloying elements lost in emissions was higher than the amount needed for the production. This also has adverse impact on the environment apart from the economic losses.

S. Patra et al. (Eds.): METCENT 2023, *Proceedings of the International Conference on Metallurgical Engineering and Centenary Celebration*, pp. 13–18, 2024.
https://doi.org/10.1007/978-981-99-6863-3_2

Table 1. Amounts of alloying elements required annually and the amount lost to slag and dust 1.

Metal	Annual demand/t	Amount lost annually as waste emissions /t
Chromium	100 000	180 000
Manganese	50 000	70 000
Zinc	15 000	33 000
Nickel	25 000	17 500
Molybdenum	10 000	8 000

There is a strong need to optimize steelmaking processes and this requires an urgent need to understand the chemistry of the elements like Cr and V in steel slags. The present studies involved experimental studies of the thermodynamic activities of the oxides of Cr and V in as well as the valence states of these elements in the slags under steelmaking conditions by novel techniques. The emission of CrO_3 and V_2O_5 from thin films of slags at high temperature has been studied using a SHTT equipment. The present paper summarizes the efforts made at the Royal Institute of Technology, Stockholm in this direction. A part of the work was carried out in collaboration with the Institute for Iron and Steel Technology, Bergakademie, Freiberg, Germany.

Thermodynamic Activities of the Oxides of Cr^2 and V^3 in Slags

The thermodynamic activities of chromium and vanadium oxides were measured in the temperature range 1803−1923 K using an equilibration technique, where the slags containing low amounts of Cr or V (CaO-MgO-Al_2O_3-SiO_2-CrO_x (or VO_x)) kept in Pt crucibles were equilibrated with a gas mixture of CO, CO_2 and Ar corresponding to well-defined oxygen partial pressures, which were kept very low. The experimental set-up is shown in Fig. 1

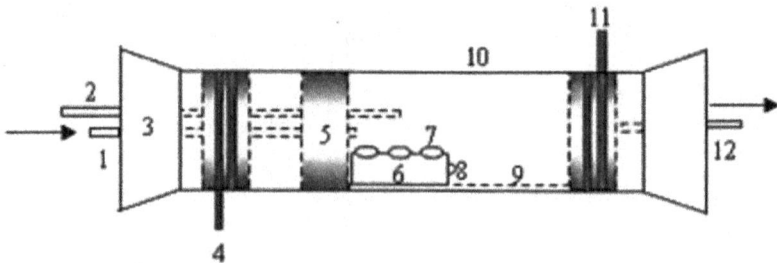

Fig. 1. The sketch diagram of the experimental equipment. 1. Gas inlet; 2. Thermocouple; 3. Silicon stopper; 4. Cooling water inlet; 5. Refractory; 6. Pure alumina holder; 7. Hole in the alumina holder for Pt crucibles; 8. Ceramic ring for pulling; 9. Alumina runners; 10. Reaction tube; 11. Cooling water outlet; 12. Gas outlet

The gases were purified suitably so that impurity oxygen levels were kept below 10^{-18} atm. The equilibration was carried out for a maximum of 20 h. The transition metal oxides were to be at their lowest oxidation states (Cr as Cr^2+ and V as V^{3+}) due to the low oxygen potentials involved.

From the results, the activities of CrO and $VO_{1.5}$ were calculated with a knowledge of the thermodynamics of Pt-Cr and Pt-V alloys respectively. The activities showed positive deviation from Raoult's law in the case of CrO and negative deviation in the case of $VO_{1.5}$ as shown in Fig. 2. The standard states are pure CrO and $VO_{1.5}$ in liquid state.

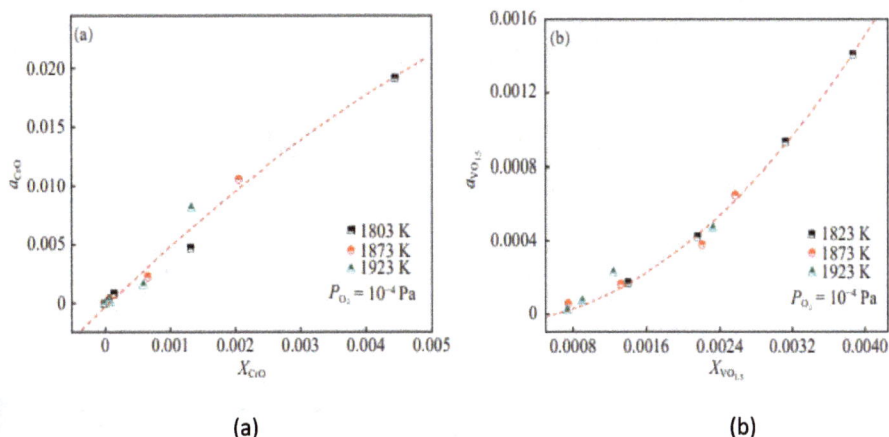

Fig. 2. Thermodynamic activities of CrO (2a)[2] and VO1.5 (2b)[3] at various temperatures. The standard states are pure liquids at the experimental temperatures.

Valence States of the Oxides of Cr and V in Slags

An understanding of the chemistry of alloy steel slags under the operation conditions with varying temperatures and oxygen partial pressures, it is very important to have a knowledge of the valence states of the transition metals in the slag, specifically the oxides of Cr and V. The data available in literature were contradictory. The main problem is to analyze the slags after the experimentation without any oxidation during the chemical analysis, which otherwise would distort the results. Two new novel techniques were employed, one involving the X-ray Analysis Near Edge Spectroscopy (XANES) method, used mostly by geologists. The slags with suitable compositions were premelted and the compositions were determined accurately. The samples were subjected to XANES investigations [4, 5]. The other method was a new approach using the results of High Temperature Knudsen Cell Mass Spectrometric method [4, 6]. The mass spectrometry method which is a common method for measuring the vapour species was used for the first time to determine the valence states of transition metals in slag. The principles involved have been explained in the original publication 4. The results of the XANES studies are presented in Fig. 3.

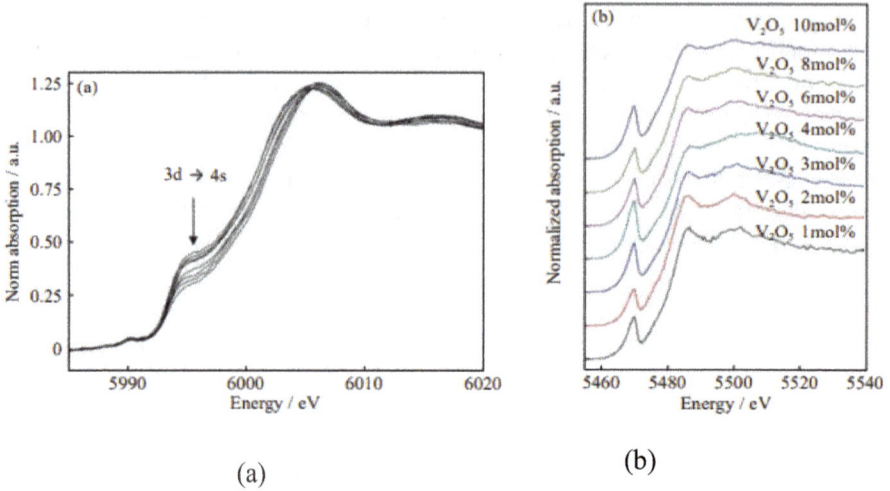

Fig. 3. XANES investigation results in the case of Cr (3a) and V (3b) slags [4, 5].

The ratio of Cr^{2+}/Cr^{3+} as functions of basicity as obtained by both the techniques is summarized in Fig. 4.

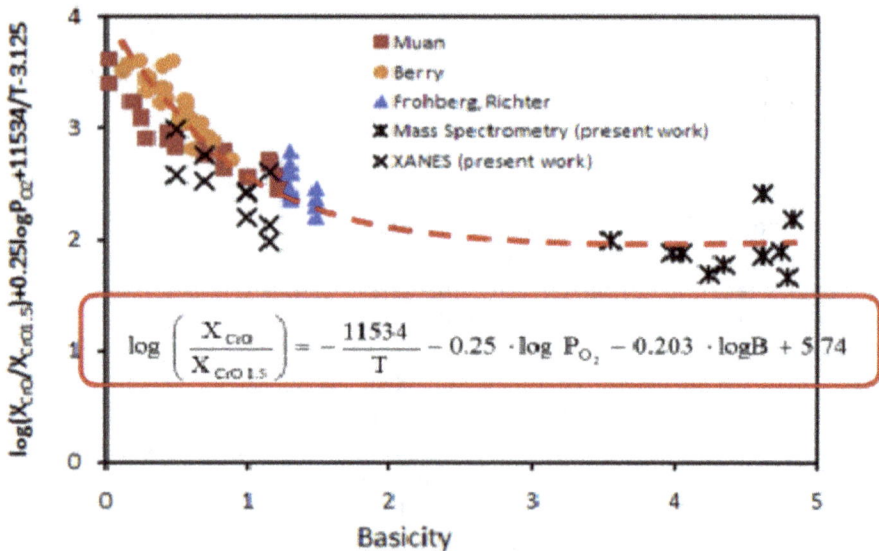

$$\log\left(\frac{X_{CrO}}{X_{CrO_{1.5}}}\right) = -\frac{11534}{T} - 0.25 \cdot \log P_{O_2} - 0.203 \cdot \log B + 5.74$$

Fig. 4. The Cr^{2+}/Cr^{3+} ratio as a function of the basicity in Cr-containing steel slags 4.Similar results were ported in the case of vanadium-containing slags.

With a knowledge of the valence states of Cr in the slags as functions of temperature, oxygen partial pressure and basicity, it was now possible to understand the sulphide capacities of of these slags. As Cr_2O_3 is a surface-active oxide in slags, there would be an accumulation of this oxide in slag surfaces. It was found by XPS analysis that conventional sulphide capacity measurement provide only an average value of the sulphide capacities, whereas the oxidation state of sulphur varied from $+8$ close to the gas/slag interface to -2 in the bulk of the slag.

Cr and V Emissions From Slags

It is known that both CrO_3 and V_2O_5 have high vapour pressures at steelmaking temperatures. Since CrO_3 is extremely carcerogenic, it is important to understand the emissions of CrO_3 when steel slags are tapped. The Single Hot Thermocouple Technique (SHTT) available at Bergakademie, Freiberg Germany was employed to study the evaporation from thin slag films. These slag films in the loop of the thermocouple (which is used both as a heater as well as for temperature measurement) were heated in air or oxygen to steelmaking temperatures and the loss of Cr or V was estimated after quenching using SEM/EDS method. The loss of Cr from chromium slags and V from vanadium slags are presented in Fig. 5(a) and (b) respectively.

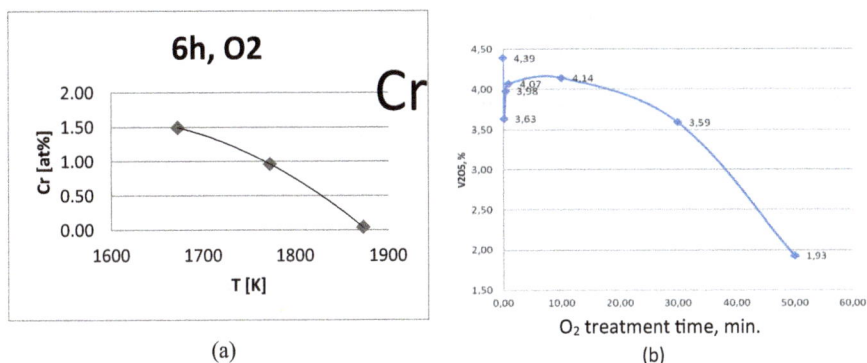

(a)   (b)

Fig. 5. Cr (5a)[7] and V (5b)[8] losses from slags after heating in air at steelmaking temperatures. T $= 1873$ K for vanadium experiments

This method has further been developed in the case of vanadium to recover this valuable metal from converter slags and other secondary resources 9.

Summary

The thermodynamic activities of Cr and V in steel slags were measured by a gas equilibration method involving CO-CO_2-Ar mixtures at molten steel temperatures.

The valence states of Cr and V in steel slags were measured by XANES method as well as high temperature Knudsen cell mass spectrometry.

The emission of CrO_3 and V_2O_5 from thin films of molten steel slags were studied by SHTT method. The method is now developed for commercial extraction of vanadium from lean sources and steel slags.

Acknowledgements. My sincere thanks to the organizers of METCENT for the opportunity to present our results. The contributions by various co-workers in obtaining the results reported are gratefully acknowledged.

References

1. Final Report, Eco Steel Production, Swedish Steel Producers Association (2012)
2. Dong, P., Wang, X., Seetharaman, S.: Thermodynamic activity of chromium, vanadium oxide in CaO-SiO2-MgO.Al2O3-CrOX slags. Steel Res. Int. **80**, 202–208 (2009)
3. Dong, P., Wang, X., Seetharaman, S.: Activity of VO1.5 in CaO-SiO2-MgO-Al2O3 slags at low vanadium contents and low oxygen pressures. Steel Res. Int. **80**, 251–255 (2009)
4. Wang, L.J., Seetharaman, S.: Experimental studies on the oxidation states of chromium oxides in slag phase. Metall. Mater. Trans. B **41**, 946–954 (2010)
5. Wang, H.J.: Investigations on thevoxidation of iron-chromium and iron-vanadium molten alloys, Ph. D. thesis, Royal Institute of Technology, Stockholm (2010)
6. Wang, L.J., Teng, L.D., Chou, K.C., Seetharaman, S.: Determination of vanadium valence state in CaO-MgO-Al2O3-SiO2 system by high temperature mass spectrometry. Metall. Mater. Trans. **44B**, 948–953 (2013)
7. Seetharaman, S., Albertsson, G., Scheller, P.R.: Studies of vaporization of chromium from thin slag films at steelmaking temperatures in oxidizing atmosphere. Metall. Mater. Trans. B **44**, 1280–1286 (2013)
8. Seetharaman, S., Shyrokykh, T., Schroeder, C., Scheller, P.R.: Vaporization studies from slag surfaces using a thin film technique. Metall. Mater. Trans. B **44**, 783–788 (2013)
9. Shyrokykh, T., Wei, X., Seetharaman, S., Volkova, O.: Vaporization of Vanadium Pentoxide from CaO-SiO2- VOx slags during alumina dissolution. Metall. Mater. Trans. B **52**, 1472–1483 (2021)

Solid State Reduction of Hematite Ore Using Hydrogen at Moderate Temperatures

Devendra Nama and Rahul Sarkar[✉]

Department of Materials Science and Engineering, Indian Institute of Technology Kanpur, UP 208016 Kalyanpur, Kanpur, India

{dnama20,rsarkar}@iitk.ac.in

Abstract. This work is carried out to investigate the reduction of hematite ore fines in a pure hydrogen atmosphere using thermogravimetric analysis (TGA). The effect of temperature (973-1173K) and hydrogen flow rate (1–2.5 L per minute) on the reduction kinetics was examined. The phases, morphological features, and chemical composition of hematite ore were identified using X-ray diffraction (XRD), Scanning electron microscopy (SEM), and X-ray fluorescence (XRF), respectively. The reduction kinetics were investigated using global reduction methods. The results show that the rate of reduction increases with temperature. An increase in hydrogen flow rate after the critical flow rate 2.5 LPM does not affect the rate of reduction. For global reduction, the reduction kinetics analysis indicates that the reduction was controlled by chemical mechanism. The apparent activation energy and preexponential factor for the global step are $20.4137 \pm .22$ kJ/mol and 0.7202 respectively.

Keywords: Hematite ore fines · hydrogen · TGA · reduction kinetics

1 Introduction

The BF-BOF routes is primarily responsible for producing steel. The coke is used as a reductant and source of energy in the blast furnace. It releases an enormous amount of carbon dioxide. The Iron and steel industry accounts for 7–9% of global carbon emissions [1]. Major economies are shifting towards net zero carbon emissions, so the Iron and steel industry is facing a lot of challenges to reduce its carbon footprint. To mitigate CO_2 emissions, several methods are suggested such as the partial substitution of fossil fuel with biomass [2], carbon capture and storage [3]. However, the drastic reduction in the emission is not possible from the existing technologies. The most promising way to reduce the carbon emissions from this sector is through the application of hydrogen-based direct reduction and hydrogen produced from electrolysis with renewable energy [4]. The new ironmaking technology needs to be developed that can reduce carbon emissions and take any type of iron ore in the form of fines as a raw input directly without the need for additional agglomeration procedures. It further decreases carbon emissions and energy consumption due to the exclusion of the agglomeration step [5].

© The Author(s), under exclusive license to Springer Nature Singapore Pte Ltd. 2024
S. Patra et al. (Eds.): METCENT 2023, *Proceedings of the International Conference on Metallurgical Engineering and Centenary Celebration*, pp. 19–27, 2024.
https://doi.org/10.1007/978-981-99-6863-3_3

A new ironmaking process will be developed which can convert the fines of iron ore concentrates into the iron in moving bed reactor by using reducing gas such as hydrogen or natural gas or coal gas at low temperatures. A schematic diagram of process is shown in the Fig. 1.

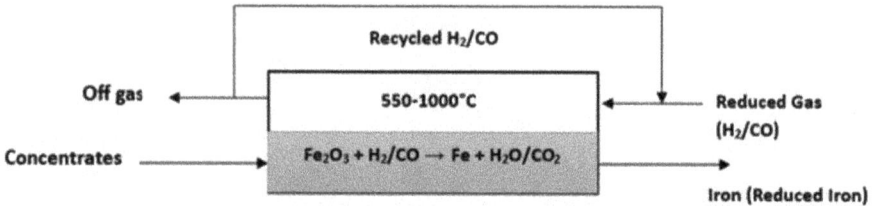

Fig. 1. Schematic diagram of possible ironmaking technology

When the hydrogen is used as a reductant then the conversion of hematite to iron occurs through the formation of intermediate oxides [6].

Fe_2O_3 (Hematite) $\rightarrow$ Fe_3O_4 (Magnetite) $\rightarrow$ Fe_xO (Wustite) $\rightarrow$ Fe (Iron) above 570 °C.

Fe_2O_3 (Hematite) $\rightarrow$ Fe_3O_4 (Magnetite) $\rightarrow$ Fe (Iron) below 570 °C [7].

These consecutive reactions, along with structural modifications in solids during reactions, increase the complexities. Many studies reported different activation energies and rate-controlling steps. The rate-controlling step varies with operational conditions and characteristics of solid particles such as shape, size, and mineralogical composition [8, 9].

In this study, the intrinsic kinetics of hematite ore reduction with hydrogen is experimentally determined. It is imperative to ensure that reduction must be carried out under such a condition in which the overall rate is controlled by the chemical kinetics. The influence of external mass transfer can be minimized by increasing the flow rate of gas past the sample. The influence of interparticle diffusion can be reduced by taking a thin layer of powder particles.

2 Experimental Methods

The isothermal reduction was carried out in the Thermo gravimetric analyzer (TGA). The schematic diagram of the setup is given in Fig. 1. It consisted of the vertical tubular furnace equipped with a micro weighing balance (accuracy 0.1mg) which is connected with a suspended crucible via a Kanthal wire. The Hematite concentrates were placed on the alumina crucible and the initial weight is recorded. The furnace was heated to set the temperature in the Nitrogen atmosphere. When the desired temperature is reached, hydrogen gas replaces the inert gas, and a computer-connected balance is used to continuously record the sample's weight. When there was no discernible change in the weight then reducing gas was replaced by nitrogen gas. Once the furnace had cooled to ambient temperature under the nitrogen atmosphere, the sample was removed and characterised (Fig. 2).

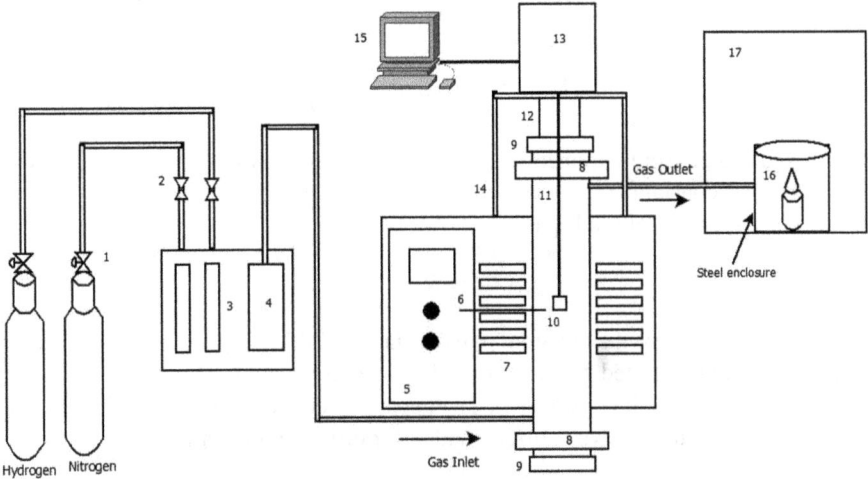

Fig. 2. Schematic diagram of the experimental setup

1. Regulator	2. Ball valve	3. Flowmeter	4. Mixing chamber	5. Furnace controller
6. Thermocouple	7. Heating element	8. Water jackets	9. Flange	10. Crucible
11. Kanthal wire	12. Connecting tube	13. Micro Balance	14. Stand	15. Computer
16. Gas burner	17. Fume hood			

2.1 Kinetics Analysis

The degree of reduction can be written as follow.

$$\text{The degree of reduction } X(t) = \frac{W_i - W_t}{W_o} \tag{1}$$

Where W_i is the initial weight of the sample, W_t is weight at t time and W_o weight of oxygen present in the iron oxide.

The differential reaction rate for a gas solid reaction can be written as

$$\frac{dX}{dt} = K_{app}(T)f(X) \tag{2}$$

where t is time for a reaction, $f(X)$ is the differential function and k_{app} is the apparent rate constant.

The apparent rate constant can be expressed as

$$K_{app}(T) = A\exp\left(-\frac{E}{RT}\right) \tag{3}$$

where A and E are the pre-exponential factor and apparent activation energy respectively, and R is the gas constant.

The integrate form of the Eq. (2)

$$g(X) = K_{app}(T)t \tag{4}$$

where K_{app} (T) and t are the apparent rate constant and time respectively. The apparent constant can be determined by best fitting between g(X) and t. The different reaction models are given in the literature [10]. The apparent activation energy can be obtained by plotting the diagram between $\ln K_{app}$ and $\frac{1}{T}$ for different temperatures.

3   Results and Discussion

3.1   Characterization of Hematite Concentrates

The Hematite ore used in this study was collected from Iron ore mine, India. The lump ore was crushed using roll crusher and sieved to get the average particle size 56 μm. The results of XRF and XRD analysis are given in the Table 1 and Fig. 3 below. The microstructure of hematite ore is given in the Fig. 4.

Table 1. The chemical composition of Hematite ore (wt%)

Fe_2O_3	Al_2O_3	SiO_2	MnO	TiO_2
84.20	2.80	12.8	0.02	0.05

3.2   Effect of Flow Rate on the Reduction Rate

The isothermal reduction experiment was carried out with different flow rates 1, 1.5, 2, and 2.5 L per minute at 973K. The conversion vs time diagram was shown in Fig. 5 which shows that an increase in the flow rate increases the reduction rate, but the effect of the flow rate diminishes above 2 L per minute. It means that the concentration of hydrogen at the particle surfaces is as same as in the bulk.

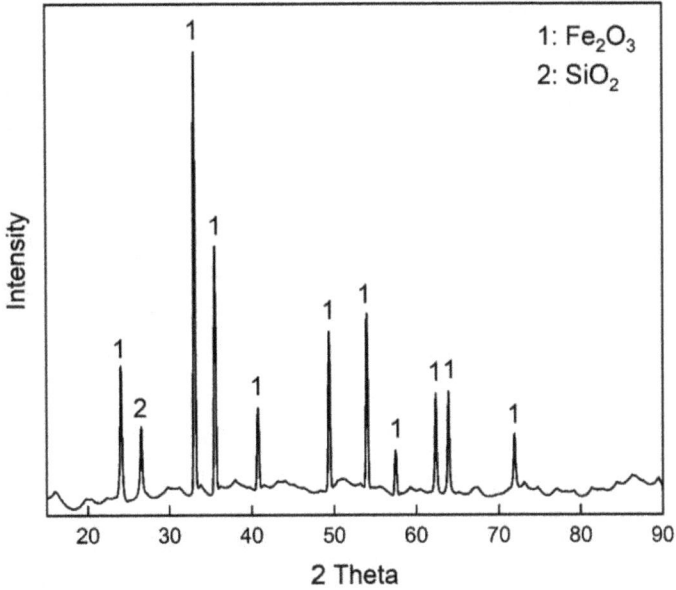

Fig. 3. XRD image of Hematite ore

Fig. 4. SEM image of hematite ore where 1 denotes 1- Fe_2O_3

3.3 Effect of Temperature

The isothermal reduction experiments were carried out at three different temperature 973,1073 and 1173K at a flow rate of 2.5 LPM. Figure 6 show the graph plotted between

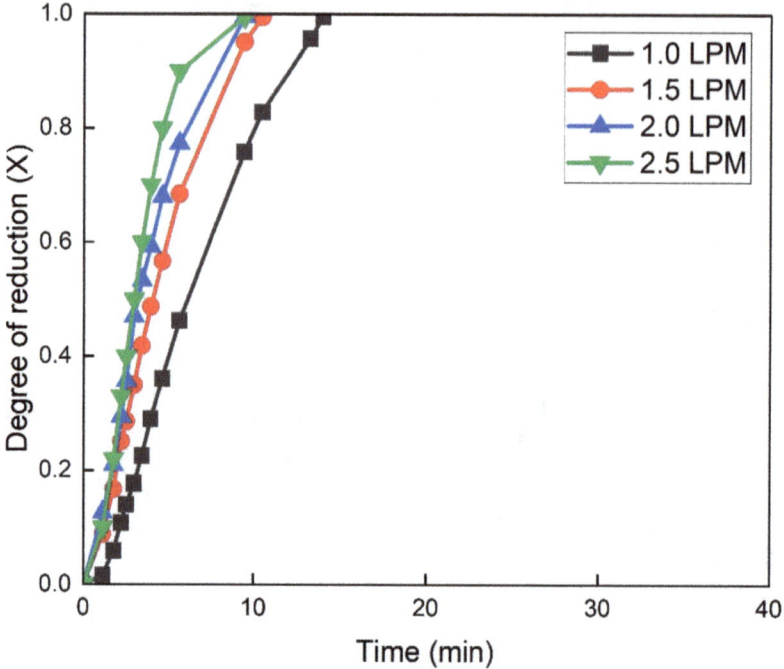

Fig. 5. The Degree of reduction (X) vs time (t) diagram for different flow rates at 973K.

conversion and time. It shows that the reduction rate increases with increase in the reduction temperature due to increase in the diffusivity.

3.4 Kinetics Analysis

The model fitting method was used by substituting different integral forms of reaction mechanisms. Figure 7(a) shows that chemical reaction mechanism was found to be the best-fitting model to describe the reaction kinetics of the global step ($Fe_2O_3 \rightarrow Fe$). The apparent activation energy and preexponential factor were determined using the slope and intercept of the linear curve mentioned in Fig. 7(b). The values of activation energy and preexponential factor were $20.4137 \pm .22$ kJ/mol and 0.7202 respectively.

3.5 Characterization of reduced hematite

Figure 8 presents the microstructure of reduced hematite at 973K. EDS analysis shows that the reduction of hematite increases the iron content in the particles while the gangue particle (SiO_2 and Al_2O_3) remains unreduced. The microstructure shows the pore holes on the surface of the reduced particle but the particle which contains high amount of silica and alumina does not show pore holes on the surface. It was also confirmed by XRD peaks which showed only peaks of Fe and SiO_2.

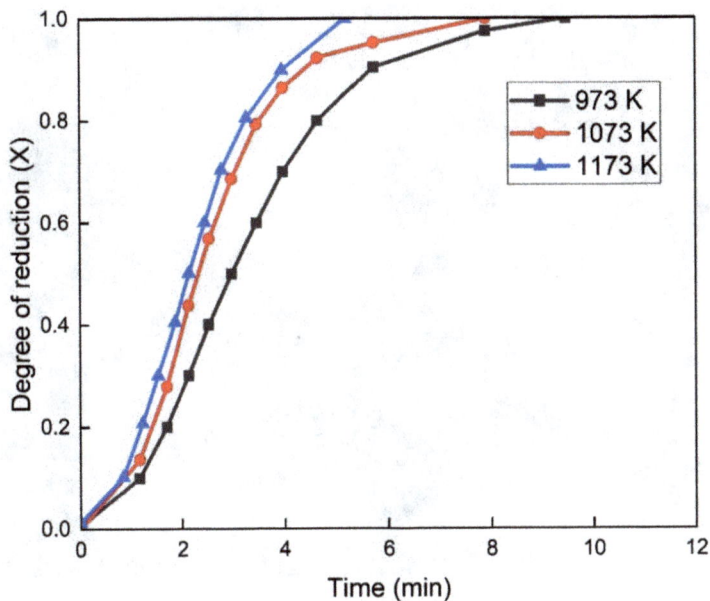

Fig. 6. The graph between the degree of reduction (X) vs time (t) for a different temperature 973, 1073 and 1173K.

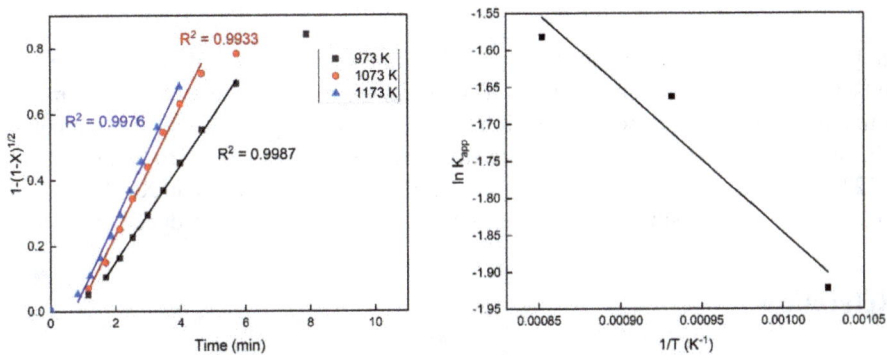

Fig. 7. The model fitting curve (a) and Arrhenius plot (b)

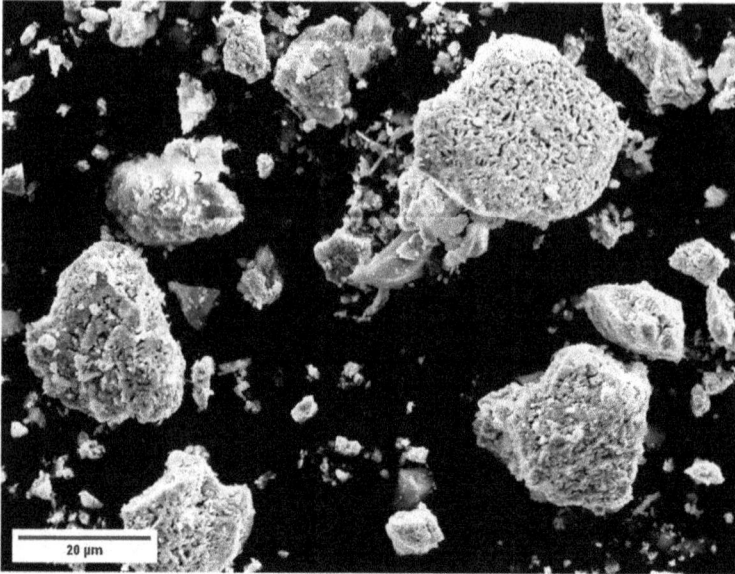

Fig. 8. SEM and EDS analysis of reduced hematite ore at 973K where different phases are 1-Fe, 2-SiO$_2$ and 3-Al$_2$O$_3$.

4 Conclusion

In this work, the effect of temperature and flow rate was investigated on the isothermal reduction of hematite ore using TGA. The kinetics analysis was done by using model fitting method. It was found that reaction rate is controlled by the chemical mechanism. The apparent activation energy and preexponential factor was found to be 20.4137 $\pm$ 0.22 kJ/mol and 0.7202 for the global step (Fe$_2$O$_3$ $\rightarrow$ Fe). The SEM and XRD was done to analyse the composition and morphology of hematite ore and its reduced products.

References

1. J. Kim, B.K. Sovacool, M. Bazilian, S. Griffiths, J. Lee, M. Yang, J. Lee, Decarbonizing the iron and steel industry: A systematic review of sociotechnical systems, technological innovations, and policy options, Energy Res Soc Sci. 89 (2022). https://doi.org/10.1016/j.erss.2022.102565
2. Mousa, E., Wang, C., Riesbeck, J., Larsson, M.: Biomass applications in iron and steel industry: An overview of challenges and opportunities. Renew. Sustain. Energy Rev. **65**, 1247–1266 (2016)
3. H. Ding, H. Zheng, X. Liang, L. Ren, Getting ready for carbon capture and storage in the iron and steel sector in China: Assessing the value of capture readiness, J Clean Prod. 244 (2020). https://doi.org/10.1016/j.jclepro.2019.118953
4. K. Rechberger, A. Spanlang, A. Sasiain Conde, H. Wolfmeir, C. Harris, Green Hydrogen-Based Direct Reduction for Low-Carbon Steelmaking, Steel Res Int. 91 (2020). https://doi.org/10.1002/srin.202000110

5. Hasanbeigi, A., Arens, M., Price, L.: Alternative emerging ironmaking technologies for energy-efficiency and carbon dioxide emissions reduction: A technical review. Renew. Sustain. Energy Rev. **33**, 645–658 (2014). https://doi.org/10.1016/j.rser.2014.02.031
6. D. Wagner, O. Devisme, F. Patisson, D. Ablitzer, A LABORATORY STUDY OF THE REDUCTION OF IRON OXIDES BY HYDROGEN, 2006
7. Pineau, A., Kanari, N., Gaballah, I.: Kinetics of reduction of iron oxides by H2. Part I: Low temperature reduction of hematite, Thermochim Acta. **447**, 89–100 (2006). https://doi.org/10.1016/j.tca.2005.10.004
8. D. Spreitzer, J. Schenk, Reduction of Iron Oxides with Hydrogen—A Review, Steel Res Int. 90 (2019). https://doi.org/10.1002/srin.201900108
9. Shimokawabe, M., Furuichi, R., Ishii, T.: Influence of the Preparation History of A-Fe 203 on its Reactivity for hydrogen reduction (1979)
10. Khawam, A., Flanagan, D.R.: Solid-state kinetic models: Basics and mathematical fundamentals. J. Phys. Chem. B **110**, 17315–17328 (2006). https://doi.org/10.1021/jp062746a

Influence of Silicon Source on the Steel Cleanness

Sanjay Pindar🄳 and Manish M. Pande$^{(\boxtimes)}$ 🄳

Department of Metallurgical Engineering and Materials Science, Indian Institute of Technology Bombay, Mumbai, Maharashtra 400076, India
manish.pande@iitb.ac.in

Abstract. Silicon is commonly used as a deoxidizer in the steel industry. The addition of silicon in the steel melt during refining is carried out in the form of ferrosilicon and commercial pure silicon. Ferrosilicon is widely used due to economic reasons. The non-metallic particles form when silicon is added into the steel melts, termed non-metallic inclusions. The impurities of the silicon source could lead to a change in the composition of the inclusion. In the present study, high-temperature experiments were carried out in the liquid steel at 1873 K at a laboratory scale. The effect of both silicon sources, i.e., high-purity silicon and ferrosilicon, on the steel cleanness were comparatively analysed. Pure silica inclusions were observed in the solidified steel samples at all silicon concentrations when high-purity silicon was added. However, apart from the silica, Si-Al-Ti-O complex inclusions were also observed in the case of the ferrosilicon addition, owing to its impurities. Thermodynamic analysis was carried out to support the experimental results.

Keywords: Silicon deoxidation · Ferrosilicon · Clean steel

1 Introduction

Silicon is a common deoxidizer in steelmaking practices. Even though much lower oxygen content can be achieved with a few ppm of aluminum [1], the inclusions formed due to aluminum deoxidation (alumina and spinel) are hard and non-deformable [2]. These inclusions are problematic to the steel processing (e.g., causing nozzle clogging during continuous casting) and harm the steel properties [3]. Therefore, silicon (or silicon-manganese) deoxidation is preferred in many steel grades. However, ferrosilicon is usually used in industries to meet the silicon requirement of the steel composition. Ferrosilicon contains many impurities, such as aluminum, and calcium, which have a significant affinity towards the oxygen contents [4].

A pure silica phase can be expected on the ferrosilicon addition; however, complex oxide inclusions could also form due to impurities in the ferrosilicon source. Therefore, the influence of the silicon source (ferrosilicon impurities) on the inclusion characteristics is of considerable interest [4–8]. The presence of the various intermetallic phases and inclusions in the ferrosilicon have been reported [4, 5, 8].

© The Author(s), under exclusive license to Springer Nature Singapore Pte Ltd. 2024
S. Patra et al. (Eds.): METCENT 2023, *Proceedings of the International Conference on Metallurgical Engineering and Centenary Celebration*, pp. 28–32, 2024.
https://doi.org/10.1007/978-981-99-6863-3_4

Mizuno et al. [8] investigated the influence of ferrosilicon impurities (aluminum and calcium) on the MnO-SiO$_2$ inclusions. The impurities present in the silicon source play a vital role in changing the inclusion scenario or steel cleanness, which eventually affects steel properties [4, 9]. A more careful consideration/control of impurities (of silicon source) is essential in steel grades requiring high silicon addition. Therefore, we conducted high-temperature experiments in liquid steel at 1873 K at a laboratory scale. The characteristics of the non-metallic inclusions were evaluated using SEM-EDS analysis. The effect of both silicon sources, i.e., high-purity silicon and ferrosilicon, on the steel cleanness were comparatively analysed.

2 Materials and Methods

High-purity electrolytic iron and ferrosilicon were commercially purchased. High-temperature experiments have been carried out in the vertical tube furnace under a purified argon atmosphere. Details of the experimental setup are reported in our previous study [3].

The predetermined amount of electrolytic iron and ferrosilicon (batch size: 90 g) was kept in the furnace using an alumina crucible. After initially flushing the purified argon for several hours, the sample was heated to 1873 K, held at that temperature for 120 min, and then cooled. The inert atmosphere of the furnace was maintained throughout the experiments using purified argon.

The sampling methodology was similar to our previous study [3]. The sample for inclusion composition analysis was taken from the central middle portion for inclusion characteristics. The characteristics of the inclusions were evaluated using scanning electron microscope energy dispersive spectroscopy (SEM-EDS, 15 kV) analysis.

3 Results and Discussion

Six high-temperature experiments were carried out to study the influence of silicon sources. Silicon was added through high-purity silicon [3] and ferrosilicon in the high-temperature experiments. The silicon ([Si]) concentration of the steel sample varied from 0.1–6 wt%.

Aluminum content ([Al]) in steel samples for high-purity silicon addition was less than 0.01 ppm [3]. On the other hand, the aluminum and titanium impurities were inherited from the ferrosilicon alloy to steel melt. Therefore, the concentration of [Al] in the steel sample from ferrosilicon addition experiments was up to 60 ppm, whereas [Ti] concentration was up to 138 ppm. Aluminum and titanium in the steel sample indicate that complex inclusions could form.

SiO$_2$, mullite, and alumina can form in the steel melt depending on the aluminum and silicon concentrations. The titanium addition in the steel melt can result in phases like Ti$_3$O$_5$ and ilmenite. The oxide stability diagram can predict the thermodynamic stability of the different phases. FactSage8.1 was used to construct the oxide stability diagram for the Fe-Si-Al-O and Fe-Si-Ti-O systems, as shown in Fig. 1(a) and (b), respectively. Thermodynamic databases used for these calculations are FactPS, FTOxid, and FTmisc.

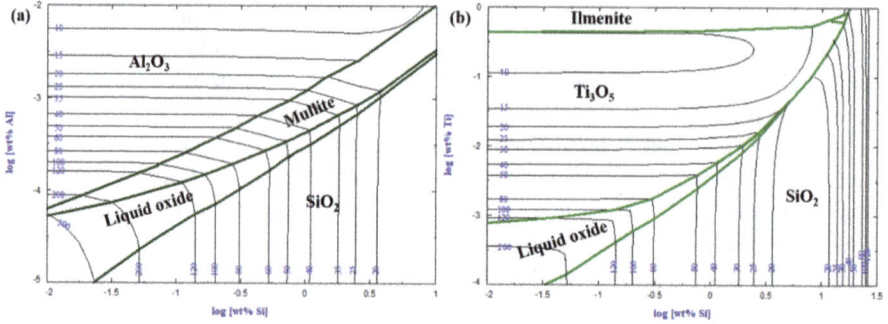

Fig. 1. Oxide stability diagram for (a) Fe-Si-Al-O and (b) Fe-Si-Ti-O system, constructed using FactSage8.1 (Color figure online)

The green line indicates the boundaries between the phases. The black lines with the numbers are iso-oxygen lines (ppm).

All the inclusions were pure silica (SiO_2) at all silicon concentrations in case of the high-purity silicon addition. A typical silica inclusion observed for high-purity silicon addition is shown in Fig. 2(a).

At lower silicon concentrations, the amount of the inherited impurities was negligible, and their effect was insignificant on the inclusion composition. Thus, only pure silica inclusions were observed. Pure silica was observed almost at all silicon concentrations in the case of the ferrosilicon addition. A typical silica inclusion in the case of ferrosilicon addition is shown in Fig. 2(b).

Most of the SiO_2 inclusions were round. Few inclusions with an irregular morphology were also observed at higher silicon concentrations in both cases, i.e., high-purity silicon and ferrosilicon addition. The melting point of silica (1996 K) and steelmaking temperature (1873 K) are closer, so the expected morphology for the silica particles is mostly spherical. The growth of silica particles during cooling probably results in a non-spherical morphology [3].

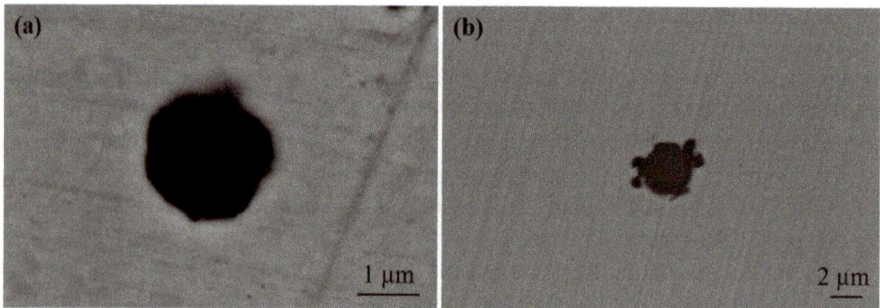

Fig. 2. Typical pure silica inclusions in solidified steel for (a) high-purity silicon and (b) ferrosilicon addition

These silica non-metallic particles are inevitable to form due to the supersaturation of the steel melt. The formation of silica inclusion can be expressed as per Eq. (1). The element in the square bracket represents the dissolved element in the steel melt, and a similar notation is used throughout the manuscript.

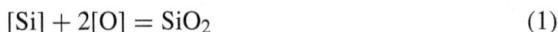

$$[Si] + 2[O] = SiO_2 \tag{1}$$

Furthermore, complex oxide inclusions were also observed at higher silicon concentrations apart from the pure silica inclusions. Typical Si-Al-Ti-O complex inclusions observed in the solidified steel sample obtained from the ferrosilicon addition experiments are shown in Fig. 3(a-b).

On increasing the silicon concentration, the impurities such as aluminum and titanium, which have a good affinity towards oxygen, were also inherited to the steel melt. The dissolved aluminum and titanium can react with the oxygen present in the steel melt when (local) supersaturation is available, forming Al_2O_3 and TiO_x as per Eqs. (2) and (3), respectively [10].

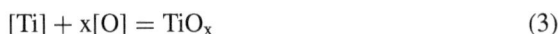

$$2[Al] + 3[O] = Al_2O_3 \tag{2}$$

$$[Ti] + x[O] = TiO_x \tag{3}$$

The collision of these Al_2O_3 and TiO_x, and SiO_2 inclusions can lead to complex inclusions. The joining of alumina and silica will result in mullite formation. Dual-phase inclusions were also observed in SEM-EDS analysis, as shown in Fig. 3(b). These inclusions have two regions one is silicon-rich inclusion, and the other is aluminum-rich inclusions. The silica inclusion will form immediately due to plentiful available saturation. Then, the presence of aluminum in the steel melt could reduce these silica inclusions, resulting in the formation of complex Si-Al-O inclusions. The reduction of SiO_2 by [Al] can be represented by Eq. (4).

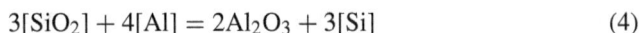

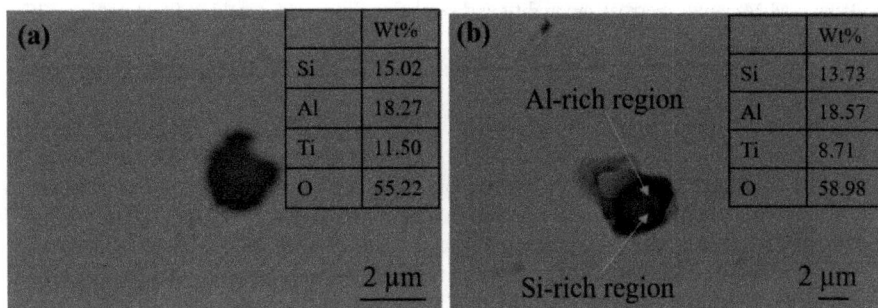

$$3[SiO_2] + 4[Al] = 2Al_2O_3 + 3[Si] \tag{4}$$

(a)

	Wt%
Si	15.02
Al	18.27
Ti	11.50
O	55.22

2 µm

(b) Al-rich region

	Wt%
Si	13.73
Al	18.57
Ti	8.71
O	58.98

Si-rich region   2 µm

Fig. 3. (a-b) Typical Si-Al-Ti-O complex inclusions formed during ferrosilicon addition

The commercial ferrosilicon alloys thus have sufficient trace impurities to change the composition of the inclusions. Therefore, carefully considering or controlling the impurities is strictly required to produce high-quality steel.

4 Conclusions

The influence of ferrosilicon on the inclusion characteristics was investigated in the present study. The effect of both silicon sources, i.e., high-purity silicon and ferrosilicon, on the steel cleanness were comparatively analyzed. Based on the present study, it can be deduced that the silicon source has a considerable effect on the composition of the inclusions. The pure silica inclusions were formed on the high-purity silicon addition. However, Si-Al-Ti-O complex inclusion was also observed apart from the pure silica inclusions. The aluminum and titanium inherited from the ferrosilicon can directly react with the dissolved oxygen and/or interact with existing inclusions, resulting in complex inclusions.

Acknowledgments. The financial support for preparing this manuscript has been provided by (1) the Industrial Research and Consultancy Center (IRCC) IIT Bombay, Mumbai (project no. RD/0518-IRCCSH0-011), and (2) the Science and Engineering Research Board (SERB), Department of Science and Technology, Government of India (project no. CRG/2019/000086) (3) Centre of Excellence in Steel Technology (CoEST), IIT Bombay. The authors would like to thank Ms. Neelam Bhatkar for her help during sample polishing. Thanks to Mr. Amit Joshi for the help during the experimental setup.

References

1. Ishfaq, M., Pande, M.M.: Application and assessment of aluminium deoxidation equilibria in liquid steel using various formalisms based on some statistical thermodynamic models. Ironmaking Steelmaking **50**(1), 44–54 (2023)
2. Silva, A.L.V.D.C.E.: Non-metallic inclusions in steels – origin and control. J. Market. Res. **7**(3), 283–299 (2018)
3. Pindar, S., Pande, M.M.: Assessment of Si–O equilibria and non-metallic inclusion characteristics in high silicon steels. Steel Res. Int. (2023). https://doi.org/10.1002/srin.202300115
4. Pande, M.M., et al.: Ferroalloy quality and steel cleanliness. Ironmaking Steelmaking **37**(7), 502–511 (2010)
5. Wijk, O., Brabie, V.: The purity of ferrosilicon and its influence on inclusion cleanliness of steel. ISIJ Int. **36**, S132–S135 (1996)
6. Li, M., Li, S., Ren, Y., Yang, W., Zhang, L.: Modification of inclusions in linepipe steels by Ca-containing ferrosilicon during ladle refining. Ironmaking Steelmaking **47**(1), 6–12 (2020)
7. Park, J.H., Kang, Y.B.: Effect of ferrosilicon addition on the composition of inclusions in 16Cr-14Ni-Si stainless steel melts. Metall. Mater. Trans. B **37**(5), 791–797 (2006)
8. Mizuno, K., Todoroki, H., Noda, M., Tohge, T.: Effect of Al and Ca in ferrosilicon alloy for deoxidation on inclusion composition in type 304 stainless steel. Iron Steelmaker (USA) **28**(8), 93–101 (2001)
9. Pande, M.M., Guo, M., Devisscher, S., Blanpain, B.: Influence of ferroalloy impurities and ferroalloy addition sequence on ultra-low carbon (ULC) steel cleanliness after RH treatment. Ironmaking Steelmaking **39**(7), 519–529 (2012)
10. Pindar, S., Pande, M.M.: Formation mechanism, evolution and modification of non-metallic inclusions in Al-killed, Ti-alloyed steel melt: a review. Trans. Ind. Inst. Met. (2023). https://doi.org/10.1007/s12666-023-02947-9

Utilization of Biomass Pellets in the Iron Ore Sintering Process

Dhanraj Patil[1]([✉]), Ashwin Appala[2], Rameshwar Sah[1], and Ganesh Shetty[2]

[1] Research and Development, JSW Steel Ltd., Bellary, Karnataka 583275, India
dhanraj.patil@jsw.in
[2] Sinter Plant, JSW Steel Ltd., Bellary, Karnataka 583275, India

Abstract. The Iron and Steel industry plays a vital role in global economic growth, consuming a significant portion of the annual industrial energy, primarily sourced from fossil fuels. Sinter making is a crucial process in Ironmaking, having 10% energy consumption share of the entire Steel industry with 78% of it coming from coke breeze thus contributing significantly to greenhouse gases, SOx and NOx emissions. Substituting fossil fuels with biomass, a clean and renewable energy source, is an attractive option for carbon-neutral iron ore sintering. In the present study, an attempt was made to utilize biomass pellets (called Biopellet) for sintering process by varying it from 0 to 4% of the total raw mix blend as a solid fuel substitution. The resulting sinter was evaluated based on productivity, product yield, sintering time, and tumbler index, while the exhaust gas emissions were analysed for NOx and SOx, emissions. With an increase in biopellet the sintering time increased from 25.30 to 31 min, with a drop in sinter productivity from 1.78 to 1.09 $t/m^2/hr$. The calorific values of biopellet, was around 66% of coke breeze. The maximum sintering temperature reduced due to evaporation of volatile matter from the biopellet ahead of the flame front. The increased biopellet proportion reduced the sinter product yield from 82 to 70%, owing to lower heat input from the biopellet. The strength of the sinter however was more or less maintained at around 65% for all the trials. The presence of biopellet affected the exhaust gas composition, with a reduction in overall NOx and SO_2 emissions. Despite the trade-offs, such as decreased sinter yield, biomass pellets are an attractive alternative for achieving sustainable steel production due to their ability to reduce emissions.

Keywords: Iron Ore sintering · Biopellet · Emissions · Fossil fuels · Sustainability

1 Introduction

Global climate change has initiated huge attempts to reduce carbon dioxide emissions. The European Union is leading efforts to combat climate change by developing policies for a low carbon economy. Their goal is to reduce greenhouse gas emissions to 80% below 1990 levels by 2050, with intermediate targets of 40% cuts by 2030 and 60% by 2040 [1].

© The Author(s), under exclusive license to Springer Nature Singapore Pte Ltd. 2024
S. Patra et al. (Eds.): METCENT 2023, *Proceedings of the International Conference on Metallurgical Engineering and Centenary Celebration*, pp. 33–44, 2024.
https://doi.org/10.1007/978-981-99-6863-3_5

The iron and steel industry is crucial for global growth and economy but is also a major consumer of energy and carbon. It consumes around 20% of the annual industrial energy, mainly from fossil fuels, and contributes significantly to global CO_2 emissions. Approximately 6.7% of total world CO_2 emissions come from the iron and steel sector, according to the International Energy Agency [2]. It heavily relies on fossil fuels mainly coke for heating, melting, and reducing iron ores. Sinter making is a crucial process in Ironmaking, and its energy consumption represents around 10%of the entire industry. It is the second largest energy-consuming process in the steel industry with 78% of energy coming from coke breeze [3]. The combustion of coke breeze produces heat, which helps agglomerate iron ore granules for the blast furnace. These emissions, including NOx, SOx, dioxins, contribute to 50% of the total emissions from steel industry [4].

Environmental regulations have imposed limitations on the use of coal/coke in sintering due to its negative environmental impact and high greenhouse gas emissions. Reducing emissions of CO_2, NOx, and SOx is essential for environmental preservation and addressing global concerns [5]. The search for alternative fuel sources has become crucial to ensure future energy security, given the environmental damage caused by traditional energy sources and their diminishing availability. The adoption of innovative technologies depends on factors like fuel and energy costs, resource availability, and carbon pricing [1].

The use of biomass in iron and steelmaking is seen as an economically and technically viable solution to reduce CO_2 emissions from fossil fuels. Biomass offers several advantages over coal/coke, including its abundant availability in nature, lower greenhouse gas emissions, renewability, carbon neutrality, low sulfur content, low ash content, high reactivity, high specific surface area, and stable pore structure. Biomass is also advantageous due to its widespread availability, renewable nature, reliable ignition, lower pollutant emissions, and lower production costs when compared to coal/coke [2, 4].

Several studies have explored the potential use of biomass pellets in the iron ore sintering process. Zandi et al. [6] suggested that it is possible to replace up to 10% of coke breeze with biomass to produce blast furnace grade sinters. Zhang et al. [7] conducted an investigation where they co-utilized 5% biomass with coke, resulting in a reduction in sintering indices. According to Cheng et al. [8] higher combustion rate of Biomass shortens the holding time at high temperature, and low combustion efficiency resulting in weaker sinter.

The current study aims to examine the feasibility of using biomass pellets (called Biopellet) as a substitute fuel in the iron ore sintering process. Five sets of lab-scale sinter pot trials were conducted, varying the proportion of biopellet from 0 to 4% in the sinter raw mix. The study focused on evaluating the impact of different biopellet dosages on sintering process parameters, as well as the physical and metallurgical properties of the resulting sinter.

2   Materials and Methods

2.1   Material Selection for Sintering Process

For the current work, 5 sets of sinter pot trials were planned wherein the biopellet were varied from 0 to 4%. Table 1 shows the mix proportion of different raw materials used for the trials.

Table 1. Raw material mix proportion.

Raw material, %	Exp-1	Exp-2	Exp-3	Exp-4	Exp-5
Iron Ore Blend	48.20	47.82	47.43	47.43	47.05
Blast Furnace Return Fines	12.50	12.40	12.30	12.20	12.10
Imported Limestone	5.00	4.96	4.92	4.88	4.84
Local Dolomite	6.00	5.95	5.90	5.86	5.81
Micro Pellets	3.70	3.67	3.64	3.61	3.58
Anthracite Coal	2.00	1.88	1.76	1.65	1.53
Coke Breeze	2.00	1.88	1.76	1.65	1.53
Internal Return Fines	16.00	15.87	15.75	15.62	15.49
Calcined lime	2.60	2.58	2.56	2.54	2.52
KR Slag	2.00	1.98	1.97	1.95	1.94
Biopellet	0	1	2	3	4
Total	100.00	100.00	100.00	100.00	100.00

Biopellets were grinded to below 3 mm size to be used as a fuel substitution for sintering process. Table 2 shows the proximate and ultimate analysis along with the gross calorific value (GCV) for the fuels consumed during the trials. A large proportion of the biopellet is comprised of Volatile matter (VM), which is approximately 3 times more than that in the coke breeze. It is also visible that the Nitrogen (N), Sulphur (S) contain and the GCV for the biopellet is much lower compared to other solid fuels used for the trials. For all the trials the fixed carbon intake through the fuels was kept between 3–3.5%.

Table 2. Ultimate and Proximate analysis of the fuels consumed.

Sample	Coke breeze	Anthracite coal	Biopellet
Moisture, %	1.86	5.78	6.69
Ash, %	4.23	14.74	16.08
VM, %	23.29	4.66	65.52
Fixed Carbon, %	72.48	80.6	18.4
C, %	73.2	81.1	38.8
N, %	1.9	1.55	0.55
H, %	0.59	1.54	4.72
S, %	0.52	0.3	0.17
GCV (cal/gm)	6794	7552	4475

2.2 Sintering Conditions

Sinter pot trials consists of two parts as shown in Fig. 1, first is preparation of green mix with an optimum moisture levels and second is conducting the actual pot trials using the green mix.

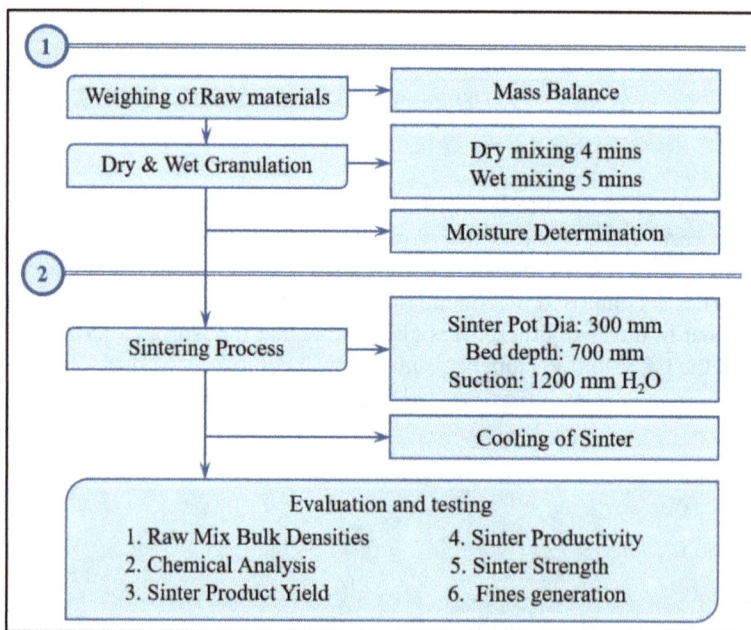

Fig. 1. Sintering pot trials workflow.

Preparation of Green Mix

First, a blend of required raw materials (Table 1) is prepared and transferred to a granulation drum. The total granulation duration is 9 min with final moisture levels maintained at around 8%. During granulation, fine particles adhere to each other or to the coarser particles to form grains of appropriate size [9]. Table 3 shows the specification for the granulation drum used in the granulation process.

Table 3. Specification for the Granulation drum.

Drum diameter, m	1.00
Drum length, m	0.50
Rotation speed, rpm	15

Sintering Pot Trials

Experiments used downsized suction lab scale sintering machine with cylindrical bed, suction fan, exhaust pipe, and burning hood. Figure 2 shows the layout for the lab scale sinter pot setup. Thermocouples inserted at 4 locations within the sinter bed monitored temperatures and flame front movement. Granulated mixture is ignited for 2 min and burned downward with running suction fans. Table 4 shows the test conditions that were applied for all the trials. At the end of experiments, sinter is produced as a result of fusion of iron ore, limestone and coke.

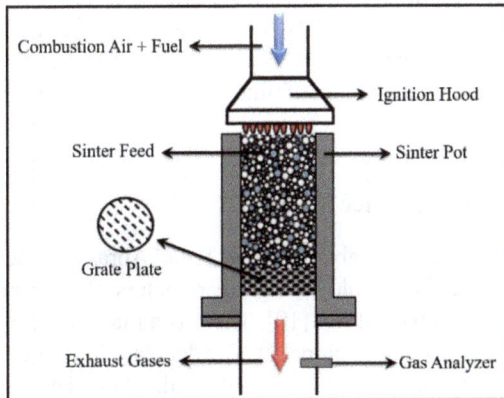

Fig. 2. Lab sinter pot layout

The sinter was studied for its different process parameter along with its metallurgical and physical properties including sinter strength [10], product yield, fine generation, sintering time, productivity and exhaust gas emissions.

Table 4. Sintering parameters for all the trials.

Ignition time	120 s
Air suction (during ignition)	400 mm of H_2O
Air suction (after ignition)	1200 mm of H_2O
Total bed height	700 mm
Hearth layer	60 mm
Ignition flame temp	1300 °C
Sinter pot diameter	300 mm

3   Results and Discussions

3.1   Impact of Biopellet on Sintering Temperature Profiles

Figure 3 shows the temperature profile observed during iron ore sintering process at different proportions of biopellet. With an increase in biopellet the maximum temperature within the sinter bed reduces. In the presence of biomass, the water content required to granulate the sinter raw mix to a moisture levels of 8% increased from 8 to 12 L which can be a reason for the drop in maximum temperature within the sinter bed [7]. The calorific values of biopellet is around 66% of coke breeze which can be another reason for drop in peak temperatures within the sinter bed. A large proportion of biopellet comprises of volatile matter which are burned-out at a relatively lower temperatures, which reduces the potential of biopellet to increase the temperature of the sinter bed above 1100 °C. This also decreases the holding time at high temperature, and lowers the combustion efficiency which reduces the overall heat release [6, 11]. The drop in peak sintering temperatures would in turn result in lower quality of sinter since the temperatures are not high enough for a sufficient amount of time to form the require melt within the sinter bed [12].

3.2   Impact of Biopellet on Sinter Strength

Figure 4 shows the variation in Tumbler Index (TI) and Abrasive Index (AI) of the sinter with the use of biopellet. These cold strength parameters of the sinter are determined by the Tumbler test (TI) IS 6495:1984 [10]. There is a marginal drop in sinter TI from 66.27 to 63.73% with an increase in biopellet. The decrease in strength can be associated with the lower peak temperature due to lower fuel utilization. The volatiles in solid fuel thermally decompose at low temperature. The amount of volatile matters going in the sinter bed increases with an increase in biopellet. These volatile matters are partially oxidized in the downstream of combustion zone. Their low oxidization temperature is not sufficient for the iron ore melting. Sintering quality is highly dependent on the amount of melt phase which begins to form at a high temperature of 1100 °C. With most of the volatile matters burning at lower temperatures there is drop in maximum sintering temperature associated with the sinter bed which hamper the quality and quantity of

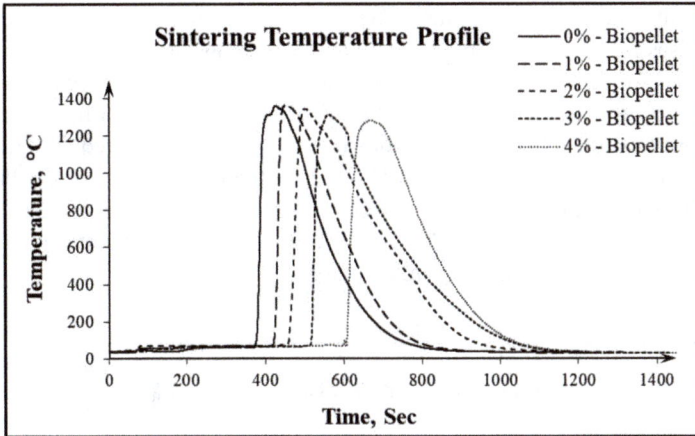

Fig. 3. Impact of biopellet on sintering bed temperature profiles at the middle portion.

melt being formed which adversely affects the quality of the sinter being produced thus reducing its strength [8, 13].

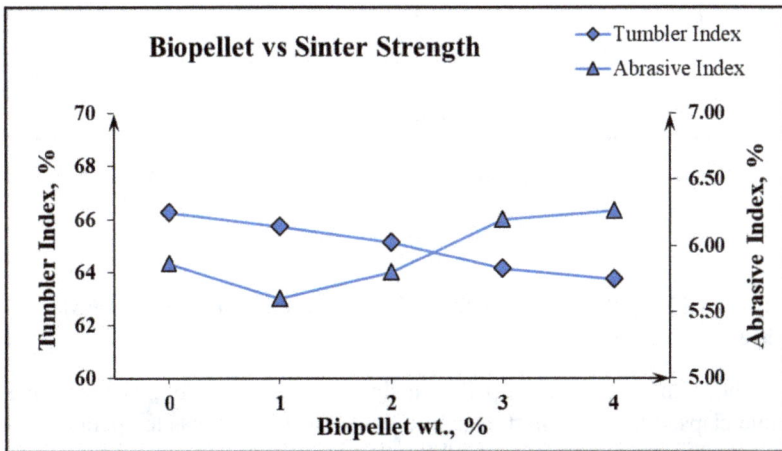

Fig. 4. Impact of biopellet on sinter strength.

3.3 Impact of Biopellet on Sinter Product Yield (+5 mm) and Fine Generation (−5 mm)

Figure 5 shows the impact of biopellet on Sinter Product yield (+5 mm) and fines generation (−5 mm). The sinter product yield and fine generation are defined as the fraction of sinter produced with +5 mm and −5 mm size fractions respectively. The level of return fines produced within the sinter is considered to be the most direct indicator of sinter quality and strength parameter. With an increase in biopellet the sinter product yield decreases from 81.58 to 70.38% with an increase in fines generation from 18.42 to 29.62%. The lower holding time at high temperature, and low combustion efficiency of the biopellet reduces the overall the heat release during the sintering process. Thus, the temperatures and heat released are not adequate to produce the required quantity of melt to bind the sinter properly [8, 11]. This can be the potential reason for the overall drop sinter product yield and increase in sinter return fines with an increase in biopellet. However further microstructure studies needs to be conducted to understand the impact of biopellet on the bonding mechanism of sinter.

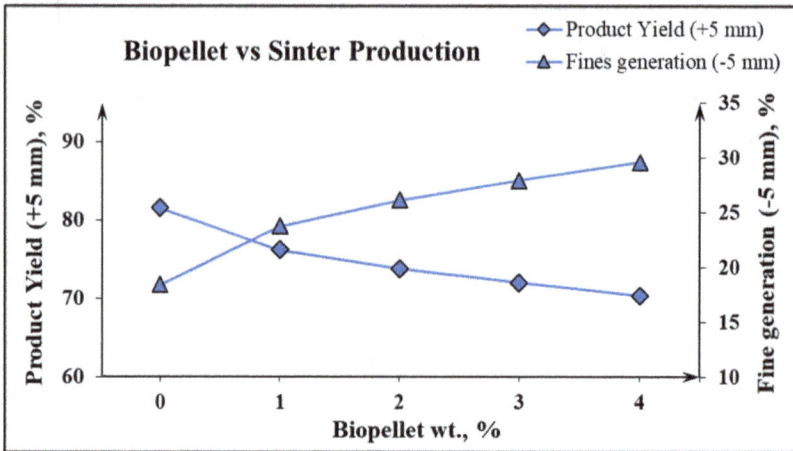

Fig. 5. Impact of biopellet on sinter product yield and fine generation.

3.4 Impact of Biopellet on Sintering Time, Flame Front Speed (FFS) and Productivity

Figure 6 shows the impact of biopellet on sintering time and FFS. The sintering time is the time elapsed from the start of ignition until the exhaust gas temperature reached a maximum. The frame front speed (FFS) can be defined as the ratio of the sinter bed height to the duration it takes for the burning zone to reach the lowest point of the sinter bed after ignition. With the increase in use of biopellets upto 4% the sintering time increases from 25.30 to 31 min.

As already mentioned, the presence of biopellet led to an increase in water addition within the sinter green mix to get the required moisture levels. Due to the absence of pre-heat treatment, biopellet led to an excessive water absorption thus dropping the sinter bed permeability. This in turn resulted in increased bed resistance, which reduced vertical sintering speed from 32.21 to 23.86 mm/min, and increased the overall sintering time [7].

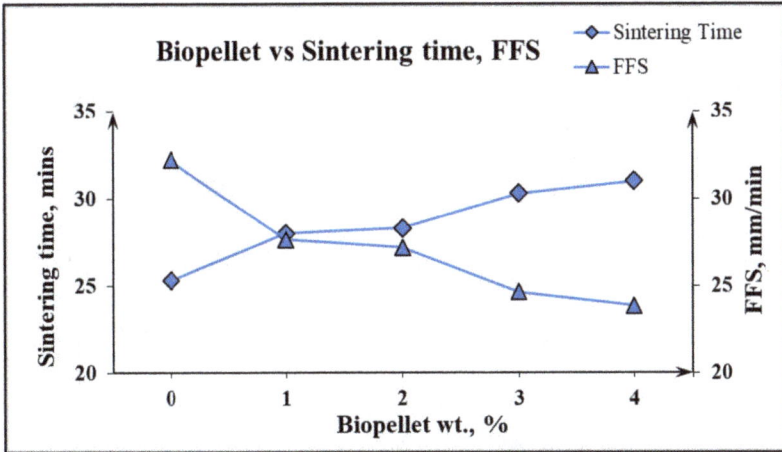

Fig. 6. Impact of biopellet on sintering time and FFS.

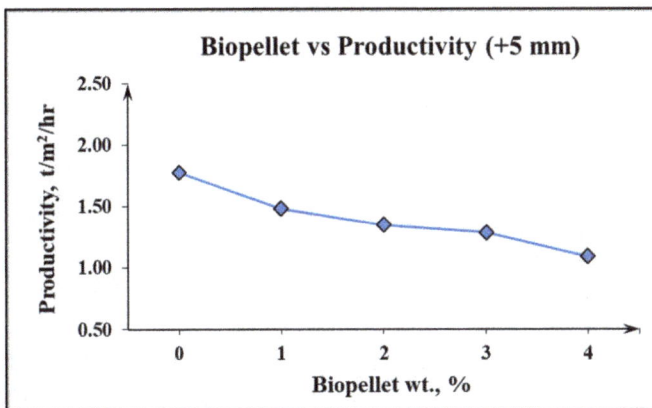

Fig. 7. Impact of biopellet on sinter productivity.

Figure 7 shows the impact of biopellet on sinter productivity. The sinter productivity is defined as the sinter product yield (+5 mm) per unit sintering time per unit area of the sinter pot. The increase in biopellet proportion is associated with a drop in sinter productivity from 1.62 to 1.45 $t/m_2/hr$. The increasing sintering time and drop in sinter product yield reduces the sinter productivity.

3.5 Impact of Biopellet on Exhaust Flue Gases

Figures 8 and 9 shows the concentration of NO_X and SO_2 in the exhaust gases emissions during sintering trials. It is found that NO_X and SO_2 emissions decreases with increasing use of biopellet. Owing to the lower share of S and N in the biopellet (Table 2), the flue gases coming from the sintering process has a lower concentration of SO_2 and NO_X [13].

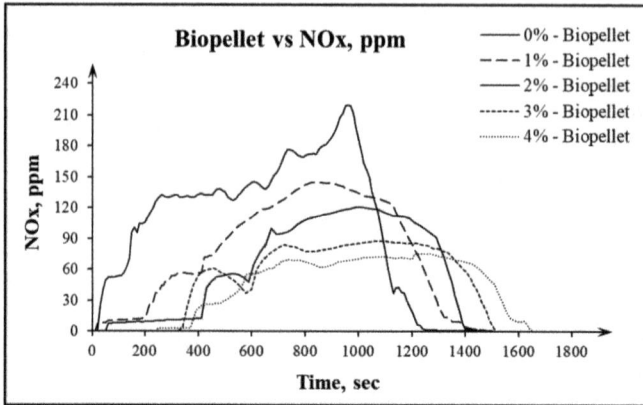

Fig. 8. Impact of biopellet substitution on NO_X emission in the sintering flue gas.

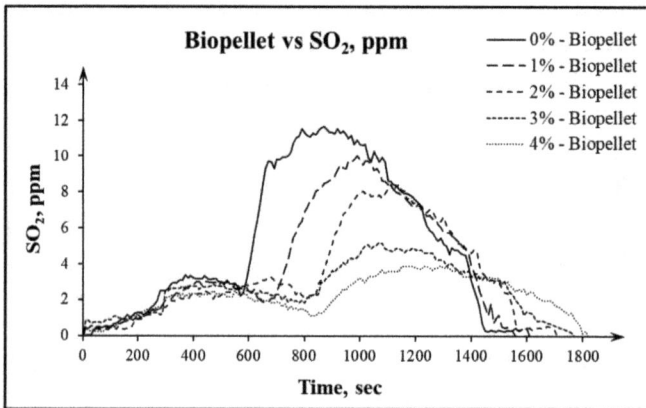

Fig. 9. Impact of biopellet substitution on SO_2 emission in the sintering flue gas.

In the whole reaction system for the sintering process, the Sulphur (S) in iron ore accounts for about 75%, whereas that in the fuel accounts for 25% [3]. When the substitution ratio of the biomass increases, the S content in the whole fuel feedstock decreases, and thus the generation of SOx also decreases. Thus, the substitution of solid fuel with biopellet shows a significant effect on reducing SO_2 concentration. According to the Zeldovich theory [3], there are three types of NO_X: thermal NO_X, prompt NO_X and fuel NO_X. Among them, thermal NOx is just produced in minute quantities when the temperature is below 1500 °C, and the prompt NOx relating to the air excess coefficient is also produced in minute quantities at almost any temperature [14]. So, the majority of NO_X produced in the sintering process is fuel NO_X that comes from the burning of the fuel. With an increase in substitution of biopellet, the N content in the whole fuel feedstock decreases, which in turn lowers the NO_X generation.

4 Conclusions

The utilization of biomass in the iron and steel industry remains restricted and faces significant competition from fossil fuels. The incorporation of biomass in this sector presents challenges that encompass both technical and economic aspects. The low energy density values, varying composition, low calorific values, high moisture, and volatile contents are some of the major concerns related to biopellet utilization within the industry. With the use of biopellet the peak temperature achieved in the bed is lower than that with the use conventional fossil fuels alone. With the use of biopellets upto 4% the sintering time increases from 25.30 to 31 min, with a drop in productivity from 1.62 to 1.45 t/m^2/hr. The product yield is also found to be decreasing from 81.58 to 70.38% with an increase in fines from 18.42 to 29.62%. However, Biomass brings benefits by reducing the SOx and NOx emission.

Looking at the overall trend of the outcomes, even though there are some positives in terms of emissions during sintering process, substitution of biopellet i.e. raw biomass only on the carbon basis alone will not be sufficient to achieve the required quality of sinter. Recently, there are several pre-process that are being developed to make the use of raw biomass fuel, making it more compatible with the process for which it is being used. Despite being in its early stages, the research field for biomass-based reducing agents is rapidly evolving. In order to utilize the biomass fuels to its full potential, there is a need to develop an economic competitiveness of biomass-based reducing agents compared to fossil-based reducing agents.

References

1. Suopajarvia, H., et al.: Use of biomass in integrated steelmaking – status quo, future needs and comparison to other low-CO2 steel production technologies. Appl. Energy **213**, 384–407 (2018)
2. Mousa, E., Wang, C., Riesbeck, J., Larsson, M.: Biomass applications in iron and steel industry: an overview of challenges and opportunities. Renewable Sustainable Energy Rev. **65**, 1247–1266 (2016)
3. Zhou, M., et al.: Thermodynamic analysis of iron ore sintering process based on biomass carbon. Energies **13**(22), 5988 (2020)
4. Jha, G., Soren, S., Deo Mehta, K.: Partial substitution of coke breeze with biomass and charcoal in metallurgical sintering. Fuel **278**, 118350 (2020). https://doi.org/10.1016/j.fuel.2020.118350
5. Kawaguchi, T., Hara, M.: Utilization of biomass for iron ore sintering. ISIJ Int. **53**(9), 1599–1606 (2013). https://doi.org/10.2355/isijinternational.53.1599
6. Zandi, M., Martinez-Pacheco, M., Fray, T.A.T.: Biomass for iron ore sintering. Miner. Eng. **23**(14), 1139–1145 (2010). https://doi.org/10.1016/j.mineng.2010.07.010
7. Zhang, J., He, Z., Jin, Y.: Utilisation of biomass fuel in sintering process. Mater. Res. Innov. **19**(5), 1140–1143 (2015). https://doi.org/10.1179/1432891714Z.0000000001265
8. Cheng, Z., Yang, J., Zhou, L., Liu, Y., Wang, Q.: Characteristics of charcoal combustion and its effects on iron-ore sintering performance. Appl. Energy **161**, 364–374 (2016). https://doi.org/10.1016/j.apenergy.2015.09.095
9. Gan, M., et al.: Optimising method for improving granulation effectiveness of iron ore sintering mixture. Ironmaking Steelmaking **42**(5), 351–357 (2015)

10. Indian Standard 6495:1984: Method of Tumbler Test for Iron Oxides: Lump Ores, Sinter and Pellet. Indian Standard Institution. IS: 6495-1984, pp. 1–8 (1984)
11. Lovel, R.R., Vining, K.R., Dell 'Amico, M.: The influence of fuel reactivity on iron ore sintering. ISIJ Int. **49**(2), 195–202 (2009). https://doi.org/10.2355/isijinternational.49.195
12. Ooi, T.C., et al.: The study of sunflower seed husks as a fuel in the iron ore sintering process. Miner. Eng. **21**(2), 167–177 (2008). https://doi.org/10.1016/j.mineng.2007.09.005
13. Suopajärvi, H., Kemppainen, A., Haapakangas, J., Fabritius, T.: Extensive review of the opportunities to use biomass-based fuels in iron and steelmaking processes. J. Clean. Prod. **148**, 709–734 (2017). https://doi.org/10.1016/j.jclepro.2017.02.029
14. Schwerdt, C.: Modelling NO_X-Formation in Combustion Processes, pp. 1–42. Department of Automatic Control, Lund University (2006)

Carbothermic Reduction and Kinetics of a Lean Grade Multimetallic Magnetite Ore

Biswajit Mishra[✉], Amit Kumar Singh, and Girija Shankar Mahobia

Department of Metallurgical Engineering, Indian Institute of Technology (BHU), Varanasi, India
biswajitmishra.rs.met19@itbhu.ac.in

Abstract. Reducibility and reduction kinetics of hardened pellets prepared from a novel, lean-grade multimetallic magnetite ore (MMO) was investigated in the present study. This lean-grade ore contains a significant amount of chromium and nickel, which can lead to a reduction in the cost of steelmaking. The hardened pellets were reduced at 950–1150 °C for 30–120 min in a bed of high ash coke. The optimum reducibility and metallisation of 76% and 85% were obtained at 1100 °C at a holding time of 120 min. The kinetic model "$[(1-\alpha)^{-1/3}-1]^2 = kt$" was found to be the best fit for the reduction of pellets. The activation energy calculated for the pellets was 220.19 kJ/mole. Pellets reduced with high ash coke fines at 1100 °C for 120 min and above are thus found necessary to achieve the industrially desired metallization of greater than 85% .

Keywords: Direct Reduction · Lean grade ore · Multi-metallic Magnetite ore · Reduction kinetics

1 Introduction

Both hematite and magnetite are sometimes associated with ores of other precious metals like titanium, chromium, nickel, vanadium, etc. Such ores, called multimetallic ores (MO), are very important in iron and steelmaking. External additions of alloying elements, like ferrochrome, ferronickel, etc., would be decreased as some of the elements required will already be present in the produced melt from the ores. The reduction behavior of Cr-V-Ti-magnetite ores has been studied extensively, and appreciable reduction was achieved with H2-CO-CO2 (%R = 95) and methane gas [1, 2]. Magnetite in pre-oxidized vanadium-titanium magnetite was 95% reduced by coal at 1050 for 60 min in another study by Liu et al. [3]. An MO found at Pokhpur in the Kiphere district of Nagaland (India) is of particular interest as it contains significant amounts of chromium (about 4%) and nickel (0.7–0.8%) [4]. About 97–98% Fe, 91–93% Cr, 95–96% Ni, 89–91% Co, 16–22% Si, and 27–34% Mn in the Nagaland multimetallic magnetite ore (MMO) was extracted in 1650 °C [5]. MMO was 80% reduced at 1200 °C with low ash coke [6, 7]. Mining the MMO will lead to the generation of fines and needs to be appropriately utilized. Pelletization is one of the many methods available wherein fines

S. Patra et al. (Eds.): METCENT 2023, *Proceedings of the International Conference on Metallurgical Engineering and Centenary Celebration*, pp. 45–51, 2024.
https://doi.org/10.1007/978-981-99-6863-3_6

can be effectively utilized to prepare input material for SR or DRI-EAF processes. During the induration of the pellets, the magnetite usually transforms into hematite which leads to the enhancement of reduction of magnetite [8]. The studies on the reduction behavior of pellets made from MMO fines have not been carried out yet. The author has already investigated the hardening properties of pellets made from MMO fines [9]. The current article contains details of the study conducted to determine the optimum reduction parameters (i.e., reduction time and temperature) for carbothermic reduction of those pellets and investigate its reduction kinetics.

2　Materials and Methods

Hardened MMO pellets (-19 + 12 mm dia) prepared in a previously published study have been used in this study for reduction studies [9]. Pellets had been prepared from MMO fines (55 wt.% below 63 μm) with calcined colemanite powder (<75μm) as a binder. The MMO and colemanite's chemical composition was determined by XRF and is given in Table I. The oven-dried pellets had been indurated at a temperature of 1523 K. Magnetite, and chromite in the green pellets had converted to hematite and sesquioxide ($Fe_{1.25}Cr_{0.75}O_3$), respectively. The current study used high ash coke fines as a reducing agent. The proximate analysis of non-coking coal is given in Table 1.

Table 1. Details of Raw materials used

Chemical composition of Multimetallic Magnetite Iron ore (wt.%)

Fe_3O_4	Al_2O_3	SiO_2	CrO_3	MgO	TiO_2	NiO	MnO	CaO	P_2O_5	LOI
62.2	14.7	14.4	4.95	1.41	0.62	0.59	0.45	0.14	0.18	Rest

Chemical composition of Colemanite Binder (wt.%)

Fe_2O_3	Al_2O_3	SiO_2	MgO	SO_4	B_2O_3	Na_2O	SrO	CaO	B	LOI
0.08	0.40	4.60	3.0	0.60	40.0	0.35	1.50	27.0	12.5	Rest

Proximate analysis of coke used (wt.%)

Moisture	Volatile matter	Ash	Fixed Carbon
1.73	2.1	37.4	58.77

The pellets were placed in a reductant bed in a galvanized iron crucible (60 mm height and 25 mm inner dia). Each crucible contained one pellet in the reductant bed. Five crucibles were placed inside the furnace at six temperatures of 950 °C, 1000 °C, 1050 °C, 1100 °C, and 1150 °C for 30, 45, 60, 90, and 120 min per temperature. Each reduction was carried out a minimum of three times to ensure the reproducibility of the results. XRD analysis of the reduced samples was carried out using a Rigaku Miniflex-600 with a Dtex ultra detector (Cu-Kα radiation λ = 0.154 nm, acceleration voltage = 40 kV, current = 15 mA). A Scanning Electron Microscope (SEM) (model ZEISS EVO-18, Oxford Instruments with software INCA Energy 300, acceleration voltage = 15 kV, and beam current ~ 5 mA) was used to study the microstructure of the reduced pellets.

3   Results and Discussion

The degree of reduction (% R) was defined as

$$\%R = (\text{change in mass of sample/change in mass expected if all of iron oxides, nickel oxides and chromium oxides are reduced}) * 100 \quad (1)$$

The percentage metallization of Fe ($\%Fe_M$) was defined as:

$$\%Fe_M = (\text{Mass of Metallic Fe in Reduced sample/Total mass of Fe in the sample}) * 100 \qquad (2)$$

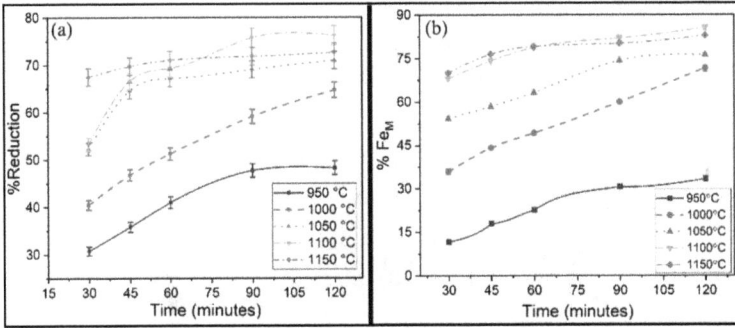

Fig. 1. (a) % Reduction and (b) % FeM for different time-temperature combinations

Metallic Fe in the reduced pellet was deduced by chemical methods according to IS 15774. Highest reducibility of 76% and metallization of 85% were obtained at 1100 °C and a reduction time of 120 min (Fig. 1). The %R was generally found to increase on increasing temperature and reduction time. The %R and $\%Fe_M$ at 1100 °C was more than the reduction at 1150 °C after about 90 min. On a similar note, except for pellets reduced at 1150 °C, metallization increased with increasing reduction time and temperature. The $\%Fe_M$ for pellets reduced at 1150 °C increased initially from 30 min to 45 min, after which it became almost constant. To study the observed anomaly, the pellets reduced for 60 min, and 90 min for both temperatures (1100 °C, 1150 °C) were characterized using XRD and SEM. Figure 2 shows the XRD analysis of the pellets reduced for 60 and 90 min for 1050 °C, 1100 °C, and 1150 °C, respectively. Iron is the major phase observed for all the pellets, followed by magnetite, cristobalite, di aluminium silicate, and chromium oxide. It can be seen that the relative intensity of the Fe phase has increased, and magnetite has decreased increasing the reduction time for each temperature. However, there was neither a phase change on increasing temperature or reduction time nor any slag phases detected. The presence of slag phases could have indicated a possible reduction in the pellets' porosity, which would have lowered the penetration of reducing gases into the pellet core and decreased reducibility.

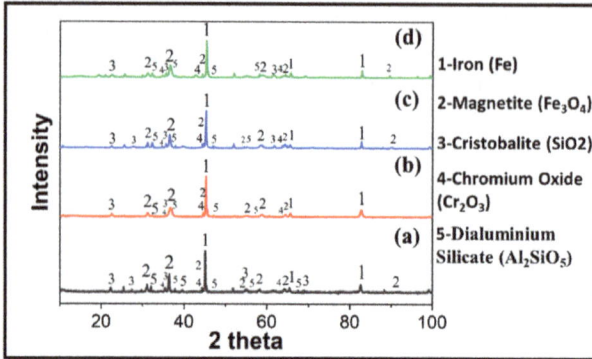

Fig. 2. XRD analysis of pellets reduced at (a) 1100 °C & 60 min (b) 1100 °C & 90 min (c) 1150 °C & 60 min (d)1150 °C & 90 min

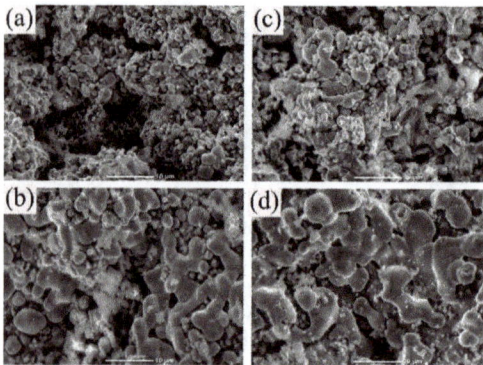

Fig. 3. SEM images of pellets reduced at (a) 1100 °C & 60 min (b) 1100 °C & 90 min (c) 1150 °C & 60 min (d)1150 °C & 90 min

Figure 3 shows the SEM images of the pellets reduced for 60 and 90 min for 1100 °C and 1150 °C, respectively. The grain size in Fig. 3(b) has increased slightly thangrain size in Fig. 3(a), which indicates that sintering has happened. The number of individual smaller grains seem to be larger in number in Fig. 3(b) than in Fig. 3(d). The grain size has also increased significantly in Fig. 3(d), indicating that the sintering and fusion of grains have increased with an increase in reduction time at 1150 °C. In Fig. 3(c), it can be seen that grains have already started to sinter, which was not the case with pellets reduced at 1100 °C. The grains in Fig. 3(d) have enlarged and fused to fill the pores, thereby reducing porosity. As the pellets at 1100 °C still possess sufficient porosity even at 90 min, the reduction was less hindered than in the case of pellets reduced at 1150 °C. The %Fe$_M$ values for pellets reduced at 1150 °C for the initial 60 min are much higher than that for pellets reduced at 1100 °C. Thus, the amount of metallic species that would have formed during the initial 60 min in the pellet reduced at 1150 °C could have been more than that formed at 1100 °C. The formed metal could have been sintered together, filled up the pores, and formed an impervious layer around the unreduced core.

The enlargement and fusion of grains in Fig. 3(e) show that the resistance to the flow of reducing gases had already started at 60 min for pellets reduced at 1150 °C. Thus, beyond a reduction time of 60 min, the reducibility and the consequent metallization of pellets reduced at 1100 °C became more than that of pellets reduced at 1150 °C.

The various equations used to fit the experimental data are as follows [10]

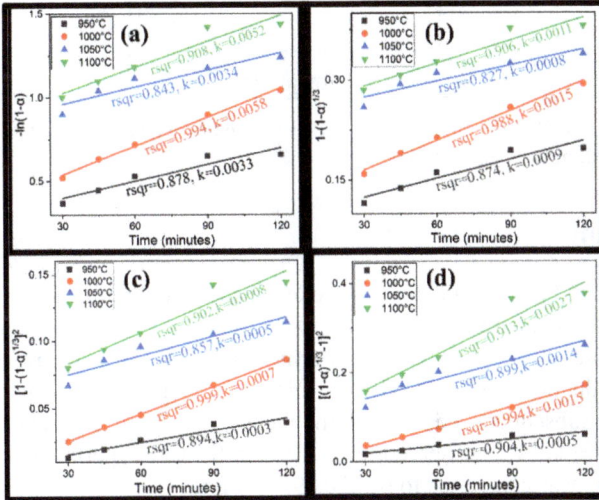

Fig. 4. Plots of model vs. time along with R^2 and slope values

$$- \ln(1 - \alpha) = kt \text{(Gasification control model)} \tag{3}$$

$$1 - (1 - \alpha)^{1/3} = kt \text{(Contracting sphere model)} \tag{4}$$

$$\left[1 - (1 - \alpha)^{1/3} \right]^2 = kt \text{(Three - dimensional diffusion: Jander equation)} \tag{5}$$

$$\left[(1 - \alpha)^{-1/3} - 1 \right]^2 = kt \text{(Three - dimensional diffusion: Zhuravlev - Lesokhin - Tempel' man (ZLT) diffusion)} \tag{6}$$

where α = fraction of reduction, t = reduction time, k = rate constant. Figure 4 shows the plots of the various models. The R^2 (rsqr) and the slope of the lines are mentioned in the graphs themselves. Equation (6) was the best fit model as the rsqr values were closest to 1. The good fitting between the experimental data and the diffusion-controlled model indicates that the reduction of MMO could be controlled by three-dimensional diffusion through the product layer. The % R of the pellets had decreased at higher temperatures as the metallic iron formed had fused and filled up the porosities. Thus, further reduction was hindered, which shows that the diffusion of the gas into the pellet controls the reduction reaction, which is in line with the mechanism predicted by the fitted model. Rate constants (k values) at different temperatures were attained from the best-fit model. Their logarithm data were plotted versus the reciprocal of absolute temperature to yield the apparent activation energy (Fig. 5) of 220.195 kJ/mole.

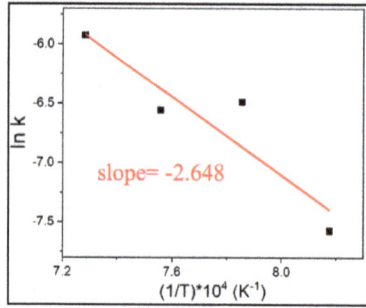

Fig. 5. Plots of (a)-ln(1-α) vs time (b) 1-(1-α)1/3 vs time (c) [1-(1- α)1/3]2 vs time (d) [(1-α)-1/3–1]2 vs time; along with R2 and slope values

4 Conclusion

The reducibility and kinetic behaviour of pellets made from multimetallic magnetite ore was studied in the temperature range of 950–1150 °C. A maximum reduction of 76% and maximum metallization of 85% were achieved at 1100 °C and a reduction time of 120 min. The reduction at 1100 °C after 60 min was found to be more than 1150 °C due to sintering and subsequent reduction in the porosity of the pellets. The kinetic model "$[(1-\alpha)^{-1/3}-1]^2 = kt$" was observed to be the best fit with experimental data, and the activation energy was calculated to be 220.195 kJ/mole.

References

1. Tang, J., Chu, M.-S., Li, F., Tang, Y.-T., Liu, Z.-G., Xue, X.-X.: Reduction mechanism of high-chromium vanadium–titanium magnetite pellets by H2–CO–CO2 gas mixtures. Int. J. Miner. Metall. Mater. **22**(6), 562–572 (2015)
2. Halli, P., Taskinen, P., Eriç, R.H.: Mechanisms and Kinetics of Solid State Reduction of Titano Magnetite Ore with Methane. Journal of Sustainable Metallurgy **3**(2), 191–206 (2017)
3. Liu, S.-s., Guo, Y.-f., Qiu, G.-z., Jiang, T., and Chen, f., Solid-state reduction kinetics and mechanism of pre-oxidized vanadium–titanium magnetite concentrate. Transactions of Nonferrous Metals Society of China, 24(10): p. 3372–3377, (2014)
4. B. Nayak, A.K.V., S. D. Singh and K. K. Bhattacharya, Petrography, Chemistry and Economic Potential of the Magnetite Ores of Pokphur Area, Nagaland. Memoir Geological Society of India, 75: p. 341–348, (2010)
5. Vaish, A.K., Singh, S.D., Minj, R.K., Gupta, R.C.: Exploration and Exploitation of Multi-Metallic Magnetite Ore of Nagaland for Value Added Product. Trans. Indian Inst. Met. **66**(5), 491–499 (2013)
6. Vaish, A., Nayak, B., Goswami, M.C., Singh, S.D., Singh, D.P., and Gupta, R.C., Magnetite ore of Nagaland - it's mineralogy and reduction kinetics. XI International Seminar on Mineral Processing Technology: p. 1064–1072, (2010)
7. Vaish A. K., G.R.C., Mehrotra S. P., Thermodynamic and kinetic aspects of the smelting reduction of multimetallic Indian magnetite ore. Journal of Metallurgy and Materials Science, 48(1): p. 1–12, (2006)
8. Kumar, M., Baghel, H., Patel, S.K.: Reduction and Swelling of Fired Hematite Iron Ore Pellets by Non-coking Coal Fines for Application in Sponge Ironmaking. Miner. Process. Extr. Metall. Rev. **34**(4), 249–267 (2013)

9. Mishra, B., Dishwar, R.K., Omar, R.j., and Mahobia, G.S., Hardening Behaviour of Pellets Prepared from a Novel Combination of Rare Multimetallic Magnetite Ore and Binder. Transactions of the Indian Institute of Metals, 74(8): p. 2049–2055, (2021)
10. Dickinson, C.F., Heal, G.R.: Solid–liquid diffusion controlled rate equations. Thermochim. Acta **340–341**, 89–103 (1999)

Reduction Kinetics of Composite Steel Slag-Coke Pellets

Charwak Ambade[1], Sheshang Singh Chandel[1,2], and Prince Kumar Singh[1(✉)]

[1] Department of Metallurgical and Materials Engineering, Indian Institute of Technology Ropar, Rupnagar 140001, Punjab, India
princeks@iitrpr.ac.in

[2] Metal Extraction and Recycling Division, CSIR-National Metallurgical Laboratory, Jamshedpur 831007, India

Abstract. Generation of slag during steel making is a process consequence but it produces in high volume and management of such wastes is a primary environmental concern and a challenge also for the steel making industries. Steel slag is rich in mineralogy, such as iron oxides, calcium oxide, silicon oxide, magnesium oxide, manganese oxide, and aluminum oxide etc. Out of these, the major minerals in the steel slag are iron and calcium oxides. The iron content in such slag is 25–35%, which varies with steel production processes. Iron recovered from the slag can be used for secondary steel industry replacing a handsome amount of scrap. In this experiment, the kinetics has been investigated during the solid reduction of iron oxide from the slag are studied at different coke-to-slag ratios, different reduction temperatures for different reduction times. Here, the coke-to-slag ratio was taken as 0.5, 1, and 1.5 for reduction temperature 1000, 1100, and 1200 °C with 15, 45, and 120-min reduction time. The kinetic and reduction parameters like activation energy and pre-exponential factor for solid stage reduction of EAF slag are calculated with the help of Coats-Redfern methods.

Keywords: Reduction Kinetics · Dynamic Models · Electric Arc Furnace

1 Introduction

Steel is an essential material in early human civilization. It finds application in various fields, from safety pins to bridges, machine equipment, construction, and beyond. Steel plays a vital role in global construction. This alloy supports structural frames in various construction projects, from skyscrapers to highways [1]. The rising demand for steel is a direct result of the expansion of the world economy. A significant quantity of impurities must be extracted from the steel during manufacturing. Steel slag is the name given to the impurities removed during the smelting process. Currently, a significant amount of steel slag is being utilized in landfills. Because steel slag is alkaline, it ultimately releases toxic substances when dumped into the environment. Steel slag contains iron, calcium, silicon, magnesium, manganese, and alumina oxides. Iron and calcium oxides

S. Patra et al. (Eds.): METCENT 2023, *Proceedings of the International Conference on Metallurgical Engineering and Centenary Celebration*, pp. 52–57, 2024.
https://doi.org/10.1007/978-981-99-6863-3_7

predominate in the composition of steel slag. The iron content in steel slag from the DRI-EAF route ranges from 25–35%. The fact that there is such a large quantity of iron in the slag makes it possible to recycle it for use in the manufacturing process.

Recovery of iron from the steel slag can be done using two approaches direct and indirect reduction. Much research is already done on the reduction behavior using coke, coal, CO gas, natural gas, and many more [2]. The reduction kinetics and mechanism are the way to predict the degree of reduction for various iron ores. One of the approaches to explore the kinetics are by following an Isothermal study [3]. The influence of temperature was observed in various other processes, including steelmaking [4].

The extraction of iron from steel slag and its subsequent use as a secondary resource in the steelmaking industry are the primary focuses of this research project. The main purpose of the research was to investigate the kinetics and mechanisms behind the reduction of iron from slag. A reduction is achieved with the help of the coke. This experiment contributed to improving our ability to predict the reduction behavior for a variety of carbon-slag ratios.

2 Experiment Procedure and Result Discussion

The raw material utilized in this study was the electric arc furnace, EAF slag, sourced from Aarti Steel International Limited, located in Ludhiana, Punjab. Figure 1(a) displays the steel slag's chemical composition and the EAF slag's physical appearance. The chemical composition of slag revealing the total Fe present in the slag is 35.801%, shown in Fig. 1(b), and Fig. 1(c) tell the presence of phase in the slag sample. The coke is utilized as a reduction agent. Molasses is used as a binder to improve the binding property between slag particles during pellet production.

Figure 2(a) shows the experimental procedure. The Electric Arc Furnace (EAF) slag was initially subjected to crushing using a steel hammer. Subsequently, the sample was introduced into a ball milling apparatus, operating at the specified revolutions per minute (RPM) for three hours. This process facilitated the transformation of the sample into a uniform and finely powdered state, with an approximate particle size of 150 mesh. Subsequently, a mixture was prepared by combining 10 g of steel slag with varying concentrations of coke (0.5, 1, and 1.5 g) using a ceramic mortar. Subsequently, molasses, a binding agent, and water were incorporated into the mixture to enhance its binding characteristics. The 3 g of prepared samples are placed into Hydraulic Pellet Press and operated at 150 psi of pressure for preparing a pellet of 5 cm dia and 3 cm of height. The pellets were weighed before being placed in the oven to eliminate the moisture content. The pellets were re-weighed after removal from the oven and before their introduction into the tube furnace. The pellets were subjected to thermal treatment in a tube furnace at temperatures 1000, 1100, and 1200 °C for 15, 45, and 80 min, respectively. The pellets were subjected to a final weighing to determine the extent of iron reduction from the oxide.

The primary aim of this study is to comprehend the kinetics of the reduction of iron oxides found in steel slag. This investigation holds significance as it provides crucial insights into iron extraction. The iron that has been extracted has the potential to serve as

(a)

(c)

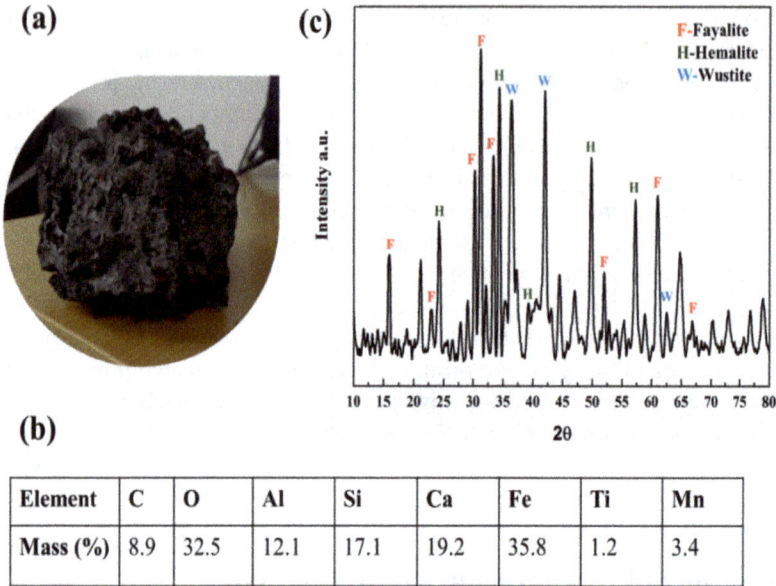

(b)

Element	C	O	Al	Si	Ca	Fe	Ti	Mn
Mass (%)	8.9	32.5	12.1	17.1	19.2	35.8	1.2	3.4

Fig. 1. (a) Physical appearance of electrical arc furnace slag, (b) Elemental analysis of EAF slag, and (c) XRD analysis of EAF slag.

a secondary resource in the steelmaking industry. The reduction process was investigated under isothermal conditions, wherein the weight loss occurring within the composite pellets was measured. The experiment was conducted at three different temperatures: 1000, 1100, and 1200 °C. The findings indicate a positive correlation between temperature and the rate of reduction, whereby an increase in temperature leads to a corresponding increase in the rate of reduction, given a constant amount of time. At a temperature of 1200 °C, the reduction rate exhibited a significantly greater magnitude than all other tested temperatures.

The increase in temperature leads to a notable enhancement in the mass transfer coefficient, consequently resulting in an augmented reduction rate [3]. Another potential factor contributing to this phenomenon could be the positive correlation between temperature and the number of moles engaged in interactions, thereby increasing the adsorption rate. This phenomenon can be. All experiments were conducted under vacuum conditions. Based on the provided information, a set of hypotheses has been formulated. The impact of any external gas that had permeated the composite pellets through the gas phase boundary layer was found to be insignificant. Furthermore, uniform dispersion of iron oxide and carbon particles was observed throughout the entire sample. The Avrami–Erofeev approach was utilized to elaborate on the experimental findings. $1 - \alpha = \exp(-kt)$ and $\ln(-\ln(1 - \alpha)) = \ln(k) + m (\ln(t))$; where is α reaction fraction calculate from [2]; t is unit time; m is slope and k is rate constant. The value of variable m is determined by calculating the slope of various straight lines fitting obtained from different reaction model equations, as presented in Table 1 below.

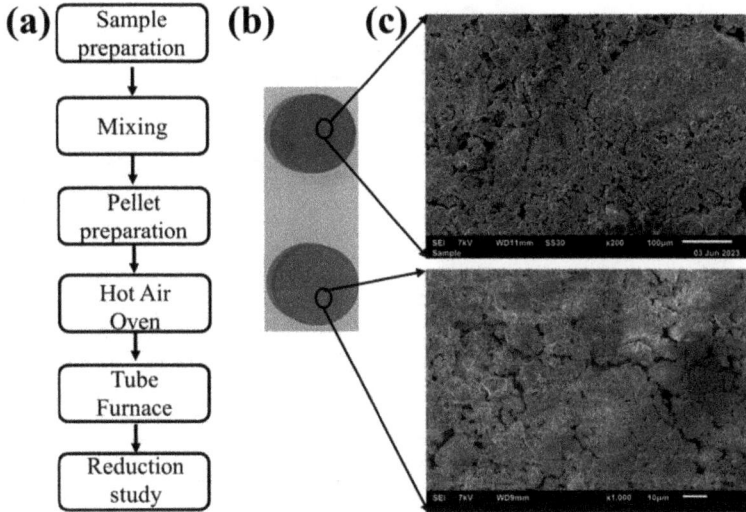

Fig. 2. (a) Flow chart of Experimental procedure, (b) Pellets obtained before and after reduction experiment, and (c) SEM image.

Table 1. Several different kinetic models for reactions in solid state [5]

Kinetic reaction models	Symbol	Equation
One-dimensional formation and growth of nuclei	A	$-\ln(1-\alpha) = kt$
Three-dimensional phase-boundary- controlled reaction	B	$1-(1-\alpha)^{1/3} = kt$
Three-dimensional diffusion	C	$[1-(1-\alpha)^{1/3}]^2 = kt$
One-dimensional diffusion	D	$\alpha^2 = kt$

Based on the data depicted in Fig. 3(a), it is evident that there is a significant rise in the rate of reduction within the initial 15 units of measurement. Nevertheless, the reduction rate decreases approximately 45 min following the initial period. This phenomenon could be attributed to achieving optimal diffusion of carbon monoxide (CO) gas from the pellet at the onset, subsequently leading to a subsequent decline in the level of reduction. A higher carbon-to-oxygen (C/O) ratio can lead to a more significant reduction. The reduction data are utilized to investigate the kinetic parameters, as depicted in Fig. 3(b). The data fitting process was conducted for version 1.0 of the C/S software. The data about the fitting is presented in Fig. 3(c). The three-dimensional diffusion model was identified as the suitable model. The determined activation energy was determined to be 4.56 kJ/mol.

(a)

(b)

(c)

Kinetics process controlling		R^2
One-dimensional formation and growth of nuclei	A	0.447
Three-dimensional phase-boundary- controlled reaction	B	0.653
Three-dimensional diffusion	C	0.653
One-dimensional diffusion	D	0.652

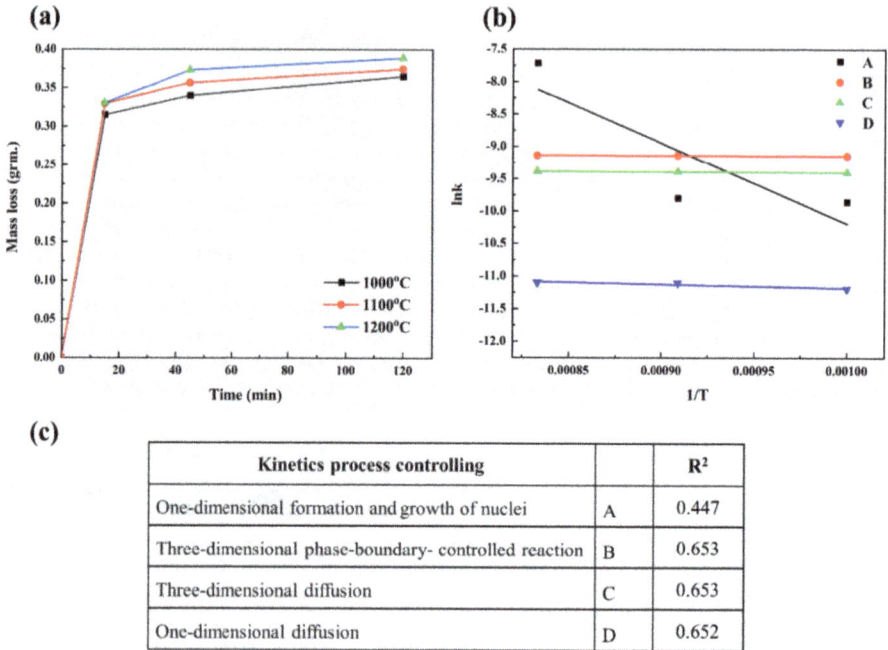

Fig. 3. (a) Mass loss of sample concerning reduction time at different reduction temperatures, (b) Arrhenius plots for the determination of activation energy, and (c) Data fitting parameters.

3 Conclusion

The current research used an isotropic experiment to investigate the recovery of iron from EAF slag. This experimental study maintains temperatures ranging from 1000 to 1200 °C for 15 to 80 min. The obtained results are used in a range of models, and the reduction kinetic parameter is calculated using these results. Three distinct phases could be distinguished within the reduction reaction process: the accelerated reduction phase, the stable reduction phase, and the decelerated reduction phase. The activation energy raises itself to a higher level as the reaction continues.

Acknowledgments. The Ministry of Education of India and the Council of Scientific & Industrial Research supported the research. The Author thanks to Aarti Steel International Limited for providing raw materials and providing help in conducting high temperature experiments.

References

1. Teo, P.T., et al.: Assessment of electric arc furnace (EAF) steel slag waste's recycling options into value-added green products: a review. Metals **10**(10), 1347 (2020)
2. Chandel, S.S., Randhawa, N.S., Singh, P.K.: Thermodynamic and kinetic aspect of solid state reduction of Electric Arc Furnace slag through coke: an experimental study. Mater. Today Proc. (2023)

3. Man, Y., Feng, J.X., Li, F.J., Ge, Q., Chen, Y.M., Zhou, J.Z.: Influence of temperature and time on reduction behavior in iron ore–coal composite pellets. Powder Technol. **256**, 361–366 (2014)
4. Chandel, S.S., Maurya, A., Singh, P.K.: Numerical investigation of quenching technique for steel alloy hardening process using twins liquid jets. Mater. Today Proc. **62**, 7348–7352 (2022)
5. Kowitwarangkul, P., Babich, A., Senk, D.: Reduction kinetics of self-reducing pellets of iron ore. AISTech Proc. **1**, 611–622 (2014)

Investigating the Suitability of Local Riverbed Sand as a Mold Material for Foundry Industry: A Comparative Study

Dheerendra Singh Patel[✉] and Ramesh Kumar Nayak[✉]

Department of Materials and Metallurgical Engineering, Maulana Azad National Institute of Technology, Bhopal 462003, Madhya Pradesh, India
rameshkumarnayak@gmail.com

Abstract. The present study aims to assess the viability of using locally available sand (Narmada Riverbed sand (NRS)) for sand casting molds, which can lead to cost and time savings compared to commercial grade silica sand. Physical and mechanical tests of the molds were performed on selected sand samples, and aluminum alloy (A356) was casted. The as cast properties such as porosity, surface roughness (SR), microstructure and phase content was evaluated. The results showed that commercial grade silica sand (CGSS) had the finest quality with the highest grain fineness number (GFN) of 64, while NRS sand was rougher with the lowest GFN of 53. However, all sands exhibited good bonding ability with sodium silicate binders. The study also investigated the impact of different sodium silicate content (6–12%) and time of supply of CO_2 gas into the mold on its properties. It is observed that with the increase in the sodium silicate content decreases the mold permeability and increases the strength. The optimum mold properties were observed at 8 wt.% sodium silicate and passing of CO_2 gas for a duration of 18 s. Casting of A356 alloy was performed using CGSS and NRS mold. It is observed that NRS can be used for non-ferrous casting in Indian foundries.

Keywords: Narmada Riverbed Sand (NRS) · Commercially Graded Silica sand(CGSS) · Sodium Silicate · CO_2 · surface roughness · Microstructure

1 Introduction

Commercial grade silica sand has been used as a primary molding material in ferrous and nonferrous foundry industry. However, the growing industrialization and mechanized production methods have led to an increased demand for sand, resulting in a scarcity of quality and quantity of it [1, 2]. The global production of silica sand varies annually, with differing levels among countries and continents [2]. In 2007, the total global production of silica sand was approximately 122.0 million metric tons [3]. However, due to the global recession starting in 2008, economic activity slowed down, leading to a decrease in both production and consumption of industrial sand. In 2008 and 2009, the production of silica sand declined to 118.1 million tons and 111.5 million tons, respectively [4].

S. Patra et al. (Eds.): METCENT 2023, *Proceedings of the International Conference on Metallurgical Engineering and Centenary Celebration*, pp. 58–71, 2024.
https://doi.org/10.1007/978-981-99-6863-3_8

This scenario, there is a pressing need to explore suitable alternatives to silica sand for foundry applications [5] Finding viable substitutes for silica sand can help alleviate the strain on its availability and ensure the continued progress of the foundry industry [6, 7]. Nural et al. [8] they have used six different types of sand with different type of binder for mold making and they have found that all sand has excellent flowability. Yekinni et al. [9] investigated the foundry characteristics of silica sands obtained from six different places in Lagos, Nigeria. They likely examined properties such as grain fineness, permeability, and strength, shedding light on the suitability of locally available silica sands for foundry applications. Aweda et al. [10] evaluated the molding properties of sand deposits in Ilesha and Ilorin, Nigeria. They assessed the mold properties such as moldability, compatibility, and collapsibility of the sands. Their research contributes to the understanding of the local sand deposits and their potential for foundry use. Ugwuoke [11] investigation focused on the suitability of Adada river sand bonded with Nkpologwu clay for foundry use. The study likely examined properties such as bonding ability, permeability, and strength of the sand-clay mixtures, providing insights into alternative foundry materials in Nigeria. Atanda et al. [12] inspected the influence of cassava starch and bentonite on the molding characteristics of silica sand. Some researcher also used industrial waste such as stone dust, granulated blast furnace slag fly ash and red mud as a molding material and they found that good mold properties compared to the silica sand [1, 4, 6, 7, 12–16]. The study likely investigated properties such as moldability, strength, and permeability of the sand mixed with these additives. This research highlights the potential of incorporating natural additives for improved molding properties. Previous researchers have examined the foundry properties of locally available sands and binders, but few have analyzed the casting quality obtained from these sand molds. In India, there are numerous untested sand deposits, leading to reliance on sand procurement from other states, increasing production costs and time. By exploring the casting quality of locally available sand molds, India can reduce production costs and enhance efficiency in the foundry industry. Utilizing these untapped sand resources will contribute to sustainability and economic growth while reducing dependence on external suppliers. The present study aims to assess the viability of using locally available sand (Narmada Riverbed sand) for sand mold casting process, which can lead to cost and time savings compared to procured sands. The researchers focused on Narmada Riverbed Sand (NRS) and conducted extensive tests to evaluate its properties for metal casting. Aluminium alloy (A356) was cast using NRS and CGSS molds to analyze casting properties such as porosity and surface roughness (SR), measured as Ra value, microstructure and phase content in it.

2 Experimental Process

2.1 Material Collection

The CGSS used in the study was sourced from RSB Metal Tech Pvt. Ltd, located in Odisha, India and mined Narmada riverbed sand was collected near Bhopal, Madhya Pradesh. The binder employed in the process was sodium silicate, which was obtained from DECG International Private Limited, Mandideep, Bhopal. Additionally, carbon dioxide (CO_2) was acquired from the nearest gas refilling agency in Bhopal, India.

2.2 Chemical Composition

The chemical composition of CGSS and NRS was analyzed using the wet chemical analysis method. The results is presented in Table 1. It is revealed that CGSS contains 98.45% SiO_2 by weight, while NRS contains 79.71% SiO_2. The loss on ignition (LOI) for CGSS was measured to be 0.1% and for NRS, it was found to be 0.31%. These LOI values fall within the acceptable range for foundry applications, which is below 0.5% [17].

Table 1. Chemical composition of Commercial grade silica sand and NRS

Type of sand	Chemical composition (wt.%)									
	SiO_2	Al_2O_3	Fe_2O_3	MnO	K_2O	CaO	MgO	TiO_2	P_2O_5	Na_2O
CGSS	98.45	1.1	0.2	–	0.1	–	–	0.05	–	–
GBFS	79.71	8.23	4.52	O.34	0.89	2.31	1.44	0.47	0.11	1.52

2.3 Grain Size and shape Analysis

The weight percentages of different sand samples, preserved on each of the eight sieves, along with their corresponding sieve numbers and Grain Fineness Number (GFN), are presented in Table 2. It is important to note that as the sieve number increases, the aperture size decreases. The GFN values were calculated for the sands, resulting in values of 63, and 53 for CGSS, and NRS, respectively. This confirms that CGSS possesses the finest texture with the highest GFN, while NRS sand exhibits a rougher texture with the lowest GFN among the sand samples.

Table 2. The particle size distribution of CGSS and GBFS

AFS sieve no	CGSS Retained (%)	NRS Retained (%)
30	0	0
40	0	0
50	2.4	0.1
70	39.4	90.8
100	53.4	8.3
140	3.2	0.5
200	1.4	0.4
Pan	0.2	0.6
AFN (GFN) number	63	53

Sand mold properties depend on physical and mechanical characteristics. Sand grain shape is crucial. SEM analysis showed CGSS (Fig. 1a) and NRS (Fig. 1b). NRS particles are a mix of sub-round and angular shapes. Round grains have point contact, requiring less binder for permeability but weaker strength. Angular grains need more binder, offering greater strength but reduced permeability. A combination of sub-round and round-angular grains achieves optimal mold permeability and strength. Mold properties are governed by sand size, distribution, and binder content. Smaller particles narrow gaps, hindering and gas flow from the mold cavity.

Fig. 1. SEM micrographs of (a) CGSS and (b) NRS

2.4 Density

Density plays a critical role in foundry mold making as it directly affects the mold's structural integrity, porosity, thermal conductivity, flow of molten metal, dimensional accuracy, and overall casting quality. High-density sand has several effects on mold materials used in foundry processes. Firstly, high-density sand contributes to improved compaction and stability of the mold, reducing the risk of deformation or collapsing during casting. It enhances the mold's structural integrity and overall strength, which is crucial for holding the shape and withstanding the pressure exerted by molten metal. Secondly, high-density sand results in a lower level of permeability in the mold material. This reduced permeability restricts the flow of gases and moisture, preventing their escape from the mold cavity. As a result, high-density sand molds can provide better dimensional accuracy, minimizing the risk of defects such as gas porosity or water vapor-related defects The density of CGSS and NRS was determined using Archimedes' principle after subjecting the samples to a drying process at 150 °C for 2 h to remove moisture. NRS shows a density of 2.64 g/cm^3, whereas CGSS exhibited a density of 2.5 g/cm^3. The 5% higher density of NRS compared to CGSS can be attributed to the presence of Al_2O_3 and Fe_2O_3.

2.5 pH Value

The pH analysis determines the acidity or alkalinity of CGSS and NRS used in foundry industries. The pH values were determined as per AFS 5113-00-S, [18]. A 25 g sample mixed with 100 ml distilled water was stirred for 5 min in a beaker. The pH values were measured using a Lab man (LMPH-10) pH meter. NRS and CGSS exhibited pH values of 7.8 and 9.01, respectively, indicating alkaline nature.

2.6 Fusion Point

Fusion point of molding sand play crucial role for maintaining the integrity of mold during the casting process. Sand having low melting point can lead to premature melting or fusion of the sand grain. Due to this collapse or deformation of the mold cavity it may detorsion of the casting surface of as cast product.

Table 3. SEM images of NRS and CGSS heated at different temperature

Temp°C	800	900	1100	1200
NRS				
CGSS				

The SEM images of CGSS and NRS is reported in Table 3. It is observed that NRS has fused at 1350 °C as shown in Fig. 2. However, at 1350 °C silica sand particles are not fused. Therefore, NRS mold cannot used for ferrous alloy casting and it is suitable for non-ferrous alloys casting which have pouring temperature is less than 1200 °C.

2.7 Preparation of Standard Mold Specimens

The CO_2 molding process is employed to create molds, and standard cylindrical specimens of size 5.08 cm × 5.08 cm are formed using a sand rammer and different combinations of sodium silicate as the binding agent and shown in Fig. 3. The mixture is hand-mixed with all the ingredients, and then compressed using three strokes of the sand rammer to achieve the desired sample size. Subsequently, CO_2 gas was introduced into the standard test samples for varying durations of 5 s, 10 s, 15 s, and 20 s. Various properties of the molds, such as shear strength, compressive strength, permeability, and hardness, are individually assessed through testing as shown in Fig. 4.

Fig. 2. Visual changes in silica sand and MRS with varying heating temperatures

Fig. 3. Standard sample of NRS mold for testing mold properties

The CO_2 process is a casting technique that utilizes carbon dioxide gas to harden a sand mix containing sodium silicate as the binder. When CO_2 gas is introduced into the sand mix, a chemical reaction occurs between sodium silicate and CO_2, resulting in the immediate hardening of the materials in the mold, such as sand or slag. While the chemical reactions involved in this process are complex, a simplified representation of the main reaction is as follows [19].

$$Na_2O * mSiO_2 * H_2O + CO_2 \longrightarrow Na_2CO_3 + mSiO_2 * H_2O \quad 1$$

$$(1)$$

In this reaction, sodium silicate reacts with carbon dioxide to form sodium carbonate (Na_2CO_3) and silica gel ($SiO_2 * H_2O$). The resulting silica gel, which contains a certain number of water molecules, provides the necessary strength to the mold. It is essential to regulate the duration of CO_2 gas exposure during the process. Regulating the CO_2 exposure time ensures that the mold achieves optimal strength while avoiding potential weaknesses or defects. The specific duration required may vary depending on

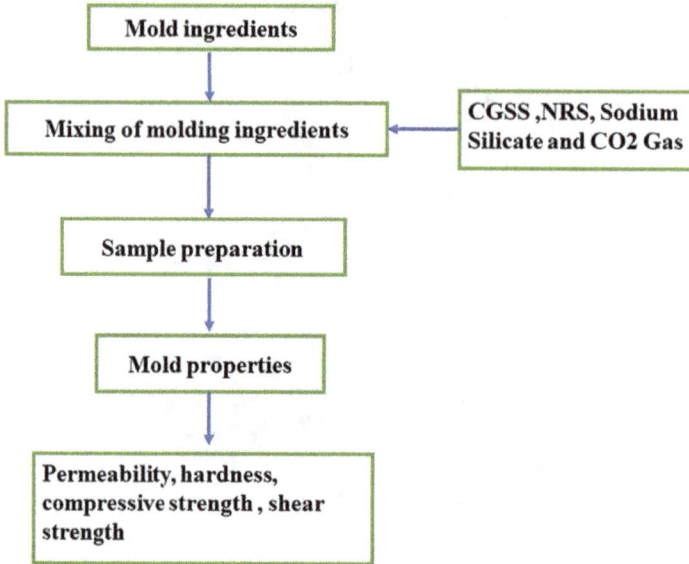

Fig. 4. Systematic flow diagram of river sand mould testing

the materials used and the desired properties of the mold. Careful control of the gassing process is crucial to ensure the quality and integrity of the final product.

3 Result and Discussion

3.1 Permeability

The results presented in Fig. 5 (a) and (b) demonstrate the relationship between the permeability of molds and the varying percentages of sodium silicate used in different types of sands. It is evident that as the content of sodium silicate increases in the molds, the permeability number decreases. Additionally, the permeability values for both samples decrease further when subjected to longer periods of CO_2 gassing and higher additions of sodium silicate, as illustrated in Fig. 5 a and b. This can be attributed to the closure of more pores in the molds due to prolonged CO_2 gassing and increased binder content. Interestingly, NRS molds consistently exhibited higher permeability values compared to sand molds for sodium silicate additions ranging from 6% to 12%. These findings suggest that the choice of mold material and the level of sodium silicate addition play crucial roles in determining the permeability and overall quality of the molds used in the casting process.

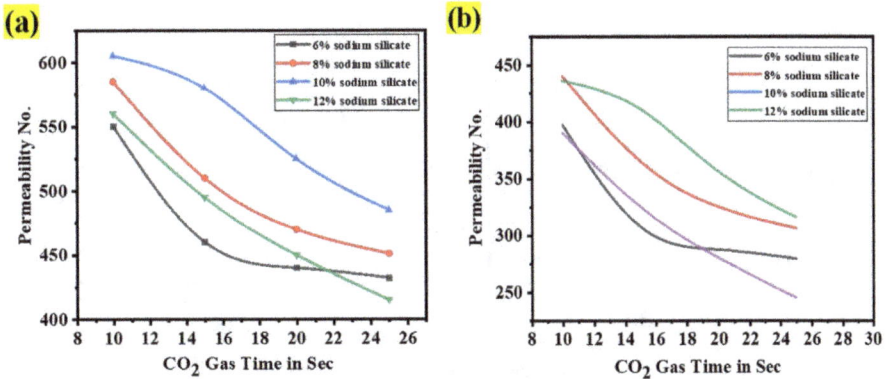

Fig. 5. Comparative mold permeability of a (a) NRS and (b) CGSS

3.2 Hardness

The hardness of the mold is an essential factor in maintaining the shape and dimensional accuracy of the mold during the casting process. If the mold becomes too soft or deformed due to the increased sodium silicate concentration and prolonged gas passing time, it can compromise the integrity of the mold. Consequently, this may lead to the production of castings with poor quality, such as inaccurate dimensions or surface defects. In Fig. 6, it is observed that when the concentration of sodium silicate increases and there is an increase in the gas passing time, leading to a higher concentration of carbon dioxide in the mold, it can have implications for the hardness of the mold. As the concentration of sodium silicate rises after certain percentage, the hardness of the mold tends to decrease. This is because higher concentrations of sodium silicate result in a more brittle and less rigid mold structure. Carbon dioxide gas is used to aid in the hardening of molds that utilize sodium silicate binders. The timing refers to the duration during which carbon dioxide gas is allowed to pass through the mold. When carbon dioxide gas is passed through the mold for an extended period, it accelerates the hardening or curing process of the sodium silicate binder. However, this accelerated curing can lead to a decrease in the overall hardness of the mold. The prolonged exposure to carbon dioxide gas promotes rapid curing, causing the sodium silicate binder to harden more quickly. This accelerated curing process can result in a mold with decreased hardness. As a consequence, the mold becomes more brittle and less rigid, which contributes to its lower hardness.

3.3 Compressive Strength

The compressive strength behavior of a mold when sodium silicate and carbon dioxide gas are combined is attributed to the chemical reactions that take place. Sodium silicate, or water glass, is utilized in mold-making due to its capacity to solidify and form a durable structure. When sodium silicate reacts with carbon dioxide, a process called carbonation occurs. This involves the reaction between sodium silicate and CO_2, resulting in the production of solid calcium carbonate ($CaCO_3$) and silica gel ($SiO_2 \cdot nH_2O$). Initially, the addition of sodium silicate enhances the compressive strength of the mold as shown

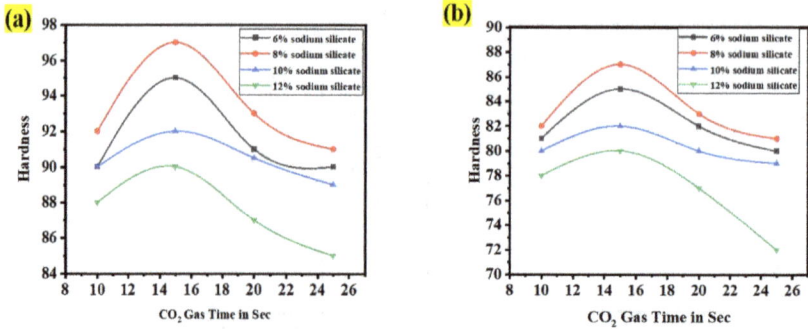

Fig. 6. Comparative mold Hardness of a (a) NRS and (b) CGSS

in Fig. 7 (a) and (b). This is due to the formation of calcium carbonate and silica gel, which contribute to strengthening and solidifying the mold. However, excessive amounts of sodium silicate can lead to a decline in compressive strength. This is attributed to the excessive formation of silica gel, which renders the mold more brittle, ultimately diminishing its compressive strength.

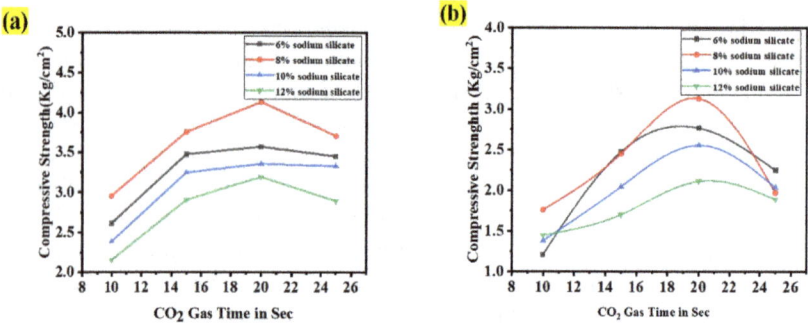

Fig.7. Comparative mold Compressive strength of a (a) NRS and (b) CGSS

3.4 Shear Strength

When the sodium silicate percentage is increased, a higher concentration of silicate ions becomes available for reaction with CO_2 gas. This results in the formation of a more extensive silica gel network within the mold, leading to enhanced interparticle bonding. As a consequence, the shear strength of the mold initially increases, making it more rigid and resistant to deformation. However, if the sodium silicate percentage continues to rise beyond a certain point, excess silicate ions can impede the formation of a well-connected and compact silica gel network. This can cause the formation of isolated clusters or uneven distribution of the gel within the mold. As a result, the shear strength of the mold starts to decrease as shown in Fig. 8 (a) and (b). This decrease occurs due to the presence of weak points created by isolated gel clusters and variations in material properties across the mold.

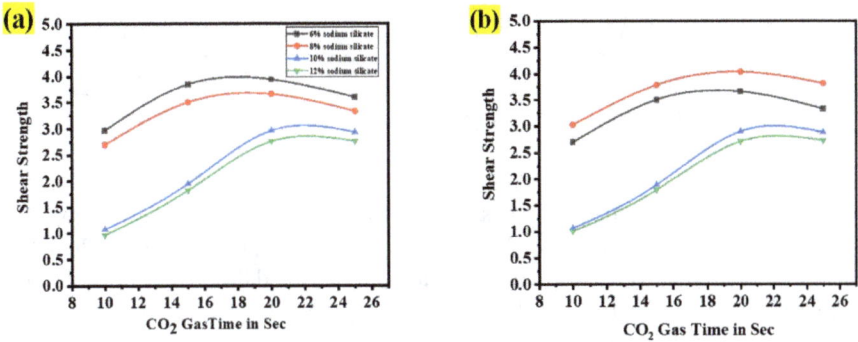

Fig. 8. Comparative mold Compressive strength of a (a) NRS and (b) CGSS

3.5 Constituent of Mold Composition

In the above study, variations in mold properties, including permeability, hardness, compressive strength, and shear strength, were observed by adjusting the sodium silicate and CO_2 parameters. Optimal mold properties were found at a sodium silicate concentration of 8% and a CO_2 gas passing time of 18 s, except for permeability. However, it should be noted that the minimum permeability requirement for sand casting, as specified by the AFS, is 75.

Table 4. Constituent of mold

	Sodium silicate	CO_2 gas time (seconds)
NRS	8	18
CGSS	8	18

After fixing the mold constituent reported in Table 4 by mixing composition of NRS and CGSS after that determine the mold properties of the NRS and CGSS. NRS show good mold properties as compared to the CGSS which are reported in Table 5.

3.6 Aluminum (A356) Alloy Casting

The A356 alloy is melted in a resistance melting furnace with a heating rate of 6 degrees per minute. The NRS mold was prepared and shown in Fig. 9. Subsequently, the ingot is placed into a graphite crucible within the furnace, and the heating process commences. The temperature is closely monitored until the ingot completely melts, at which point

Table 5. NRS and CGSS mold composition and their properties

Optimum parameters	Experimental Result (NRS)	CGSS
Permeability number	340	350
Hardness number (B Scale)	97	89
Compressive strength (kg/cm^2)	4	3
Shear strength (kg/cm^2)	1.45	0.36

any slag is removed, and the crucible is covered to prevent oxidation. Next, argon gas is introduced for 3 min to degas the molten metal and minimize porosity. The molten A356 alloy is then poured into the mold cavity, ensuring a controlled flow. After 24 h of cooling and solidification, the sand mold is broken, revealing the as-cast A356 alloy component. Finally, various analyses are conducted to assess the as-cast properties, encompassing mechanical testing, microstructure analysis.

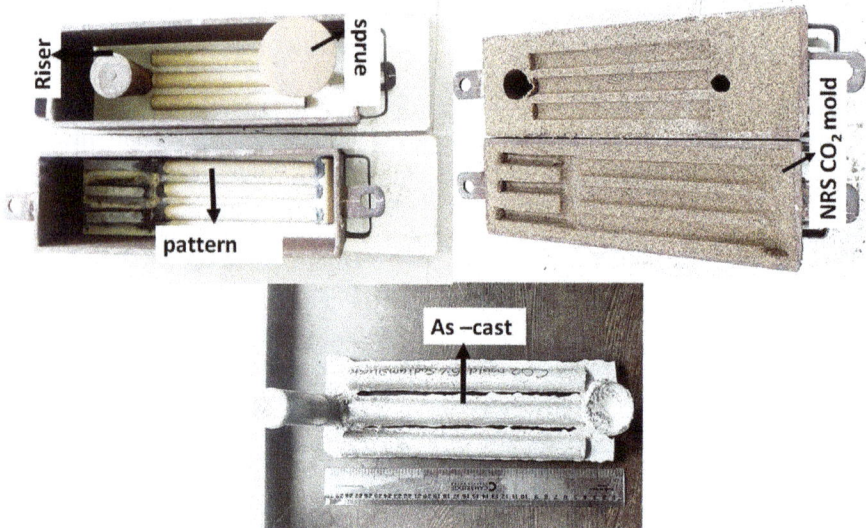

Fig. 9. CO_2 molding using NRS

3.7 As Cast Properties

To ensure direct usage of as-cast castings in component assembly without machining, achieving a superior surface finish is crucial. Mitutoyo's surface roughness tester is used with consistent parameters: a 0.8 mm cut-off length, 1.0 mm/s traverse velocity, and 4 mm traverse length. Four measurements are taken at different locations on the casting, and the average value is reported in Table 6 The surface roughness of the NRS mold is determined to be 5 ± 0.8 µm, while the surface roughness of the CGSS mold is 2.3 ± 0.9 µm. This difference can be attributed to the lower grain fineness number (GFN) of the NRS mold compared to the CGSS mold. The higher surface roughness of NRS molds compared to CGSS molds can be attributed to the irregular shapes and sizes of NRS particles, resulting in a less uniform surface. Hardness value of A-356 alloy casting in NRS and silica sand was found 80 an75 respectively reported in Table 6. NRS mold has 6.66% higher hardness as compared to CGSS mold due to high thermal conductivity in NRS mold because of high amount of Al_2O_3, and Fe_2O_3 present in the NRS mold and thermal conductivity increase the rate of cooling in NRS mold.

Table 6. As cast properties A356 Alloy

Casting mold type	Surface Roughness	Vickers Hardness	Phase % (α-aluminum and silicon)	Porosity (%)
CGSS	2.3 ± 0.9	75 ± 2	75% and 25%	2.5
NRS	5 ± 0.8	80 ± 3	70% and 30%	2.0

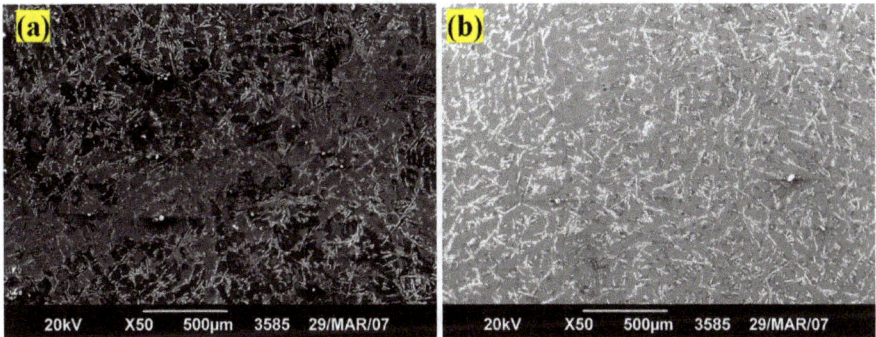

Fig. 10. SEM images of A356 alloy castings produced in (a) CGSS and (b) NRS molds

The microstructure analysis of the as-cast A356 casting produced using CGSS and NRS molds was conducted at a magnification of 50x using scanning electron microscopy (SEM). The resulting SEM image, shown in Fig. 10, reveals that eutectic silicon (Si) particles in the form of needle-shaped structures are uniformly distributed within the a-aluminum matrix. To quantify the phase content and measure the secondary dendrite arm spacing (SDAS), ImageJ software was employed. The analysis revealed that Al matrix phase constitutes approximately 79% in the CGSS casting and 74% in the NRS casting. This variation can be attributed to the higher concentration of Fe_2O_3 and Al_2O_3 in the NRS mold compared to the CGSS mold. The presence of these compounds increases the thermal conductivity of the NRS mold, leading to a higher cooling rate of the liquid metal during solidification. The accelerated cooling rate in the NRS casting results in a reduced growth rate of the solidification nuclei, leading to the formation of finer grains.

4 Conclusions

1. Permeability of NRS mold is higher than the CGSS which indicate that the NRS mold is more porous than CGSS mold. The mold hardness of NRS mold 20% higher than the CGSS mold.
2. It is found that 8% sodium silicate and 18 s duration of CO_2 supply is sufficient enough to get the required mold properties.
3. Surface roughness of as cast A356 alloy made from CGSS mold less than that of NRS mold casting. This may be due to the smaller particle size of CGSS.
4. The as casted A356 alloy made from NRS mold demonstrated higher hardness as compared to CGSS mold casting and may be attributed to its higher thermal conductivity resulting from the presence of Al_2O_3 and Fe_2O_3 in NRS.
5. Hence, NRS mold can be used for non-ferrous casting. However, its industrial trails is required for different size and shape of casting before commercialization.

References

1. Sinha, N.K., Singh, J.K.: Utilization of industrial waste as mold material and its effect on the evolution of microstructure and mechanical properties of Al–Si (A319) alloy. Int. J. Metalcast. **15**(4), 1238–1249 (2021). https://doi.org/10.1007/s40962-020-00555-7
2. Market Research Repot: Silica Sand Market Size, Share & COVID-19 Impact Analysis, By End-use Industry (Construction, Glass Manufacturing, Filtration, Foundry, Chemical Production, Paints & Coatings, Ceramics & Refractories, Oil & Gas, and Others) and Regional Forecast, 2022–2029 (2023)
3. G. Ministry of Mines: Government of India Ministry of Mines Annual Report 2021–22 (2023)
4. Nayak, R.K., Sadarang, J., Kumar, A.: Development of Fe–Cr slag mold for Al–Si alloy (A356) casting. Int. J. Metalcast. (2022). https://doi.org/10.1007/s40962-022-00876-9
5. Padmalal, D., Maya, K.: Sand mining: the world scenario. In: Padmalal, D., Maya, K. (eds.) Sand Mining. Environmental Science and Engineering, pp. 57–80. Springer, Dordrecht (2014). https://doi.org/10.1007/978-94-017-9144-1_5
6. Murthy, I.N., Rao, J.B.: Molding and casting behavior of ferro chrome slag as a mold material in ferrous and non-ferrous foundry industries. Mater. Manuf. Processes **32**(5), 507–516 (2017). https://doi.org/10.1080/10426914.2016.1257796
7. Kumar Sinha, N., Choudhary, I.N., Singh, J.K.: Influence of mold material on the mold stability for foundry use. https://doi.org/10.1007/s12633-021-01070-y/
8. Anwar, N., Sappinen, T., Jalava, K., Orkas, J.: Comparative experimental study of sand and binder for flowability and casting mold quality. Adv. Powder Technol. **32**(6), 1902–1910 (2021). https://doi.org/10.1016/j.apt.2021.03.040
9. Bello, S.K., Yekinni, A.A.: Volume 1; Issue 7 Paper-1 and Comparative Analysis of Clay Content, Grain Size and Grain Size Distribution of Foundry Moulding Sands, Lagos Area (Print) (2013). www.ijmst.com
10. Aweda, J.O., Alaro Jimoh, Y., Jimoh, Y.A.: Assessment of Properties of Natural Moulding Sands in Ilorin and Ilesha, Nigeria A Study of Performance Characteristics of Warm Mix Asphalt Modified with Dissolved Plastic Polymer View project Assessment of Properties of Natural Moulding Sands in Ilorin and Ilesha, Nigeria (2009). https://www.researchgate.net/publication/230662498
11. Ugwuoke, J.C.: Investigation on the suitability of Adada river sand and Nkpologwu clay deposit as foundry moulding sand and binder. Int. J. Sci. Res. Eng. **3**(1), 19–27 (2018)

12. Yajjala, R.K., Inampudi, N.M., Jinugu, B.R.: Correlation between SDAS and mechanical properties of Al–Si alloy made in Sand and Slag moulds. J. Mark. Res. **9**(3), 6257–6267 (2020). https://doi.org/10.1016/j.jmrt.2020.02.066
13. Narasimha Murthy, I., Babu Rao, J.: Granulated blast furnace slag: potential sustainable material for foundry applications. J. Sustain. Metall. **3**(3), 495–514 (2017). https://doi.org/10.1007/s40831-016-0111-3
14. Rao, J.B., Murthy, I.N.: Properties of silica sand and GBF slag moulds practiced by Nishiyama process. Int. J. Cast Met. Res. **31**(6), 360–372 (2018). https://doi.org/10.1080/13640461.2018.1495862
15. Nayak, R.K., Sadarang, J.: A study on the suitability of Mahanadi riverbed sand as an alternative to silica sand for Indian foundry industries. Trans. Indian Inst. Met. **75**(5), 1169–1179 (2022). https://doi.org/10.1007/s12666-021-02472-7
16. Nayak, R.K., Sadarang, J.: Feasibility study of stone-dust as an alternative material to silica sand for Al–Si (A356) alloy casting. Int. J. Metalcast. **16**(3), 1388–1396 (2022). https://doi.org/10.1007/s40962-021-00695-4
17. Holtzer, M., Kmita, A.: Mold and Core Sands in Metalcasting: Chemistry and Ecology. Springer, Cham (2020). https://doi.org/10.1007/978-3-030-53210-9
18. Heine, R.W., Loper, C.R., Rosenthal, P.C.: Principles of Metal Casting, 2nd edn. Tata McGraw-Hill Education, New York
19. Jain, P.L.: Principles of Foundry Technology. Tata McGraw Hill Education Private Limited, New York (2009)

Computational Material Science

Effect of External Magnetic Field on Grain Boundary Migration in Non-magnetic Systems: A Phase-Field Study

Soumya Bandyopadhyay, Somnath Bhowmick, and Rajdip Mukherjee[✉]

Department of Materials Engineering, Indian Institute of Technology Kanpur, Kanpur 208016, India
rajdipm@iitk.ac.in

Abstract. In this work, we propose a phase field model to investigate grain boundary (GB) migration under an external magnetic field in a non-magnetic Bismuth (Bi) system. We take into account various arrangements, such as bicrystals (rectangular and circular) and polycrystals (hexagonal grains), where we apply magnetic fields parallel and perpendicular to the c-axis. We observe that the direction of the applied magnetic field affects the GB migration direction; the grain with the c-axis parallel to the magnetic field grows preferentially at the expense of the grain with the c-axis perpendicular to the magnetic field. Moreover, our study indicates that the difference in the magnetic free energy density of grains with various orientations results in preferential grain growth.

Keywords: Grain growth · Magnetic field · Phase field · Moose Framework

1 Introduction

There is an old saying, "Microstructure defines properties," which has been immensely applicable in the field of materials engineering for several decades [1, 2]. Thus, investigating microstructural evolution and in turn, tweaking has been a key for the state-of-the-art optimized and improved materials applications. However, this tuning of materials' properties is interdependent on the varieties of external parameters, such as temperature [3], stress [4], electric field [5], magnetic field [6], etc. Grain boundaries, which are potential sites for morphological instabilities, play an important role in determining microstructural properties in metallic materials and are extremely vulnerable to external thermo-electro-magneto-mechanical load. Some classic examples of grain boundary-related phenomena are the formation of voids and hillocks, grain boundary migration, nucleation sites for secondary phases, etc. [7–9]. However, in nanocrystalline thin films, the formation and deepening of the grooves at the multi-junctions are the primary mechanisms for the breaking of the film. Furthermore, grain boundary migration caused by electric (electromigration) or thermal (thermomigration), or both, void or associated hillock formation, has always been a long-term reliability issue in the microelectronic

S. Patra et al. (Eds.): METCENT 2023, *Proceedings of the International Conference on Metallurgical Engineering and Centenary Celebration*, pp. 75–82, 2024.
https://doi.org/10.1007/978-981-99-6863-3_9

packaging and integrated circuit (IC) industries. More interestingly, some recent studies show grain boundary migration due to a magnetic field in non-magnetic alloys (Bi, Zn, Ti, etc.), especially in metals with an anisotropic crystal structure. For example, many experimental investigations report that by vigilant control of the magnetic field in Sn-Ag-Cu solder interconnects, one can tune the Sn orientation, which can pertain drastic affect in the material performance and reliability [10].

In the present work, we employ a phase field model to investigate the effect of magnetic fields on grain boundary migration in non-magnetic systems such as Bi. We also show the GB migration kinetics are related to the grain orientation, driven by the anisotropy in magnetic free energy. We also emphasize on the grain boundary migration as well as growth as a function of magnetic field for bicrystal as well as polycrystalline systems.

2 Formulation

Phase field or diffuse interface approach has been used as a potent and preferred approach to model complex mesoscopic microstructural evolution in solidification, precipitation, and strain-induced transformations for the last few decades [11]. In the general phase-field approach, we can describe the microstructure of a system considering two types of order parameters to differentiate between the compositional (conserved) as well as structural (non-conserved) heterogeneities [11].

In this study, we use a set of non-conserved order parameters $\eta_i(\vec{r})$ $(i = 1, 2 \ldots N)$ to define the N number of grains possessing different orientations, where $\vec{r}$ defines the position vector. The total free energy of the system is defined as

$$F = \int_v [u f_{(0)}(\eta_1, \eta_{(2)}, \ldots \eta_N) + \frac{\kappa_\eta}{2} \sum_{(i=1)}^{N} (\nabla \eta_i)^2 + f_m] dv, \tag{1}$$

where the parameter u and the gradient energy coefficient κ_η is obtained from the GB energy σ_{GB} and the corresponding interfacial width l_{GB} as $u = 6\sigma_{GB}/l_{GB}$ and $\kappa_\eta = 3\sigma_{GB}l_{GB}/4$.

The first term in Eq. (1) defines the bulk free energy density and for a polycrystalline material is defined as

$$f_0(\eta_1, \eta_2, \ldots \eta_N) = \sum_{i=1}^{N} \left(\frac{\eta_i^4}{4} - \frac{\eta_i^2}{2} \right) + \gamma \sum_{i=1}^{N} \sum_{i>j}^{N} \eta_i^2 \eta_j^2 + 0.25. \tag{2}$$

Here γ is considered to be 1.5. The second and third terms in Eq. (1) are related to the gradient as well as the magnetic contributions to the free energy.

Considering the contributions from differently oriented grains we define the magnetic free energy density as

$$f_m = \sum_{i=1}^{N} h(\eta_i)f_{m,i}, \tag{3}$$

where $h(\eta_i) = \eta_i^3(6\eta_i^2 - 15\eta_i + 10)$ defines the switching function corresponding to different grains. Furthermore, we define the magnetic free energy density of grain i as $f_{m,i} = \mu_0 H^2 \chi_i / 2$, where μ_0, χ_i and $\vec{H}$ are the magnetic permeability, magnetic susceptibility of grain, i and applied magnetic field, respectively. χ_i is a second-rank tensor which also depends on the grain orientations in the anisotropic systems.

In this present study, we intend to focus on a diamagnetic hexagonal material Bi; thus the susceptibility tensor can be considered as $\chi^a = \chi^b$, χ^c. Furthermore, following Liang et al. [10] we can decompose the magnetic field into its components considering χ_i^c along the c-axis and $\chi^a = \chi^b$ along the X axis:

$$\vec{H} = H_m sin\theta_i \widehat{n_x} + H_m cos\theta_i \widehat{n_z}, \tag{4}$$

where θ_i represents the angular relation between c-axis corresponding to the particular grain and the direction of the magnetic field. H_m defines the amplitude of the magnetic field. Thus, the magnetic free energy density finally takes the form as [10].

$$f_{m,i} = \mu_0 H_m^2 (\chi_a - cos^2\theta_i(\chi_c - \chi_a))/2. \tag{5}$$

Once we have obtained all the free energy densities, the spatiotemporal evolution of the order parameters is governed by the relaxation type Allen-Cahn equation:

$$\frac{\partial \eta_i}{\partial t} = -L\frac{\delta F}{\delta \eta_i}, \tag{6}$$

where $L = 4m/3l_{GB}$ defines the relaxation coefficient related to the grain boundary mobility, and $m = m_0 exp(-Q/K_bT)$. Here m_0 is the preexponential factor, Q defines the activation energy and K_b is the Boltzman constant. In this study, we choose an interfacial energy $\sigma_{GB} = 0.155\,J/m^2$ [10] and interfacial width of $l_{GB} = 1.6\,nm$, ensuring there are five grid points across the interfaces. m_0 is considered to be $1.1 \times 10^{24}\,m^4/Js$ [10, 13], $Q = 3.38\,eV$ [10, 13, 14], $\mu_0 = 1.257 \times 10^{-6}\,N/A^2$ [15], $K_b = 1.38 \times 10^{-23}\,J/K$, $\chi_a = -1.24 \times 10^{-4}$ [10, 15], and $\chi_c = -1.47 \times 10^{-4}$ [10, 13]. All the above equations are solved in the weak form using Open Source MOOSE Framework using periodic boundary conditions in all directions [12].

3   Results and Discussions

First, we try to study the migration of GB under the influence of magnetic field $H_m = 1.6 \times 10^8\,A/m$ in a bicrystal system. We start with a system size of 1024 nm $\times$ 128 nm, where initially we do not apply any external magnetic field as shown in Fig. 1a. Furthermore, we also assume two different grain orientations; $0°$ for Grain 1 and $90°$ for Grain 2 such that we obtain maximum anisotropy in the magnetic driving force. Note that the corresponding crystallographic c-axis of the grain that is oriented along $0°$ is aligned parallel to the Z-axis of the Cartesian frame of reference. On the other hand, the grain that is $90°$ oriented possesses the crystallographic c-axis perpendicular to Z-axis or in this case parallel to the X-axis. On the application of the magnetic field parallel to the Z-axis, we observe that Grain 1 increases at the expense of Grain 2 as displayed in Fig. 1b. On the contrary when the applied magnetic field is perpendicular to the Z-axis we observe the growth of Grain 2 while Grain 1 shrinks (shown is Fig. 1c). We also checked the GB position vs. time for both the cases, they show a linear trend as shown in Fig. 2a. Figure 2b displays the GB position vs. time for different magnetic fields (parallel to the Z-axis) and in all the cases we observe a linear response. We expect a very similar response if we apply the magnetic field in the other direction.

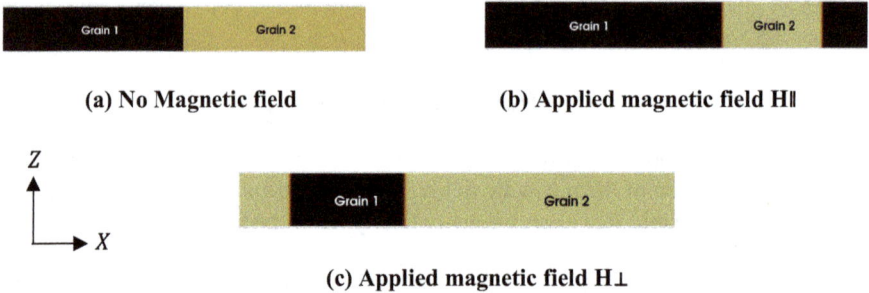

(a) No Magnetic field   **(b) Applied magnetic field H∥**

(c) Applied magnetic field H⊥

Fig. 1. Grain boundary migration in a bicrystal (a) without magnetic field (b) magnetic field applied parallel to the Z axis, (c) magnetic field applied perpendicular to the Z axis

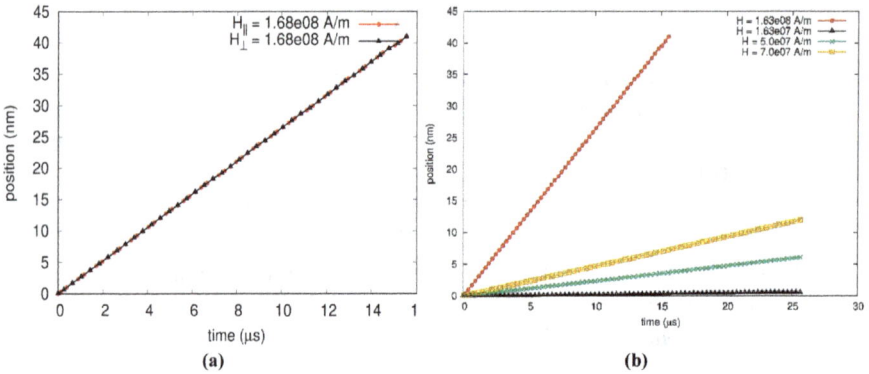

Fig. 2. (a) Comparison between grain boundary position when applied magnetic field is **H∥** and **H⊥**. (b) Comparison between grain boundary positions for different magnitudes of magnetic fields

After investigation for the bicrystal planar morphology as discussed above we simulate a system with a circular grain (Grain 1) inside Grain 2 as shown in Fig. 3a. The considered system size is 128 nm × 128 nm with a circular grain of radius 20 nm.

In this case, we require a critical magnetic field of $H_m = 8.762 \times 10^8$ A/m for the grain with 20 nm to grow. However, the amplitude of the required magnetic field varies depending on the system dimension and size which can be attributed solely due to the curvature associated with the circular morphology.

In this case, we observe a similar growth of Grain 1 as shown in Figs. 3b–c for different time steps, when we apply magnetic field parallel to the Z-axis. However, for the other direction Grain 1 shrinks (not shown). Figure 3d displays the contour plot of the temporal evolution and growth of Grain 1.

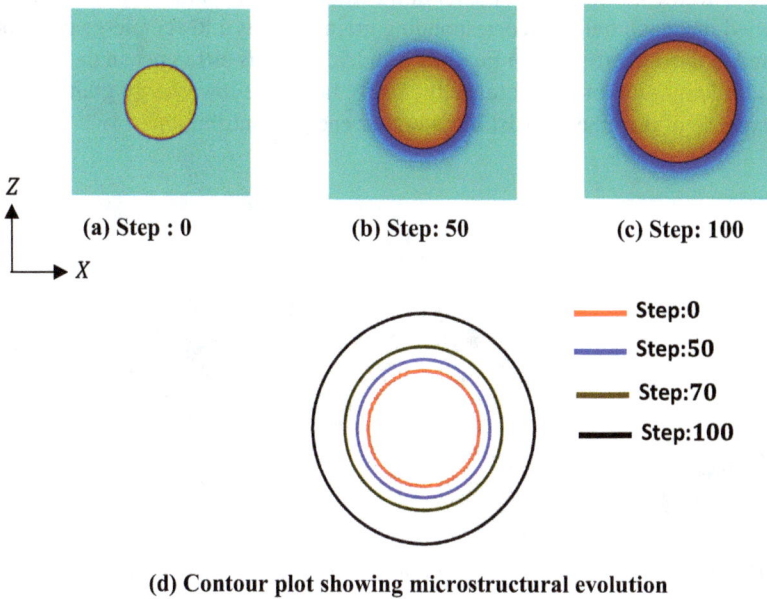

(a) Step : 0   (b) Step: 50   (c) Step: 100

(d) Contour plot showing microstructural evolution

Fig. 3. Temporal microstructural evolution for a circular grain at (a) Step: 0, (b) Step: 50, (c) Step: 100 when applied magnetic field is **H∥**. (d) Contour plots showing microstructural evolution at different time steps

We further extend our model to study a polycrystalline system where we consider a hexagonal morphology. The system size was chosen to be 300 nm × 261 nm as shown in Fig. 4a. We consider the central hexagonal grain is oriented at an angle of 90°, while the other grains are of 0°. When we apply the magnetic field perpendicular to the Z-axis we observe the central grain starts growing while the other grains shrink (Figs. 4b–c). On the other hand, when we apply the field in the other direction we observe an exact opposite phenomenon as displayed in Figs. 5a–c. Figure 4g and Fig. 5g represent the corresponding contour plots of the grain evolution as a function of the applied magnetic field.

The growth and the shrinkage of the grains in all the cases can be attributed to the change in the overall free energy (Bulk + Magnetic free energy) due to the variation in the direction of the magnetic field. Moreover, from all the above studies it is prominent that the grain with the c-axis parallel to the direction of the magnetic field grows preferentially at the expense of the other grain with c-axis perpendicular to the magnetic field direction. This is due to the fact when the c-axis of a certain grain is parallel to the direction of the applied magnetic field the corresponding grain exhibits a lower energy contribution compared to the other (Shown in Figs. 4d–f and 5d–f). This difference in the free energy triggers the migration of the more energetically favorable low-energy grain to migrate at the expense of the other consisting of higher energy contribution.

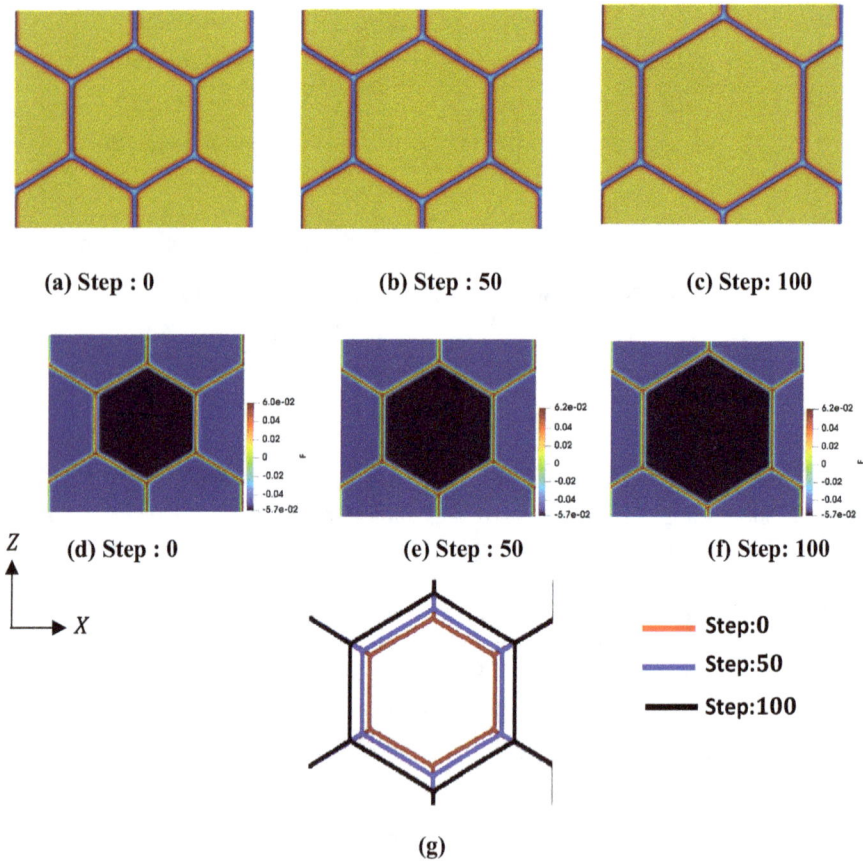

Fig. 4. Temporal microstructural evolution of hexagonal grain at (a) Step: 0, (b) Step: 50, (c) Step: 100 when an applied magnetic field is **H⊥**. Corresponding energy (bulk + magnetic) distribution at (a) Step: 0, (b) Step: 50, (c) Step: 100. (g) Contour plots showing microstructural evolution at different time steps

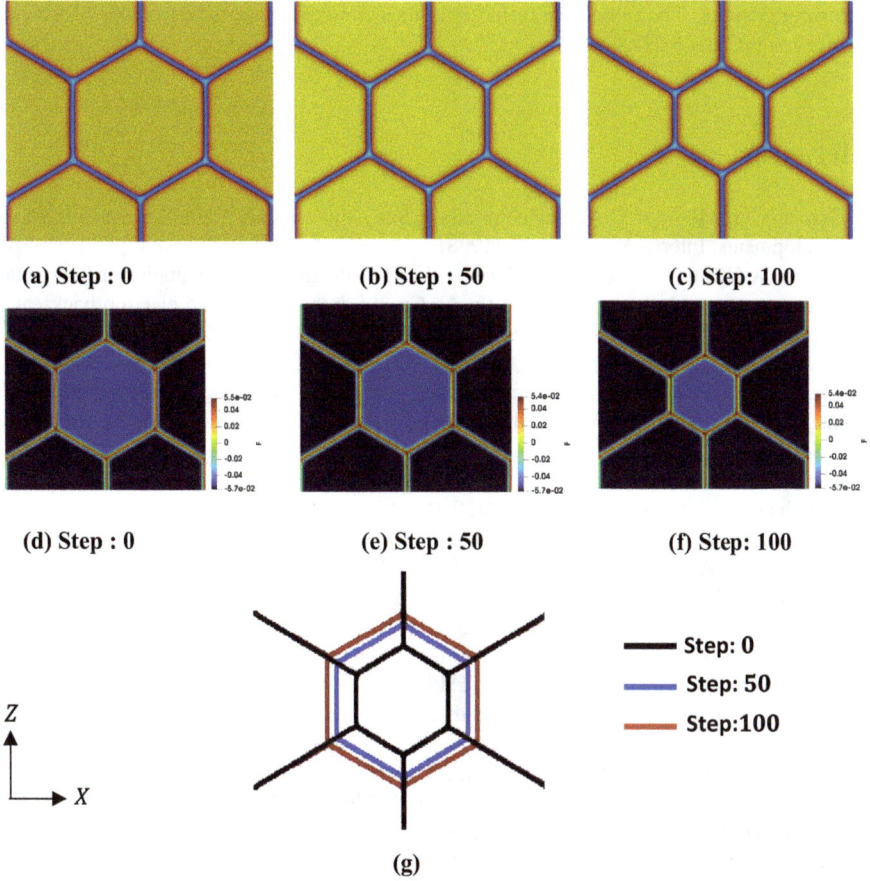

Fig. 5. Microstructural evolution of hexagonal grain at (a) Step: 0, (b) Step: 50, (c) Step: 100 when the applied magnetic field is **H**‖. Corresponding energy (bulk + magnetic) distribution at (d) Step: 0, (e) Step: 50, (f) Step: 100. (g) Contour plots showing microstructural evolution at different time steps

4 Conclusion

In this work we study the effect of magnetic field on the migration of the GBs in a non-magnetic system using a phase field approach. Although this study is motivated by the work of Liang *et al.* [10] that considers only the migration of the planar interfaces; however, in this work, in addition to the planar interface, we also studied the curvature driven magnetic field induced GB migration. Furthermore, we have also extended our model to study polycrystalline systems. Our studies suggest that the grains with *c*-axis parallel to the magnetic field direction preferentially grow at the expense of the other grains due to the lower total energy contribution. This model can further be extended to investigate abnormal grain growth under external magnetic field in different systems.

Acknowledgement. The authors are thankful for financial support received from the SERB core research grant (CRG/2021/003687).

References

1. Gottstein, G., Molodov, D.A., Shvindlerman, L.S.: Grain boundary migration in metals: recent developments. Interface Sci. **6**, 7–22 (1998)
2. Linares, X., Kinney, C., Lee, K.O., Morris, J.W.: The influence of Sn orientation on intermetallic compound evolution in idealized Sn-Ag-Cu 305 interconnects: an electron backscatter diffraction study of electromigration. J. Electron. Mater. **43**, 43–51 (2014)
3. Liang, S.B., Ke, C.B., Huang, J.Q., Zhou, M.B., Zhang, X.P.: Phase field simulation of microstructural evolution and thermomigration-induced phase segregation in Cu/Sn58Bi/Cu interconnects under isothermal aging and temperature gradient. Microelectron. Reliab. **92**, 1–11 (2019)
4. Rupert, T.J., Gianola, D.S., Gan, Y., Hemker, K.J.: Experimental observations of stress-driven grain boundary migration. Science **326**(5960), 1686–1690 (2009)
5. Jeong, J.W., Han, J.H., Kim, D.Y.: Effect of electric field on the migration of grain boundaries in alumina. J. Am. Ceram. Soc. **83**(4), 915–918 (2000)
6. Sheikh-Ali, A.D., Molodov, D.A., Garmestani, H.: Migration and reorientation of grain boundaries in Zn bicrystals during annealing in a high magnetic field. Scripta Mater. **48**(5), 483–488 (2003)
7. Chakraborty, S., Kumar, P., Choudhury, A.: Phase-field modeling of grain-boundary grooving and migration under electric current and thermal gradient. Acta Mater. **153**, 377–390 (2018)
8. Mukherjee, A., Ankit, K., Mukherjee, R., Nestler, B.: Phase-field modeling of grain-boundary grooving under electromigration. J. Electron. Mater. **45**, 6233–6246 (2016)
9. Laxmipathy, V.P., Wang, F., Selzer, M., Nestler, B.: Phase-field simulations of grain boundary grooving under diffusive-convective conditions. Acta Mater. **204**, 116497 (2021)
10. Liang, S., Liu, C., Zhou, Z.: Phase field study of grain boundary migration and preferential growth in non-magnetic materials under magnetic field. Mater. Today Commun. **31**, 103408 (2022)
11. Chen, L.-Q.: Phase-field models for microstructure evolution. Annu. Rev. Mater. Res. **32**, 113–140 (2002)
12. Gaston, D., Newman, C., Hansen, G., Lebrun-Grandie, D.: MOOSE: a parallel computational framework for coupled systems of nonlinear equations. Nucl. Eng. Des. **239**, 1768–1778 (2009)
13. Modolov, D.A., Gorkaya, T., Günster, C., Gottstein, G.: Migration of specific planar grain boundaries in bicrystals: application of magnetic fields and mechanical stresses. Front. Mater. Sci. China **4**, 219–305 (2010)
14. Modolov, D.A, Günter, G., Heringhaus, F., Shivindlerman, L.S: True absolute grain boundary mobility: motion of specific planar boundaries In bi-bicrystal under magnetic driving forces. Acta Materialia **46**, 5627–5632 (1998)
15. Asai, S., Sassa, K.S., Tahashi, M.: Crystal orientation of non-magnetic materials by imposition of a high magnetic field. Sci. Technol. Adv. Mater. **4**, 455–460 (2003)

Deformation in Metals: Insights from *ab-initio* Calculations

Albert Linda, Md. Faiz Akhtar, and Somnath Bhowmick[✉]

Department of Materials Science and Engineering, Indian Institute of Technology Kanpur, Kanpur 208016, India
{albert20,faiza21,bsomnath}@iitk.ac.in
https://sites.google.com/view/cmsmseiitk/home

Abstract. Density functional theory-based *ab initio* calculation is the most accurate tool for computational prediction of materials properties. In this paper, we present an *ab initio* study of stacking fault energy (SFE) calculation for face-centered cubic (FCC) metals using different models. One particular model allows us to calculate the generalized stacking fault energy (GSFE) curve and surface, which can be used to calculate the Peierls-Nabarro (PN) stress, facilitating the study of dislocations in metals from *ab initio* calculations. Comparing with experimental results, we find a notable overestimation compared to experimental values of Cu.

Keywords: DFT · GSFE · PN-stress

1   Introduction

The study of dislocations involves two length scales, a continuum elasticity theory to understand the elastic strain energy and dislocation interactions and an atomic description to understand the dislocation core. Peierls and Nabarro were the first to solve the problem and gave some reasonable estimates of the lattice resistance to the dislocation motion. Due to their seminal contribution, the shear stress required to make a dislocation glide in a crystal is called the Peierls-Nabarro (PN) stress [1,2], which provides insights into the deformation mechanism and mechanical behavior of metals and alloys [3–5].

The original PN model assumes sinusoidal stress, written as a function of the misfit in the displacement field across the slip plane, known as disregistry. However, instead of sinusoidal approximation, generalized stacking fault energy (GSFE), calculated directly from atomistic simulations, makes the outcome more accurate. Thus, GSFE calculation is the key to *ab initio* based deformation modeling in metals and alloys [6].

Besides input for the PN model, GSFE itself reveals the deformation behavior of metals. For example, deformation twinning is common in Ag, which has a low stacking fault energy, but rare in Al, which has a high stacking fault energy. Experimental methods such as transmission electron microscopy (TEM), X-ray diffraction (XRD), and atom probe tomography (APT) enable direct observation

S. Patra et al. (Eds.): METCENT 2023, *Proceedings of the International Conference*
it on Metallurgical Engineering and Centenary Celebration, pp. 83–92, 2024.
https://doi.org/10.1007/978-981-99-6863-3_10

and measurement of stacking faults, providing valuable data [7]. Computationally, classical molecular dynamics (MD) and density functional theory (DFT) simulations offer atomic-level insight along with generalized stacking fault energy (GSFE) by comparing energy differences between perfect and defective crystal structures [6, 8–10]. Although MD calculations are faster and can afford to handle a large system size, DFT simulations are preferred for their accuracy.

In this manuscript, we compare and discuss two SFE estimation techniques, the Axial Next Nearest Neighbor Interaction (ANNNI) [11] method and the periodic supercell [12] method. We calculate the PN stress of Cu using a semi-discrete variational Peierls-Nabarro (SVPN) model [13], which utilizes GSFE values obtained from DFT. We also investigate the differences between 1D and 2D SVPN models in determining the PN stress. Comparing with experimental results, we find a notable overestimation compared to experimental values of Cu. This study aims to provide comprehensive insights into accurate SFE estimation and the influence of different modeling approaches on predicting PN stress, starting from *ab initio* calculations.

2   Computational Methods

We perform all *ab initio* calculations using the Vienna Ab initio Simulation Package (VASP) [14] within the Density Functional Theory (DFT) framework. We use generalized gradient approximation (GGA) [15] for exchange-correlation energy functional. We take a kinetic energy cutoff of 400 eV for all our calculations and use Gaussian smearing with a smearing width of 0.03 eV. We use 1E-06 eV and 1E-04 eV convergence criteria for electronic and ionic optimization. We take a k-point mesh of $21 \times 15 \times 2$ for the GSFE calculation. For the ANNNI model, we use a k-point set of $38 \times 38 \times 38$ for the FCC (face-centered cubic) and $42 \times 42 \times 13$ for the HCP (hexagonal close-packed) and DHCP (double hexagonal close-packed) unit cell.

3   Stacking Fault Energy Estimation

Theoretically, when a dislocation is subjected to external stress, it splits into Shockley partials to reduce energy. Specifically, for FCC materials with a slip system of the type $\langle 110 \rangle 111$, the dislocation splits into the following form:

$$\frac{a}{2}[\bar{1}10] = \frac{a}{6}[\bar{2}11] + \frac{a}{6}[\bar{1}2\bar{1}] \tag{1}$$

Here, a represents the lattice parameter. As a result of splitting, the stacking sequence in an FCC crystal along the closed packed direction changes from "..ABCABCA.." to "..ABCABABCA.." for the intrinsic fault and "..ABCABACA.." for the extrinsic fault. The SFE is a measure of the additional energy per unit area due to the changed stacking sequence.

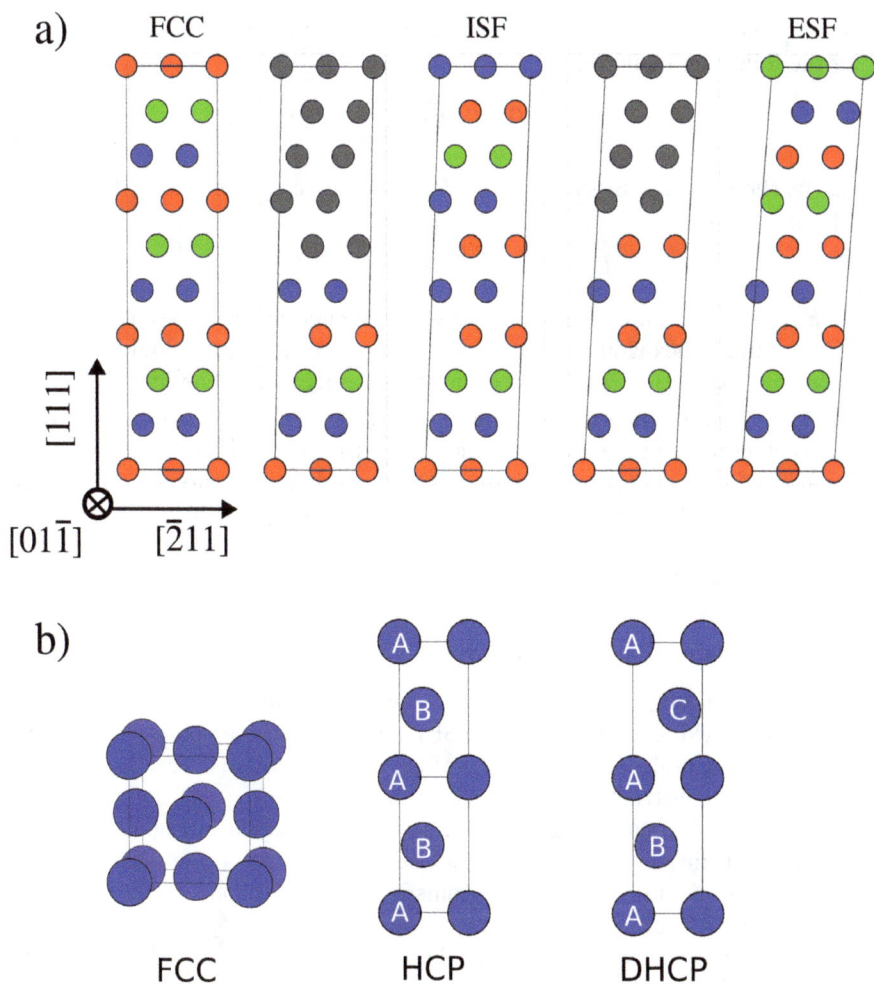

Fig. 1. a) Sequential evolution of structures for GSFE calculation using the periodic supercell approach. Red, blue, and green represent atoms in layers A, B, and C, respectively, while gray atoms represent some transition state between FCC-ISF and ISF-ESF. b) Structures used for stable SFE determination using the ANNNI model. Each structure type (FCC, HCP, and DHCP) comprises four atoms per unit cell. The FCC structure represents the conventional unit cell, while the HCP and DHCP structures are arranged in their respective stacking sequences.

3.1 Stable SFEs Using ANNNI Model

The ANNNI model provides an approximate method to determine stable Stacking Fault Energies (SFEs) by considering the energy difference among the bulk structures with differently stacked closed-packed (111) planes Fig. 1. In this approach, the energy of the Intrinsic Stacking Fault (ISF) can be estimated using the following equation:

$$E_{ISF} = \frac{E_{HCP} + 2E_{DHCP} - 3E_{FCC}}{A}. \tag{2}$$

Similarly, the Extrinsic Stacking Fault (ESF) energy can be calculated using the equation:

$$E_{ESF} = \frac{4(E_{DHCP} - E_{FCC})}{A}. \tag{3}$$

Here, E_{FCC}, E_{HCP}, and E_{DHCP} represent the energies of face-centered cubic (ABC stacking), hexagonal close-packed (AB stacking), and double hexagonal close-packed (ABAC stacking) structures, respectively. The interfacial area denoted by A can be determined using $\frac{\sqrt{3}}{4}a^2$, where a is the lattice parameter of the FCC structure. SFE values obtained from the ANNNI model are reported and compared with another model (Sect. 3.2) in Table 1. The outcomes of the two models are in good agreement with each other. Although the ANNNI approach allows for a quick estimation of stable SFEs at a relatively lower computational cost, it can not be used to calculate the GSFE curve.

3.2   GSFE Using Periodic Supercell

The GSFE curve comprises four key points: two local minima indicating intrinsic and extrinsic stacking faults (ISF and ESF) and two unstable points representing unstable stacking faults (USF) and unstable twinning faults (UTF). To trace the GSFE curve, we construct a supercell with cell vectors $[01\bar{1}]$, $[\bar{2}11]$, and $[111]$ axes, consisting of two atoms in each of the nine layers as shown in Fig. 1. To create an intrinsic stacking fault (ISF), we displace atoms in the top four layers along the $[\bar{2}11]$ direction while keeping the atoms in layers 5–9 fixed. Similarly, for the extrinsic stacking fault (ESF), we displace the atoms in the top three layers while keeping the atoms in layers 4–9 fixed. The unstable stacking fault (USF) and unstable twinning fault (UTF) are the energy barrier between the ideal lattice and the ISF and between the ISF and ESF, respectively. To maintain periodicity within the supercell, we tilt the out-of-plane cell vector aligned along the $[111]$ direction in FCC.

To determine the SFEs, we utilize the following equation:

$$\gamma = \frac{E_{defect} - E_{FCC}}{A}. \tag{4}$$

In this equation, E_{FCC} represents the energy of the FCC structure, E_{defect} corresponds to the energy of the structure with a defect, and A denotes the area of the fault plane. By applying this equation, we calculate the energies associated with each defect structure, notably the stable SFEs(γ_{ISF}, γ_{ESF}) and unstable SFEs(γ_{USF} and γ_{UTF}). Plotting these energies against the displacement vector $\frac{a}{6}[\bar{2}11]$ yields the complete GSFE curve. Figure 2 illustrates the GSFE curves for 11 FCC metals calculated along the displacement vector $\frac{a}{6}[\bar{2}11]$. Similar curves can be obtained along other displacement vectors like $\frac{a}{2}[\bar{1}10]$.

Table 1. Comparison of SFEs obtained using the periodic supercell approach with ANNNI model.

Element	Super Cell (mJ/m^2)				ANNNI (mJ/m^2)	
	USF	ISF	UTF	ESF	ISF	ESF
Ag	115.1	17.2	126.5	16.7	17.5	18.4
Al	196.1	143.6	269.4	119.8	132.0	117.4
Au	102.3	33.2	118.1	33.0	23.6	23.3
Ca	52.4	11.8	56.8	17.0	11.5	15.5
Cu	188.0	43.4	216.3	45.7	48.7	53.3
Ir	761.3	362.4	982.7	342.3	348.3	334.3
Ni	313.6	138.6	386.6	137.8	140.8	135.0
Pd	241.1	141.3	307.9	136.3	146.6	139.5
Pt	339.4	292.5	445.1	302.3	277.0	282.6
Rh	536.7	209.0	661.4	198.5	190.2	188.2
Sr	39.5	8.7	42.9	9.4	7.9	9.1

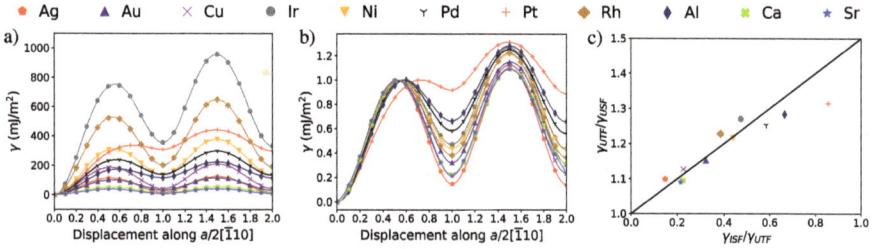

Fig. 2. a) GSFE curve along the partial direction of FCC metals, showcasing the two minima as the stable SFEs (γ_{ISF} and γ_{ESF}), while the two maxima correspond to the unstable points (γ_{USF} and γ_{UTF}). b) Normalized GSFE with respect to their γ_{USF} value. c) Verification of obtained SFEs using the universal scaling law (solid line) [16]. Metals towards the bottom left have more tendency to form deformation twins compared to the metals towards the top right.

4 PN Stress Using SVPN Model

Disregistry (or misfit) vector is defined as $\mathbf{u} = \mathbf{u}_\uparrow - \mathbf{u}_\downarrow$, where $\mathbf{u}_\uparrow$ and $\mathbf{u}_\downarrow$ are displacement field on top and bottom of the slip plane. The GSFE curve is a function of disregistry, i.e., $\gamma(\mathbf{u})$. The shear stress is given by the gradient of $\gamma(\mathbf{u})$, i.e., $\tau = -\nabla\gamma(\mathbf{u})$. The maximum of τ is the theoretical shear stress required for sliding one plane on another in a perfect crystal along a slip direction. Thus, instead of using a sinusoidal approximation directly, the GSFE curve can give a very accurate shear stress value.

The total energy of dislocation has two parts:

$$E_{total} = E_{elastic} + E_{misfit}. \tag{5}$$

Since energy is a function of disregistry vector, which is unknown, the latter is obtained by solving $\delta E_{total}/\delta u = 0$. Since the classical continuum PN model does not consider the lattice's discrete nature, it has some limitations, particularly for narrow dislocation core. Calculations are done using the semi-discrete variational Peierls-Nabarro (SVPN) model, which was initially developed by Bulatov [13, 17]. The literature presents two variants of this model: the 1D and 2D models. In order to estimate PN stress using this equation, it is necessary to provide the GSFE curve (for the 1D model) or GSFE surface (for the 2D model) as input. In the following sections, we compare the 1D and 2D SVPN models by taking the example of Cu (Fig. 3).

Table 2. Peierls stress in MPa for Cu dislocations predicted using 1D and 2D SVPN model.

	Calculated		Literature	
	1D model	2D model	MD [20]	Experiment [24]
Edge	1.39	1.87	2	~0.3
Screw	5.01	5.34	4	

4.1   1D PN Model

To generate the desired structure for 1D, we employ the same configuration and number of layers as we did in the above GSFE calculations. The key distinction

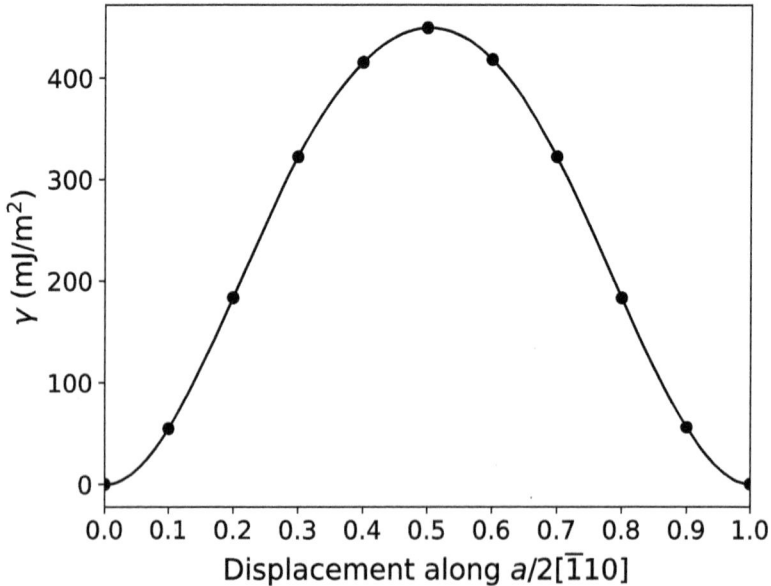

Fig. 3. GSFE plot of Cu along the direction of a perfect dislocation, specifically for utilization in 1D PN model.

lies in the direction of atom displacement, which is set as $[\bar{1}10]$ for this specific case. We apply an interval of $0.1 \times \frac{a}{2}[\bar{1}10]$ for each defect structure and ensure the periodicity of the supercell by implementing the necessary tilting. We follow the same parameter settings as the above DFT calculations to enable structural relaxation and utilize Eq. 4 to determine the SFEs. Once the results are obtained, we use the values in a 1D SVPN model using the MATLAB code PNADIS [18] to calculate the Peierls stress. We take the shear modulus and Poisson's ratio of Cu required for the model as 41 GPa and 0.37 [19] and use the burgers vector from our DFT run. The obtained results are presented in Table 2. PN stress of edge dislocation is significantly lower than that of screw dislocation, which agrees with the reported trend. However, we observe a slight difference in the values compared to the MD results present in the literature [20], which might be due to the limitations of the PN model. For example, $u(x)$ deviates from the direction of the Burgers vector across the glide plane. Thus, one ideally needs to solve a system of two coupled Peierls integral equations numerically to determine $u(x)$. Additionally, the Burgers vector of a pure dislocation in FCC splits into two non-parallel directions, as given by Eq. 1. A realistic treatment of this vital configuration is not feasible in 1D model [21,22].

4.2   2D PN Model

We now estimate the PN stress using the 2D model to get around the issues with the 1D SVPN model's estimation of the PN stress. Similar to the 1D case, we maintain a consistent number of atoms and layers for our study. To estimate the entire GSFE surface, we shift atoms in two directions: the partial direction ($[\bar{2}11]$) and the closed packed direction ($[\bar{1}10]$). To capture a comprehensive range of data, we obtained 100 data points by employing a shift of $0.1 \times \frac{a}{6}[\bar{2}11]$ and $0.1 \times \frac{a}{2}[\bar{1}10]$ for each direction. We calculated the corresponding energies using DFT, as illustrated in Fig. 4. The obtained GSFE surface is then utilized in a 2D SVPN model. The other input parameters, shear modulus, and Poisson's ratio, are kept the same as 1D. The results obtained agree well with atomistic calculations [20], but we notice a significant difference when comparing it to the experimental values of Cu. The difference in results may be because the terms for the interaction of the applied stress on the total energy of dislocation [20] are not included in the model. Also, the gradient energy term, which is responsible for displacement gradient, considerably affects PN stress [23]. This has recently been confirmed that the value of PN stress oscillates drastically by a small change in input parameters like shear modulus, Poisson ratio, and GSFE as well as the distance between the two partials of the extended dislocation [25]. These factors mentioned above may be the reason behind the difference in results from the experiment value. Nevertheless, these values are pretty close, unlike the generalized Peierls-Nabarro model [26–28] whose values are different by several orders of magnitude.

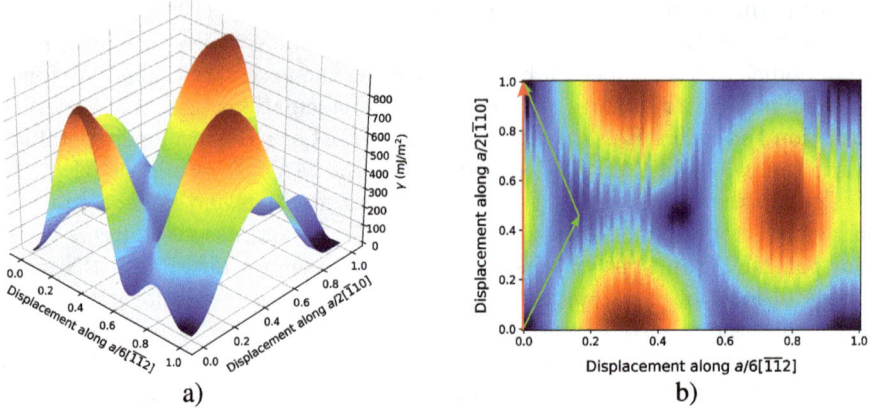

Fig. 4. a) GSFE surface of Cu plotted aganist the axis $[\bar{1}10]$ and $[\bar{2}11]$ shown as a 3D plot. b) Projection plot of same shown in 2D. The solid red and green line represents the perfect dislocation and its Shockley partial, respectively.

5   Conclusion

In conclusion, we have calculated SFE values and GSFE curves of 11 FCC metals using *ab inito* calculations. The results from the ANNNI and the periodic supercell model agree well. While the former is computationally less expensive, the latter provides more information. Based on the results, we can identify the metals in which deformation twins can easily form. We further extend the calculation to generate the entire GSFE surface and estimate the PN stress for Cu. Our work confirms the possibility of studying dislocation starting from *ab initio* calculations. The challenge would be to do similar studies for alloys, which will significantly impact the field of computational alloy designing.

Acknowledgement. We acknowledge National Super Computing Mission (NSM) for providing computing resources of "PARAM Sanganak" at IIT Kanpur, which is implemented by CDAC and supported by the Ministry of Electronics and Information Technology (MeitY) and Department of Science and Technology (DST), Government of India. We also thank ICME National Hub, IIT Kanpur and CC, IIT Kanpur for providing HPC facility. Md. Faiz Akhtar acknowledges the financial support from the Prime Minister's Research Fellows (PMRF) scheme, Government of India.

References

1. Peierls, R.: The size of a dislocation. Proc. Phys. Soc. **52**, 34–37 (1940). https://doi.org/10.1088/0959-5309/52/1/305
2. Nabarro, F.: Dislocations in a simple cubic lattice. Proc. Phys. Soc. **59**, 256 (1947). https://doi.org/10.1088/0959-5309/59/2/309
3. Anderson, P., Hirth, J., Lothe, J.: Theory of Dislocations. Cambridge University Press, Cambridge (2017)

4. Hull, D., Bacon, D.: Chapter 5 - dislocations in face-centered cubic metals. Introduction To Dislocations (Fifth Edition), pp. 85-107 (2011). https://www.sciencedirect.com/science/article/pii/B9780080966724000050

5. Joós, B., Duesbery, M.: The Peierls stress of dislocations: an analytic formula. Phys. Rev. Lett. **78**, 266–269 (1997). https://doi.org/10.1103/PhysRevLett.78.266

6. Zhang, R., et al.: First-principles design of strong solids: approaches and applications. Phys. Rep. **826**, 1-49 (2019). https://www.sciencedirect.com/science/article/pii/S0370157319303060

7. Koch, C.: Nanocrystalline high-entropy alloys. J. Mater. Res. **32**, 3435–3444 (2017)

8. Linda, A., Tripathi, P., Nagar, S., Bhowmick, S.: Effect of pressure on stacking fault energy and deformation behavior of face-centered cubic metals. Materialia **26**, 101598 (2022). https://www.sciencedirect.com/science/article/pii/S2589152922002794

9. Kumar Panda, A., Divakar, R., Singh, A., Thirumurugesan, R., Parameswaran, P.: Molecular dynamics studies on formation of stacking fault tetrahedra in FCC metals. Comput. Mater. Sci. **186**, 110017 (2021). https://www.sciencedirect.com/science/article/pii/S0927025620305085

10. Heino, P., Perondi, L., Kaski, K., Ristolainen, E.: Stacking-fault energy of copper from molecular-dynamics simulations. Phys. Rev. B **60**, 14625–14631 (1999). https://doi.org/10.1103/PhysRevB.60.14625

11. Denteneer, P., Haeringen, W.: Stacking-fault energies in semiconductors from first-principles calculations. J. Phys. C: Solid State Phys. **20**, L883 (1987). https://doi.org/10.1088/0022-3719/20/32/001

12. Kibey, S., Liu, J., Johnson, D., Sehitoglu, H.: Predicting twinning stress in FCC metals: linking twin-energy pathways to twin nucleation. Acta Materialia **55**, 6843-6851 (2007). https://www.sciencedirect.com/science/article/pii/S1359645407005927

13. Bulatov, V., Kaxiras, E.: Semidiscrete variational peierls framework for dislocation core properties. Phys. Rev. Lett. **78**, 4221–4224 (1997). https://doi.org/10.1103/PhysRevLett.78.4221

14. Kresse, G., Furthmüller, J.: Efficient iterative schemes for ab initio total-energy calculations using a plane-wave basis set. Phys. Rev. B **54**, 11169–11186 (1996). https://doi.org/10.1103/PhysRevB.54.11169

15. Perdew, J., Burke, K., Ernzerhof, M.: Generalized gradient approximation made simple. Phys. Rev. Lett. **77**, 3865–3868 (1996). https://doi.org/10.1103/PhysRevLett.77.3865

16. Jin, Z., Dunham, S., Gleiter, H., Hahn, H., Gumbsch, P.: A universal scaling of planar fault energy barriers in face-centered cubic metals. Scripta Materialia **64**, 605-608 (2011). https://www.sciencedirect.com/science/article/pii/S1359646210007918

17. Schoeck, G.: The core structure, recombination energy and Peierls energy for dislocations in Al. Philos. Mag. A **81**, 1161–1176 (2001). https://doi.org/10.1080/01418610108214434

18. Zhang, S., Legut, D., Zhang, R.: PNADIS: an automated Peierls-Nabarro analyzer for dislocation core structure and slip resistance. Comput. Phys. Commun. **240**, 60-73 (2019). https://www.sciencedirect.com/science/article/pii/S0010465519300839

19. Jain, A., et al.: Commentary: the materials project: a materials genome approach to accelerating materials innovation. APL Mater. **1** (2013). https://doi.org/10.1063/1.4812323

20. Shen, Y., Cheng, X.: Dislocation movement over the Peierls barrier in the semi-discrete variational Peierls framework. Scripta Materialia **61**, 457-460 (2009). https://www.sciencedirect.com/science/article/pii/S1359646209003224

21. Schoeck, G.: Peierls energy of dislocations: a critical assessment. Phys. Rev. Lett. **82**, 2310–2313 (1999). https://doi.org/10.1103/PhysRevLett.82.2310

22. Schoeck, G.: The Peierls model: progress and limitations. Mater. Sci. Eng.: A. **400-401**, 7-17 (2005). https://www.sciencedirect.com/science/article/pii/S0921509305002704. Dislocations 2004

23. Liu, G., Cheng, X., Wang, J., Chen, K., Shen, Y.: Peierls stress in face-centered-cubic metals predicted from an improved semi-discrete variation Peierls-Nabarro model. Scripta Materialia **120**, 94-97 (2016). https://www.sciencedirect.com/science/article/pii/S1359646216301348

24. Kamimura, Y., Edagawa, K., Takeuchi, S.: Experimental evaluation of the Peierls stresses in a variety of crystals and their relation to the crystal structure. Acta Materialia **61**, 294-309 (2013). https://www.sciencedirect.com/science/article/pii/S1359645412006921

25. Zhang, X., Cao, S., Yang, R., Hu, Q.: Drastic oscillation of peierls stress from peierls-nabarro model calculation and its remedy. J. Mater. Res. Technol. **23**, 5502-5519 (2023). https://www.sciencedirect.com/science/article/pii/S2238785423003587

26. Schoeck, G.: The Peierls energy revisited. Philos. Mag. A **79**, 2629–2636 (1999). https://doi.org/10.1080/01418619908212014

27. Schoeck, G.: The generalized Peierls-Nabarro model. Philos. Mag. A **69**, 1085–1095 (1994). https://doi.org/10.1080/01418619408242240

28. Xiang, Y., Wei, H., Ming, P., Weinan, E.: A generalized Peierls-Nabarro model for curved dislocations and core structures of dislocation loops in Al and Cu. Acta Materialia **56**, 1447-1460 (2008). https://www.sciencedirect.com/science/article/pii/S1359645407008087

A Comparative Heat Transfer Study of Gadolinium and $(MnNiSi)_{1-x}(Fe_2Ge)_x$ Alloy for Thermomagnetic Energy Generator

M. Silambarasan[1(✉)], Om Kapoor[1], Ravikiran Kadoli[1], P. Kondaiah[2], and K. Deepak[3]

[1] Department of Mechanical Engineering, National Institute of Technology, Karnataka, Surathkal, Karnataka 575025, India
{silambarasanm.207me018,omkapoor.212th014,rkkadoli}@nitk.edu.in
[2] Department of Mechanical Engineering, Ecole Centrale School of Engineering, Mahindra University, Hyderabad, Telangana 500043, India
kondaiah.p@mahindrauniversity.edu.in
[3] Department of Metallurgical Engineering, Indian Institute of Technology (BHU), Varanasi 221005, India
deepak.met@iitbhu.ac.in

Abstract. Utilizing and harvesting low grade heat energy with efficient methods has been a challenge for scientists for decades, as with several designs and materials studied the models still pose hurdles in producing usable energy, manufacturing, and affordability. The present work focuses on a comparative study for the heat transfer model of a thermomagnetic energy generator design by using a thermomagnetic material, i.e., an alloy $(MnNiSi)_{0.68}(Fe_2Ge)_{0.32}$ and a reference material taken as Gadolinium (Gd). Thermomagnetic energy generator has shown great potential in the field of low-grade energy harvesting due to their ability to convert thermal energy into electrical energy. Developed model of thermomagnetic material which is made to oscillate between a strong permanent magnet and a heat sink. The parameters considered for the transient study are the gap distance between the heat source and sink, and the shape of the thermomagnetic material used, where the results are presented as plots of temperature with respect to time for a single cycle in steady state, and for multiple cycles. These parameters are optimized to maximize the performance of the thermomagnetic harvester. Our analysis indicates the advancement of Thermomagnetic alloy $(MnNiSi)_{0.68}(Fe_2Ge)_{0.32}$ over the bench mark material Gadolinium in the enhancement of performance over the low temperature energy harvesting application.

Keywords: Heat Transfer Model · Thermomagnetic Alloy $(MnNiSi)_{0.68}(Fe_2Ge)_{0.32}$ · Thermomagnetic generator

1 Introduction

Energy is the source of life. In an ever-growing and industrialized world, sources of energy are the pillars on which the modern society stands. Thus, with an increasing population of the world, the citizen's demands and industrial needs for the energy consumption has seen a similar upsurge. This high demand has caused a serious need for an

S. Patra et al. (Eds.): METCENT 2023, *Proceedings of the International Conference on Metallurgical Engineering and Centenary Celebration*, pp. 93–107, 2024.
https://doi.org/10.1007/978-981-99-6863-3_11

increased electric supply throughout the globe. The human race still being majorly dependent on the non-renewable and exhaustible sources of energy, has led to the exploitation of natural resources, without considering any sustainable development [1]. This has in turn caused severe harm to our nature as well as other beings on the planet. Thus, a substitute way to produce cleaner and greener energy is highly imperative right now, which can be done by using either renewable energy sources or utilizing the waste heat usually expelled from various households, industries etc. This study takes into account the waste heat generated in our surroundings and to device a new way on how to utilize it to its maximum potential.

1.1 Thermal Energy Harvesting Devices

The thermal energy harvesting devices are devices that are used to convert heat energy (low grade energy) into higher grade energy such as electricity. The devices that were lately developed for converting waste heat were not efficient enough to be widely used in mass production due to their vulnerable designs, low output of power/current, high maintenance requirements [2]. This led to a decrease in interest of scientists on such topics, but in the last few years newer technologies have appeared which can be utilized for such ideas [3]. Therefore, in this study we introduce thermomagnetic materials which are based on the thermomagnetic cycle.

1.2 The Thermomagnetic Cycle

The Thermomagnetic cycle is the cycle which governs the process by which a thermomagnetic material can produce electricity. It consists of two isothermal and two iso-field processes, which is generally done in four steps, in which the heat load is cycled through the material above and below its Curie temperature [4]. As the thermomagnetic cycle progresses, there is a change in magnetization of the thermomagnetic material seen due to application and removal of heat load on it, which causes the material to oscillate between ferromagnetic and paramagnetic behaviors. This deems helpful as by Faraday's law, change in the magnetic field in a coil or a wire produces a current in the same coil, pertaining to the voltage change generated in the coil. The cycle can be explained by bifurcating it in 4 steps:

Step 1: The cycle starts with the thermomagnetic material (TMM) below its Curie temperature, where it possesses the ferromagnetic properties. At this point, when an external magnetic field is applied to it, the dipoles in the TMM are aligned in the direction of the magnetic field, and thus it is strongly attracted (or repelled depending on the polarity) to the external agent. This causes the magnetization of the TMM.

Step 2: In this step, the TMM receives heat (or waste heat) from the heat load/source such that its temperature rises above its Curie temperature. This makes the TMM go into the transition from a ferromagnetic material to a paramagnetic material.

Step 3: When the TMM has lost its magnetization being above the Curie temperature, it loses its capability to be influence by the external field, thus it starts losing contact or moves away from the source.

Step 4: Being away from the heat source, the TMM comes in contact with the heat sink, where it then cools down below its Curie temperature and thus regains its magnetization by again transitioning from a paramagnetic material to a ferromagnetic one.

1.3 Thermomagnetic Generator Design

The thermomagnetic generator model that has been used in this work is derived from the works of Ujihara (2007) [5] and remodeled as the one shown in the figure below by Joshi and Priya (2013) [6]. This is a schematic of the heat transfer model that has been used to specifically calculate the optimum temperature, gap distance and heat transfer pattern in the thermomagnetic material so as to maximize the results thus obtained. In this model (as shown in the figure below, Fig. 1), the upper plate acts as the permanent magnet, whose material has been taken of Neodymium (Nd). It also acts as the heat source as its temperature has been kept above the Curie temperature of the TMM ($T_h > T_{curie\ TMM}$). The heat sink at the bottom is maintained at a temperature below the Curie temperature of the TMM ($T_c < T_{curie\ TMM}$). The soft magnet (as shown by the middle rectangle here of t_h length) has been made of the TMM, which in this paper was taken as Gadolinium (Gd). The TMM has been supported in between the heat source and sink by the help of non-magnetic springs (shown here as two curves supporting the TMM rectangle) to help in the vibrations of the model. The TMM is then discretised in number of finite volumes to apply the FVM equations. The heat sink surface is elevated by a small offset (δ) above the spring support to ensure that the spring force acts in the opposite direction of the magnetic force, and thus facilitate in proper functioning of the model.

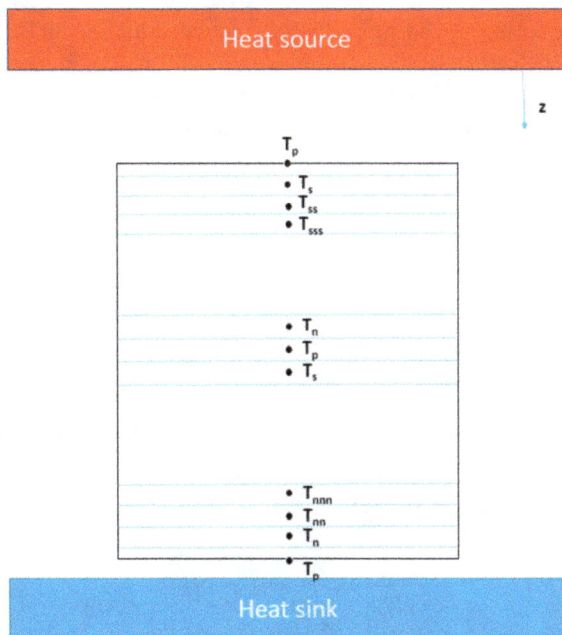

Fig. 1. Schematic diagram of the proposed model by showing the finite volume separations

2 Thermomagnetic Material

Pertaining to the literature review done before, it was found that the best suited category of materials for the job in our model were the thermo caloric materials, whose magnetic properties are dependent on their variation of temperature [7–15]. As being used in a thermomagnetic energy generator, the temperature dependence of magnetization of these materials help run the thermomagnetic cycle and thus give suitable outputs. The thermomagnetic material used is an alloy consisting of Manganese, Nickel, Silicon, Iron and Germanium, from the work done by K. Deepak et al. (2019) [16]. The composition of the elements in the alloy was best suited as $(MnNiSi)_{1-x}(Fe_2Ge)_x$, where x = 0.3 to 0.32.

3 Heat Transfer Modelling

3.1 Thermal Conductance

Thermal conductance (h) is defined as the rate of flow of heat in or out of a unit area of a material of a predefined cross section and mechanical and thermal properties. Empirically, it can be derived from the general conduction heat flux equation. In order to calculate the amount of heat transferred we just need to calculate the thermal conductance of the model, which includes several properties of the materials in contact, interstitial fluid, pressure and roughness conditions of the setup etc. Thus, to study the overall heat transfer model more accurately, we divide the overall contact conductance model into further two units, so that it can be precisely calculated and administered in the heat transfer model, which are:

a) Contact Conductance (h_c): the contact conductance arises due to the heat transfer between the two surfaces in contact through direct touching of several peaks and valleys in the surface roughness of the objects, and thus it takes place through the solid-solid conduction mechanism.

To calculate the contact conductance, we refer to Yuncu 2006 [17], which gives the expression for the conductance as:

$$h_c = 1.25 \cdot \frac{k_s \cdot \left(\frac{P}{H_{sur}}\right)^{0.95}}{\frac{\sigma}{m}} \tag{1}$$

where, $k_s = \frac{2k_1 k_2}{k_1 + k_2}$ is the harmonic mean of the thermal conductivities of the two surfaces in contact, m is the effective rms slope, and $\sigma = \sqrt{\sigma 1^2 + \sigma 2^2}$ is the effective rms surface roughness of both the surfaces in contact. The value of $\frac{P}{H_{sur}}$ can be evaluated as:

$$\frac{P}{H_{sur}} = \frac{A_r}{A_a} = \frac{1}{2} erfc(x) \tag{2}$$

where, A_r is the real area of contact, A_a is the apparent area of contact, x is the dimensionless parameter, $x = \frac{Y}{\sqrt{2} \cdot \sigma}$

And Y is the gap distance parameter and σ is the rms surface roughness.

b) Gap Conductance (h_g): the gap conductance comes into action when there are gaps in between the surfaces, and as getting a purely flat and smooth surface is nearly impossible, the gap conductance contributes enough to come in effect of overall heat transfer. This depends on the interstitial fluid in between the surfaces.

The gap conductance can be calculated by the same (Yuncu 2006) [17], in which:

$$h_g = \frac{k_g}{Y + M_c} \tag{3}$$

where, k_g is the thermal conductivity of the gas/interstitial fluid, Y is the gap distance parameter, M_c is the temperature jump at the cold surface.

3.2 Discretization

The 1-D heat diffusion discussed in is solved using the finite volume approach, with a second order central spatial discretization and implicit time marching scheme, as shown below:

For the general nodes, we take the 1-D heat diffusion equation and discretize it, i.e.:

$$\frac{\partial T_P}{\partial t} = \alpha \cdot \frac{\partial^2 T_P}{\partial z^2} \tag{4}$$

$$\Rightarrow \frac{T_P^{n+1} - T_P^n}{\Delta t} = \alpha \cdot \frac{T_S^{n+1} - 2T_P^{n+1} + T_N^{n+1}}{\Delta z^2} \tag{5}$$

$$\Rightarrow T_P^{n+1} = T_P^n + \frac{\alpha \Delta t}{\Delta z^2}\left(T_S^{n+1} - 2T_P^{n+1} + T_N^{n+1}\right) \tag{6}$$

Simplifying, we can write:

$$A_N T_N^{n+1} + A_P T_P^{n+1} + A_S T_S^{n+1} = S \tag{7}$$

Here the subscripts P, S, N denote the finite volume for which the equation is written, and its north and south neighbours respectively. The superscript (Δt) denotes the number of time step and (Δz) denotes the space discretization.

At the hot boundary, we take the second order forward difference approximation,

$$\frac{\partial T_P}{\partial t} = \alpha \cdot \frac{\partial^2 T_P}{\partial z^2} \tag{8}$$

$$\Rightarrow \frac{T_P^{n+1} - T_P^n}{\Delta t} = \alpha \cdot \frac{\left(-3\frac{\partial T_P^{n+1}}{\partial z} + 4\frac{\partial T_S^{n+1}}{\partial z} - \frac{\partial T_{SS}^{n+1}}{\partial z}\right)}{2\Delta z} \tag{9}$$

where subscripts SS and SSS correspond to second and third neighbors to the south.

Rearranging and simplifying we can write,

$$A_P T_P^{n+1} + A_S T_S^{n+1} + A_{SS} T_{SS}^{n+1} + A_{SSS} T_{SSS}^{n+1} = S \tag{10}$$

where, the subscripts A_P, A_S, A_{SS}, A_{SSS} denote the coefficients for the respective nodes, and S is the source term.

Similarly at the cold boundary, we take the second order backward difference approximation,

$$\frac{\partial T_P}{\partial t} = \alpha \cdot \frac{\partial^2 T_P}{\partial z^2} \tag{11}$$

$$\Rightarrow \frac{T_P^{n+1} - T_P^n}{\Delta t} = \alpha \cdot \frac{\left(3\frac{\partial T_P^{n+1}}{\partial z} - 4\frac{\partial T_N^{n+1}}{\partial z} + \frac{\partial T_{NN}^{n+1}}{\partial z}\right)}{2\Delta z} \tag{12}$$

Subscripts NN and NNN correspond to second and third neighbors to the north. Simplifying we get,

$$A_{NNN} T_{NNN}^{n+1} + A_{NN} T_{NN}^{n+1} + A_N T_N^{n+1} + A_P T_P^{n+1} = S \tag{13}$$

where, the subscripts A_P, A_N, A_{NN}, A_{NNN} denote the coefficients for the respective nodes, and S is the source term.

3.3 Solution Methodology

The model was bifurcated into four phases, to easily simulate and understand the working. Phase 1 was taken at the time periods when the thermomagnetic material was touching the hot surface, phase 2 and 3 when the TMM was traversing downwards and upwards in between the hot and cold surfaces, and finally phases 4 for when the TMM is touching the cold surface. The most important parameter for the numerical simulation and used in the boundary conditions was thermal conductance. It was split into two parts namely contact and gap conductance, which were calculated at each time step in all the four phases. All the values were stored in arrays for later usage in the code.

3.3.1 Fixing Constants

Several constants like the known parameters (such as temperature of the heat source and sink, thermal and physical properties of the material in use) and dimensions of the parts (gap distance, thickness of the thermomagnetic material) were initialized using fixed variables.

3.3.2 Parameter Calculation

The model was bifurcated into four phases, to easily simulate and understand the working. Phase 1 was taken at the time periods when the thermomagnetic material was touching the hot surface, phase 2 and 3 when the TMM was traversing downwards and upwards in between the hot and cold surfaces, and finally phases 4 for when the TMM is touching the cold surface. The most important parameter for the numerical simulation and used in the boundary conditions was thermal conductance. It was split into two parts namely contact and gap conductance, which were calculated at each time step in all the four phases. All the values were stored in arrays for later usage in the code.

3.3.3 Coefficient Matrix

To start with solving the equations, firstly the left-hand side coefficient matrix was created using all the coefficients of the respective temperature terms.

3.3.4 Initializing

The first-time values of the TMM were set to the room temperature and next time step values guessed as zeros for the first gauss seidel iteration.

3.3.5 Phase Calculations

The temperature values were calculated by solving the matrices by iterative gauss seidel method and the error after each iteration as well.

3.3.6 Error Check

The error after each iteration was checked for convergence and kept below a certain value for accuracy and precision. The error values were also stored separately and plotted to visualize the convergence rate and criteria.

3.3.7 Multiple Cycles

After completing the code for one cycle, loops were added so as to follow up for more number of cycles, so as to reach the steady state and analyze the products then.

3.3.8 Storing Data

The temperature values at each node and every successive time step were recorded in a single matrix, and from that the useful data was extracted to be analyzed.

3.3.9 Results

For ease of understanding and also to be used for verification and validation, the temperature values at every time step were taken only for the top-most and the bottom-most layers of the TMM. These temperature values were plotted against the time taken by the model for one full cycle, and then for multiple cycles as well to visualize the steady state of the model.

3.4 Convergence Study

To check the convergence of the code and the calculated values, error was calculated after every iteration in each phase, and it was made sure that the error obtained was of converging trend.

4   Results

The mathematical formulation once done, we have got the value of overall heat transfer contact conductance. This value plays a significant role in the heat transfer modelling as the basic heat transfer is assumed to occur as only in the form of diffusion, and thus gives the need of a diffusion constant or the heat conductance. Finally, we obtain that the model when split into three phases, gives different values of the contact conductance which are given below:

4.1   Temperature Variation Over Multiple Cycles

The code was run for multiple cycles to observe the time and number of cycles taken to achieve steady state by the model. As seen from the Fig. 1(a), the steady state was achieved after around 17 cycles or after 0.2 s from the start of the first cycle. The average temperature of the model was observed to be constant after the steady state was achieved, which conforms to the results obtained from the literature survey.

4.2   Temperature Variation Over One Full Cycle

The temperature profile of the top-most and bottom-most layers of the TMM at each time step were plotted against time to visualize the change in temperature over the phases as the model progressed. The plot (Fig. 2(b)) is taken of a full cycle in steady state of the thermomagnetic harvester model, where the blue curve shows the temperature variation in the top-most layer of TMM (one in contact with the heat source) and the red curve shows the temperature variation in the bottom-most layer of the TMM (one in contact with the heat sink).

4.3   Heat Transfer

The heat transfer study has been done to quantify the amount of heat flowing in to the TMM from the hot surface (heat source), and the amount of heat flowing out from the TMM to the cold surface (heat sink). This has been done using the following equations,

$$Q_{in} = \int_t^{t+\tau} h_{hot} A_{TMM} (T_{hot} - T_0) dt$$

$$Q_{out} = \int_t^{t+\tau} h_{cold} A_{TMM} (T_l - T_{cold}) dt$$

Here t is the time from the beginning of the cycle, which is the moment TMM leaves the cold surface (heat sink) and τ being the length of one cycle. A_{TMM} is the area of the TMM magnet perpendicular to direction of heat flow (Fig. 3).

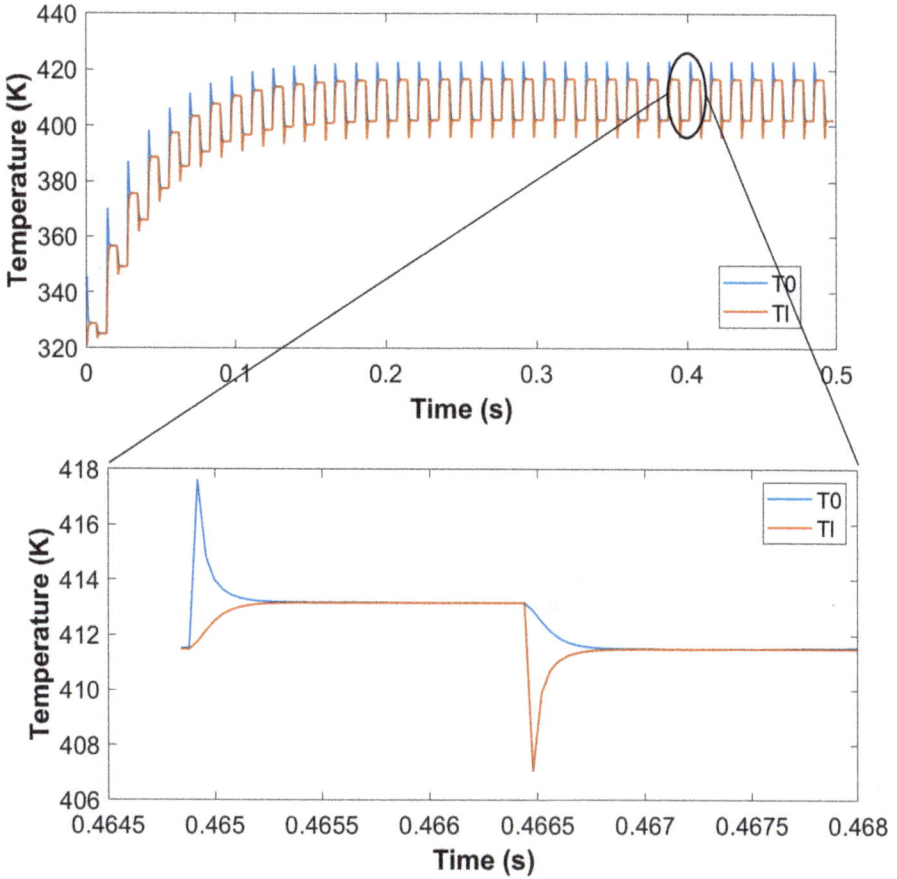

Fig. 2. a) Temperature distribution plot of the top-most and bottom-most layers of the TMM over 36 cycles for the temperature range 298 K–470 K; b) Temperature distribution plot of the top-most and bottom-most layers of the TMM in one full cycle

4.4 Heat Source and Sink Temperature for TMM with x = 0.3

The heat source and sink temperatures are to be decided so as the final stable temperature of the TMM during equilibrium cycling in the generator contains the curie temperature of the TMM. The material for the thermomagnetic harvester was selected as $(MnNiSi)_{1-x}(Fe_2Ge)_x$ with $x = 0.3$ from the study done by K. Deepak et al. (2018), and the curie temperature for this composition was taken as 417 K. Keeping practical applications in mind we keep the heat sink temperature as 298 K (room temperature). To successfully obtain oscillations from the harvester, after trial and error method we found that the required temperature for the heat source would be 470 K at the gap distance of 0.002 m. From K. Deepak et al. (2019), we take up that the required temperature to be reached at the cold side for lifting the TMM up to the heat source is 401 K, and the temperature at the hot side has to be above 421 K to make the TMM fall back to the cold side (Fig. 4).

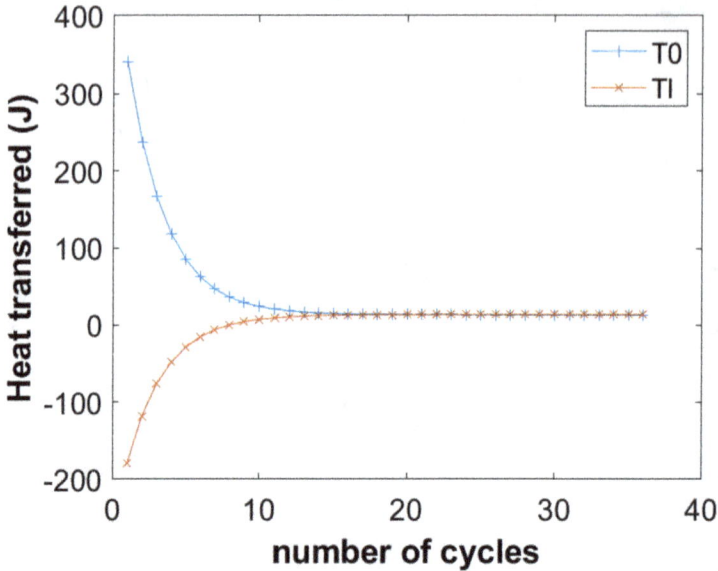

Fig. 3. Heat transferred per cycle for $(MnNiSi)_{1-x}(Fe_2Ge)_x$ with x = 0.3 at ± 10 K from the Curie temperature (417 K)

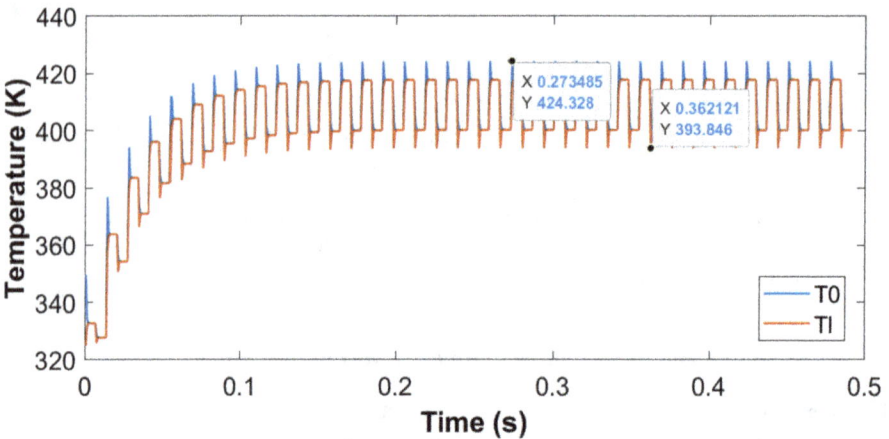

Fig. 4. Internal temperature variation in TMM alloy for working with room temperature as heat sink

4.5 Heat Source and Sink Temperature for TMM with x = 0.32

For the TMM material $(MnNiSi)_{1-x}(Fe_2Ge)_x$ with x = 0.32, it is observed from K. Deepak et al. (2019) that the temperature required to lift the TMM is 345.7 K and the temperature required for the TMM to drop from the heat source to the sink is 364 K. Hence, the Curie temperature for $(MnNiSi)_{1-x}(Fe_2Ge)_x$ with x = 0.32 is around 355 K. Now finding the best suited temperature for this variant, after trial and error method it was found that the required temperature for the heat source would be 440 K at the gap distance of 0.002 m (Fig. 5).

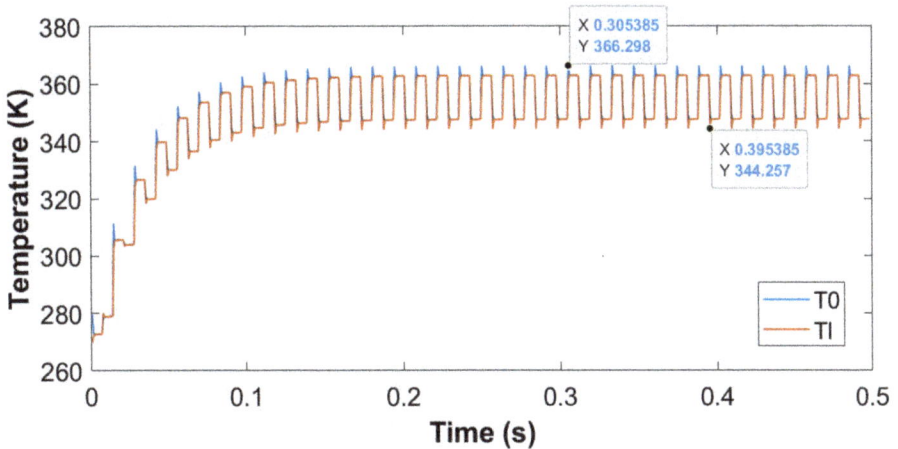

Fig. 5. Internal temperature variation in TMM alloy for working with room temperature as heat sink

4.6 Comparison Between Gadolinium (Gd) and $(MnNiSi)_{1-x}(Fe_2Ge)_x$

Gadolinium is used for TMM in harvester as a reference material, which was studied by Keyur B. Joshi et al. (2013). In this project we have chosen a thermomagnetic alloy to give us better results than gadolinium, at the same time costing less than it too. Thus to see the comparison of the two thermomagnetic materials, we compare the heat transferred, maximum and minimum temperature ranges during cycling, and the number of cycles required to achieve steady state by both the materials.

4.6.1 Heat Transferred

Now we compare our alloy $(MnNiSi)_{1-x}(Fe_2Ge)_x$ with a reference material, here taken as Gadolinium. For, Gadolinium as the TMM working material in the harvester at $\pm$ 10 K from the Curie temperature (Tcurie = 293 K) i.e., Th = 303 K and Tc = 283 K, we get,

As can be inferred from the Fig. 7.12, with Gd as the TMM the heat transferred at steady state is around 0.461 J.

Now computing the cycles with $(MnNiSi)_{1-x}(Fe_2Ge)_x$, x = 0.3, the Curie temperature is 417 K, hence keeping the heat source and sink temperatures at $\pm$ 10 K, we get,

From the Fig. 7, it can be observed that with $(MnNiSi)_{1-x}(Fe_2Ge)_x$ and x = 0.3 as the TMM the heat transferred at steady state is around 13.330 J.

Now, using $(MnNiSi)_{1-x}(Fe_2Ge)_x$, x = 0.32 as the TMM we get the Curie temperature as 355 K, hence keeping the heat source and sink temperatures at $\pm$ 10 K from the Curie temperature, we get,

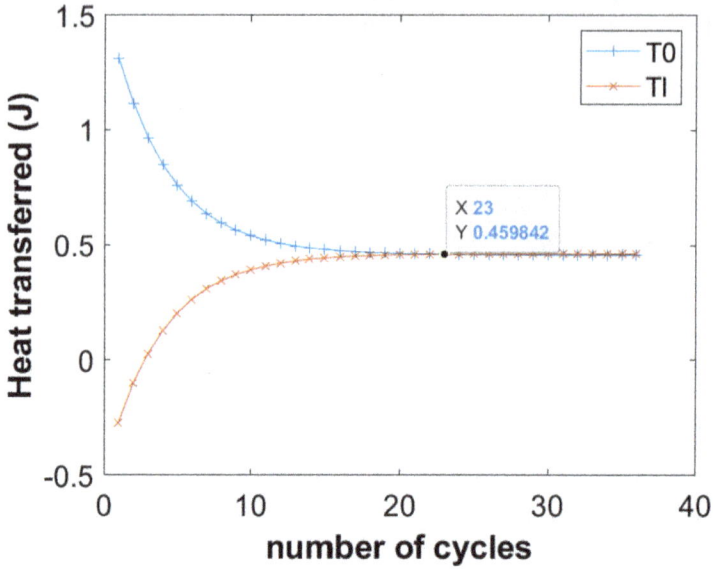

Fig. 6. Heat transferred per cycle with Gd as TMM for the harvester

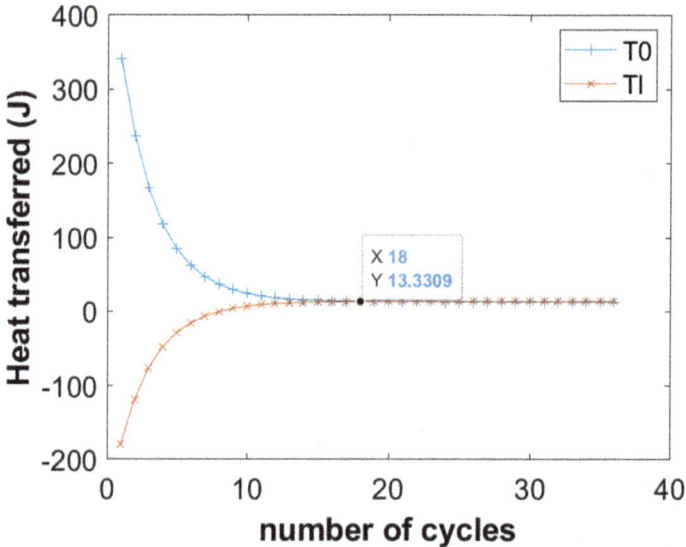

Fig. 7. Heat transferred per cycle with $(MnNiSi)_{1-x}(Fe_2Ge)_x$, $x = 0.3$ as TMM for the harvester

As seen from the Fig. 8, with $(MnNiSi)_{1-x}(Fe_2Ge)_x$ and $x = 0.32$ as the TMM the heat transferred at steady state is around 17.193 J.

From Figs. 6, 7 and 8 it can be seen that when Gadolinium is used as the TMM, the steady state of the harvester is reached in around 23 cycles; whereas when $(MnNiSi)_{1-x}(Fe_2Ge)_x$ with $x = 0.3$ is used, the steady state is achieved in 19 cycles

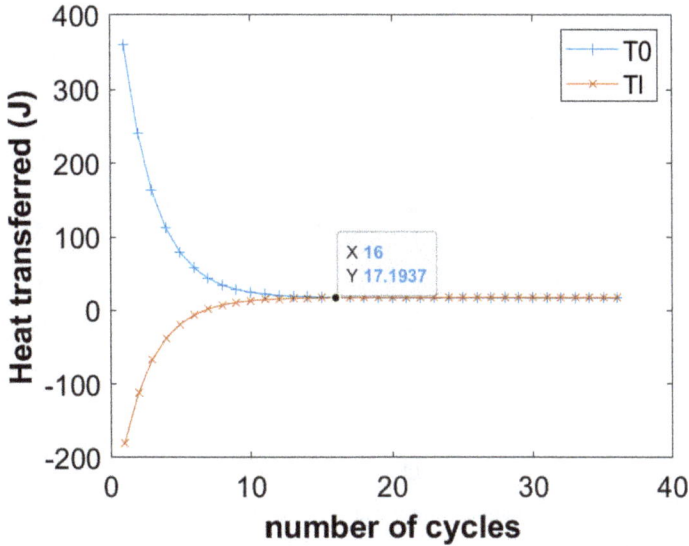

Fig. 8. Heat transferred per cycle with $(MnNiSi)_{1-x}(Fe_2Ge)_x$, $x = 0.32$ as TMM for the harvester

and with $x = 0.32$, constancy is reached in 16 cycles. This shows that when using $(MnNiSi)_{1-x}(Fe_2Ge)_x$, not only higher heat transfer is achieved, but the system achieves equilibrium in less number of cycles, hence reducing the time taken from starting point to the point where we can expect steady output of energy and cooling.

5   Verification and Validation

The plots obtained were validated from the results given by Joshi K. B. et al. (2013) in which they performed the same heat transfer modelling to a similar model of a waste heat energy harvester. The results generated in this project are almost similar to the results given by Joshi et al. (Fig. 9).

The graphs were superimposed and percentage error was quantified for three major points in the modelling, that are at the end of phase 1, end of phase 2 and end of phase 3. After each cycle, the temperature values at these points for both the plots (this project and from Joshi et al.) were recorded and used to calculate the percentage error. The error at all the points at any given cycle was very low, and diminishing as the model achieved steady state.

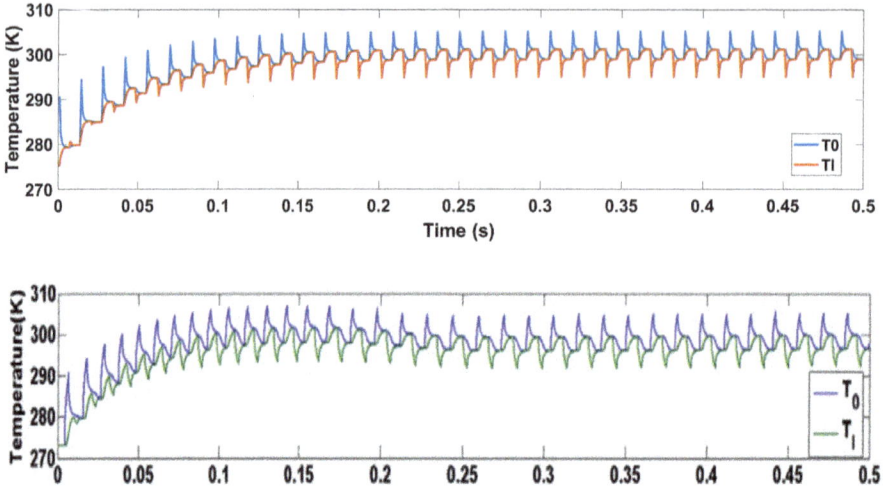

Fig. 9. Trend obtained in the temperature variation a) by the code generated and b) results given by Joshi K. B. et al. (2013)

6   Conclusion

From the heat transfer analysis done, it was observed that the fabricated alloy for the thermomagnetic energy harvester $(MnNiSi)_{1-x}(Fe_2Ge)_x$ with x = 0.3 and 0.32 is better suited than the reference material used that is gadolinium. It can be stated by giving several notions form the past chapters from this project, summarising,

1) With $(MnNiSi)_{1-x}(Fe_2Ge)_x$ as the TMM, the steady state of the harvester model was observed to be reached faster than when gadolinium was used.
2) $(MnNiSi)_{1-x}(Fe_2Ge)_x$ presented better and larger heat transfer per cycle than gadolinium, which can be useful in applications to cool down the heat source as well as producing energy simultaneously.
3) $(MnNiSi)_{1-x}(Fe_2Ge)_x$ has tuneable Curie temperature (331 K to 417K) that can be obtained by using different values of x in the empirical formula, and hence it can be used in variety of heat sources while keeping room temperature heat sinks.

After finalizing the alloy $(MnNiSi)_{1-x}(Fe_2Ge)_x$ for use in the thermomagnetic harvester, parametric study was done which gave the best suited heat source temperatures to be used when using different values of x. It was found that when using x = 0.3, i.e., $(MnNiSi)_{0.7}(Fe_2Ge)_{0.3}$ we have to set up the heat source temperature near 470 K to get proper functioning and cycling of the harvester. Whereas, if x = 0.32 is used, i.e., $(MnNiSi)_{0.68}(Fe_2Ge)_{0.32}$ we have to keep the heat source near 440 K.

This gives us an idea so as to where and at what temperatures of the source we have to use the alloy $(MnNiSi)_{1-x}(Fe_2Ge)_x$ so as to get best suited functionality. On comparing with the reference material gadolinium, it was observed that the Curie temperature of Gd being 293 K, it is best suited for working with heat sources very near or just above room temperature, and hence at times we need to keep the heat sink below the room temperature to get proper functioning of the harvester, which increases the overall cost of the setup by involving an additional chiller or heat remover. Whereas, in case of $(MnNiSi)_{1-x}(Fe_2Ge)_x$ alloy, the Curie temperatures are tuneable from 331 K to 417 K, which prove to be more fond in practical use and no chiller is required as the harvester can run perfectly with room temperature heat sink.

References

1. Garg, P.: Energy scenario and vision 2020 in India. J. Sustain. Energy Environ. **3**(1), 7–17 (2012)
2. Kishore, R.A., Priya, S.: A review on design and performance of thermomagnetic devices. Renew. Sustain. Energy Rev. **81**, 33–44 (2018)
3. Kitanovski, A., Egolf, P.W.: Innovative ideas for future research on magnetocaloric technologies. Int. J. Refrig. **33**(3), 449–464 (2010)
4. Kitanovski, A.: Energy applications of magnetocaloric materials. Adv. Energy Mater. **10**(10), 1903741 (2020)
5. Ujihara, M., Carman, G.P., Lee, D.G.: Thermal energy harvesting device using ferromagnetic materials. Appl. Phys. Lett. **91**(9) (2007)
6. Joshi, K.B., Priya, S.: Multi-physics model of a thermo-magnetic energy harvester. Smart Mater. Struct. **22**(5), 055005 (2013)
7. Wada, H., Tanabe, Y.: Giant magnetocaloric effect of MnAs1-xSbx. Appl. Phys. Lett. **79**(20), 3302–3304 (2001)
8. Spichkin, Y.I., Pecharsky, V.K., Gschneidner, K.A.: Preparation, crystal structure, magnetic and magnetothermal properties of (GdxR5-x)Si4, where R=Pr and Tb, alloys. J. Appl. Phys. **89**(3), 1738–1745 (2001)
9. Nóbrega, E.P., et al.: Theoretical investigation on the magnetocaloric effect in amorphous systems, application to: Gd 80 Au 20 and Gd 70 Ni 30. J. Appl. Phys. **113**(24) (2013)
10. Pecharsky, V.K., Gschneidner, K.A.: Giant Magnetocaloric Effect in Gd 5 Si 2 Ge 2 (1997)
11. Canepa, F., Napoletano, M., Cirafici, S.: Magnetocaloric effect in the intermetallic compound Gd 7 Pd 3. Intermetallics **10**(7), 731–734 (2002)
12. Klimczak, M., Talik, E.: Magnetocaloric effect of GdTX (T = Mn, Fe, Ni, Pd, X=Al, In) and GdFe 6Al6 ternary compounds. In: Journal of Physics: Conference Series, Institute of Physics Publishing (2010)
13. Law, J.Y., Ramanujan, R.V., Franco, V.: Tunable Curie temperatures in Gd alloyed Fe-B-Cr magnetocaloric materials. J. Alloys Compd. **508**(1), 14–19 (2010)
14. Das, S., Dey, T.K.: Magnetic entropy change in polycrystalline La1-xKxMnO3 perovskites. J. Alloys Compd. **440**(1–2), 30–35 (2007)
15. Wada, H., Taniguchi, K., Tanabe, Y.: Extremely Large Magnetic Entropy Change of MnAs 1−x Sb x near Room Temperature (2002)
16. Deepak, K., Varma, V.B., Prasanna, G., Ramanujan, R.V.: Hybrid thermomagnetic oscillator for cooling and direct waste heat conversion to electricity. Appl. Energy **233–234**, 312–320 (2019)
17. Yüncü, H.: Thermal contact conductance of nominally flat surfaces. Heat Mass Transf. **43**(1), 1–5 (2006)

Theoretical Comparison and Machine Learning Based Predictions on Li-Ion Battery's Health Using NASA-Battery Dataset

K. M. Chaturvedi$^{(\boxtimes)}$, Rohit Mathew Samuel, O. D. Jayakumar, and Aryadevi Remanidevi Devidas

Center for Wireless Networks and Applications, Amrita Vishwa Vidyapeetham, Amritapuri, India
krishnachaturvedi64@gmail.com,
amenp2wna21010@am.students.amrita.edu, aryadevird@am.amrita.edu

Abstract. This article highlights the potential of predicting cell parameters, such as potential and current, for the optimization of electrochemical processes. We used Python to make complex ML predictions. Lithium-ion (Li-ion) batteries are rechargeable batteries used for a variety of electronic devices, which range from electric vehicles, smartphones, and even satellites. A set of Li-ion batteries run through different operational profiles (charge, discharge and impedance) at room temperatures, and time. Here, input (features) parameters like, Impedance measurement, Voltage measured, Current measured, Temperature measured, Current load, Voltage load, Time, under charge and discharge cycles, for batteries evaluations and predictions. This NASA dataset is classified as a time-series dataset.

Electrochemistry in battery making is one of the prevalent and proven sciences for Efficient Batteries for the Automobile Industry. With the advent of Material Science and Technology, it also embraced the electrochemistry utility for several electrochemical parameter customizations. Not only that, electrochemistry has been a proven asset for sensor technology and budding defense technology. Meanwhile, the development of nano-/ultrafine-grain metallic materials with improved strength has improved ion trapping and high electrode-potential composites and are prominent potential candidates for future fuel cell and Hydrogen trapping based cell's technologies.

Here, we have taken the NASA Battery Dataset study that employs machine learning models, such as linear regression, decision tree, random forest, and support vector regression, to predict battery capacity based on features such as voltage, current, temperature, and time. Descriptive statistics, capacity vs. cycle plots, and capacity vs. state of health (SoH) plots are used to analyze the dataset and visualize the battery performance over time. The correlation between different features is examined using a correlation heatmap. The mean squared error (MSE), root mean squared error (RMSE), and high R2 scores are also calculated.

Keywords: Electrochemical cell · Li-ion batteries · nanomaterials · NASA dataset · electrochemical performance

S. Patra et al. (Eds.): METCENT 2023, *Proceedings of the International Conference on Metallurgical Engineering and Centenary Celebration*, pp. 108–118, 2024.
https://doi.org/10.1007/978-981-99-6863-3_12

1 Introduction

Lithium-ion (Li-ion) batteries are rechargeable batteries used for a variety of electronic devices, which range from electric vehicles, smartphones, and even satellites (Fig. 1).

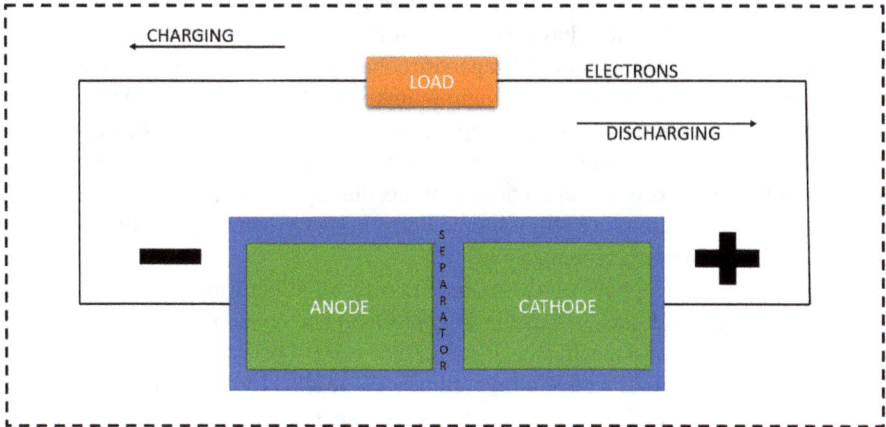

Fig. 1. The Illustration of a battery.

Batteries are electrochemical devices that store chemical energy converted from electricity. They consist of electrodes (anode and cathode) immersed in an electrolyte solution, as shown in the figure. The electrolyte, typically a non-aqueous solvent with dissolved lithium salts, facilitates charge transfer and serves as a source of lithium ions. The cathode, often an oxide material, undergoes reversible delithiation during charging, while the anode, commonly graphite or silicon-based, undergoes lithiation during discharging. Charging involves extracting electrons from the cathode and transferring Li-ions to the anode through the electrolyte, converting electrical energy into chemical energy. Discharging reverses this process, generating electrical energy as electrons flow from the anode to the cathode. A separator prevents short-circuits by allowing ion migration while maintaining charge neutrality. This cyclic charging and discharging process sustains the battery cell's operation. Li-Ion battery making is one of the prevalent and one of the oldest energy storage technologies used, and proven sciences for Efficient Batteries for the Automobile Industry. With the advent of Material Science and Technology, it also embraced the electrochemistry utility for several electrochemical parameter customizations. Not only that, electrochemistry has been a proven asset for sensor technology and budding defense technology. Li-Ions batteries are proven with an excellent life span for 15 years or more. In addition, materials based on carbon-silica composite as an electrochemical sensor encircle the applications in sensing materials [1].

A battery discharge at low temperatures affects the rate of internal chemical reactions. It is essential to predict the characteristics of battery degradation might indicate if the battery needs replacement.The indicator for battery degradation is state-of-health. TPG are becoming more accessible for applications in transportation (such as in hybrid vehicles and submarines) and in peak power generation (such as using waste heat from nuclear and fossil fuel power plants and possibly in conjunction with solar panels) [2]. The TPG is a trending technological power source for battery recharging in electric vehicles, trains and automobiles. The intricate realm of lithium-ion (Li-ion) batteries, widely recognized as the cornerstone of energy storage systems powering an array of applications, ranging from electric vehicles to portable electronic devices. The four prominent Li-ion battery chemistries that have gained significant traction in the industry: lithium iron phosphate (LFP), lithium nickel manganese cobalt oxide (NMC), lithium manganese oxide (LMO), and lithium nickel cobalt aluminum oxide (NCA). These models are scrutinized, with a particular focus on their application to the Li-ion battery chemistries [3]. These models are idealizations that account for the behavior of the liquid electrolyte. One way to achieve this is through ab initio molecular dynamics (AIMD) simulations. However, these simulations are still too time-consuming to model, to accurately model ion concentrations in electrochemistry [4].

There is lucid and emphatic evidence that micro/nano-structuring can result in better performance of electrodes as well as the solid electrolytes. A microstructural modification can result in improved mechanical, thermo-electrical (TEG) and electronic capacity of the electrode materials as well as alloys (Binary Ti–Fe alloy by adding Ga in the eutectic alloy) [5][6]. The use of graphene nanoparticles as nano-inclusions in skutterudites to enhance the electrochemical properties and the addition of graphene nanoparticles was found to increase their thermoelectric efficiency [2]. Carbon sources are known for the composite due to its low resistivity and stability under oxidizing conditions, resulting in enhanced electrochemical properties where the solid-electrolytes are used [1]. The increasing availability of computational power and means to acquire and store data at scale, machine learning has become more accessible and can be a powerful tool for analyzing electrochemical data. In some studies, a support vector machine (SVM), a deep neural network, was used to model. The physics-informed neural networks (PINNs) in electrochemical simulations, which is a new development in machine learning. Machine learning is a subset of artificial intelligence (AI) that enables computers to learn and improve using algorithms and large datasets of inputs and outputs. The goal of machine learning is to recognize patterns to make autonomous recommendations on decisions [7].

1.1 About Dataset

The time-series NASA Battery Dataset consists of data related to Li-ion batteries that were subjected to various operational profiles, including charge, discharge, and impedance measurements, at different temperatures. The impedance measurements were conducted using electrochemical impedance spectroscopy (EIS) at different frequencies. Multiple experiments were conducted on different batteries named as B0005, B0006, B0007, B0018 etc. Each battery has its own dataset containing information such as cycle count, ambient temperature, voltage, current, and time. The dataset for each battery contains approx 50,285 records. The datasets under the profile 'discharge' include the target variable, which is the battery capacity. The capacity is measured in ampere-hours (Ah) and represents the amount of charge a battery can store,

$$C_d = I \cdot t_d \tag{1}$$

where C_d is discharging capacity, I is current and t_d is discharge time. For each battery, there are 168 cycle measurements available for the capacity variable. Charging was performed in a constant current (CC) mode at 1.5A until the battery voltage reached 4.2V. The charging then continued in a constant voltage (CV) mode until the charge current dropped to 20mA. Discharging was carried out at a constant current (CC) level of 2A until the battery voltage fell to specific values depending on the battery number. Impedance measurements were conducted using a frequency sweep from 0.1 Hz to 5 kHz [8].

Cycle: A top-level structure array containing information about the charge, discharge, and impedance operations. Under the structure of discharge, fields contain voltage measured, current measured, temperature measured, current load, voltage load, time and capacity.

2 Electrochemistry Measurements

Electrochemistry is a branch of chemical science in which the potential difference due to ionic discharge is obtained and applications are derived in various devices. The field of stochastic electrochemistry involves acquiring high bandwidth amperometry and/or voltammetric data in the form of a time series, with the fluctuation of current or voltage measurements with time subject to subsequent analysis [7]. In recent years, machine learning approaches have been developed to analyze datasets more quickly and efficiently [1]. EIS is a powerful technique used in electrochemistry to measure the electrical and dielectric properties of a system over a range of frequencies. The technique involves applying a small sinusoidal perturbation to the electrochemical cell and measuring the response of the system in terms of the resulting current and voltage changes. The frequency of the perturbation is then varied over a wide range to obtain a complex impedance spectrum, which contains information about the electrochemical processes occurring in the system. The EIS models involve a series of electrical components, such as resistors, capacitors, and inductors, which mimics the behavior of the electrochemical system [11].

3 Models and Discussions

The dataset consists of repeated charge and discharge cycles resulting in accelerated aging of the batteries while impedance measurements provide insight into the internal battery parameters that change as aging progresses. We have considered the discharge cycle of the dataset to perform SoH prediction based on battery capacity. The data folder contains the voltage and current measured, voltage and current load, temperature and time. These are the parameters that are considered as independent variables in building an ML model. The discharge cycle contains the capacity of the battery for every cycle which is considered the target variable. The SoH is calculated by dividing capacity at each cycle by the initial capacity. Data from each battery is collected and merged together in a new CSV file (Fig. 2).

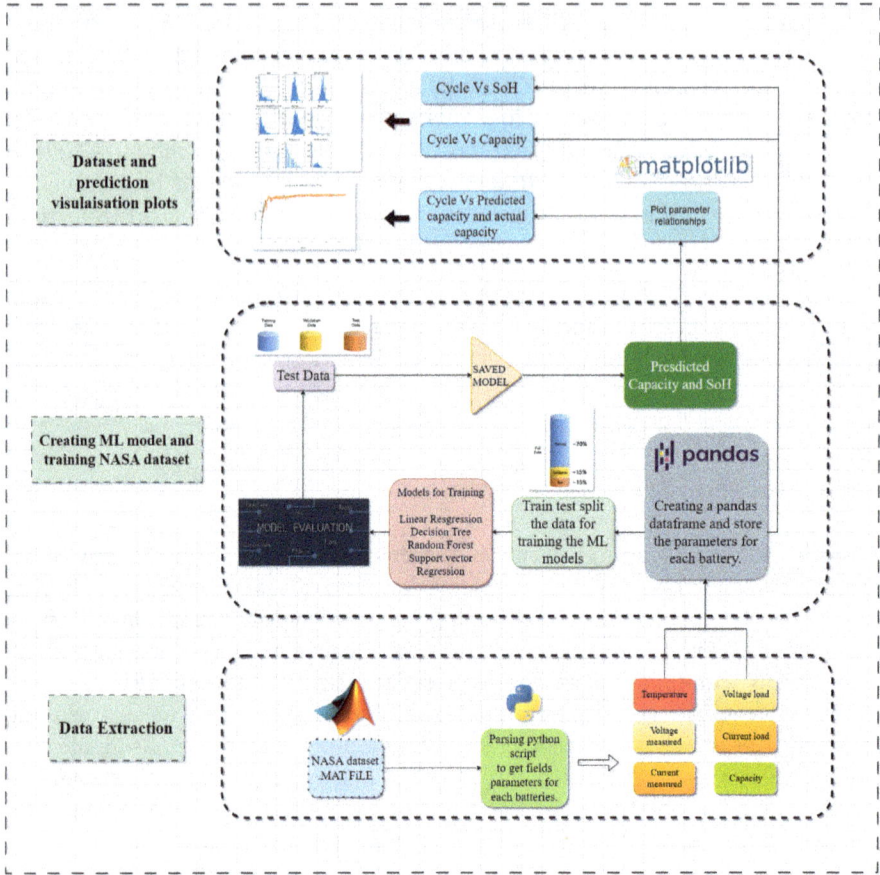

Fig. 2. Flow Chart of the ML Model and Prediction.

From cycles against Capacity plots for all batteries, we can observe that there is decrease in battery capacity over longer cycles. When the capacity is below the threshold, the battery is not usable.

The discharge capacity decreases as seen over the cycling, as shown in Fig. 3. And the lucid decreasing trend of the discharge capacity implies aging electrochemical processes occurring inside the cell. The yellow line represents threshold value, below which the below may not be usable again.

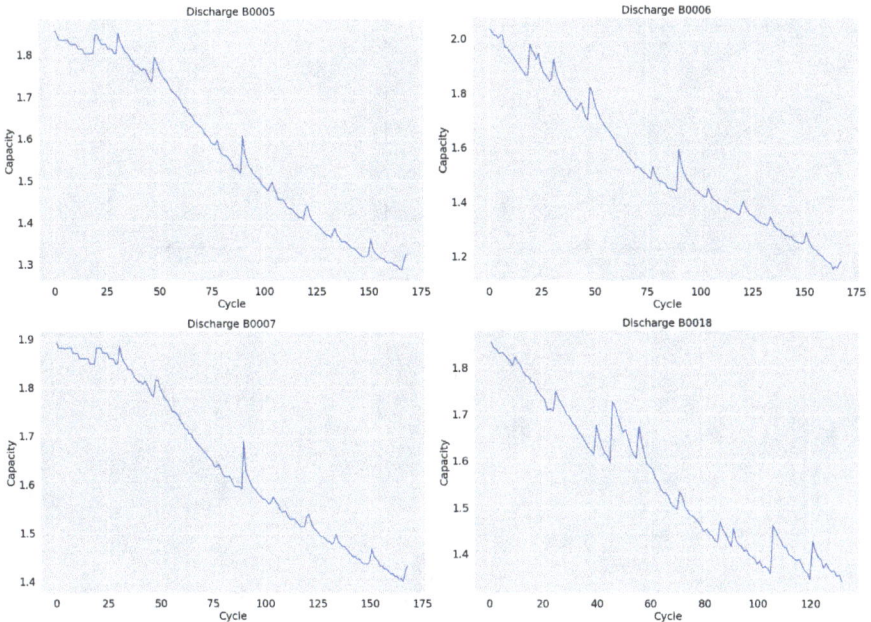

Fig. 3. Discharging: Cycles Vs Capacity Plots.

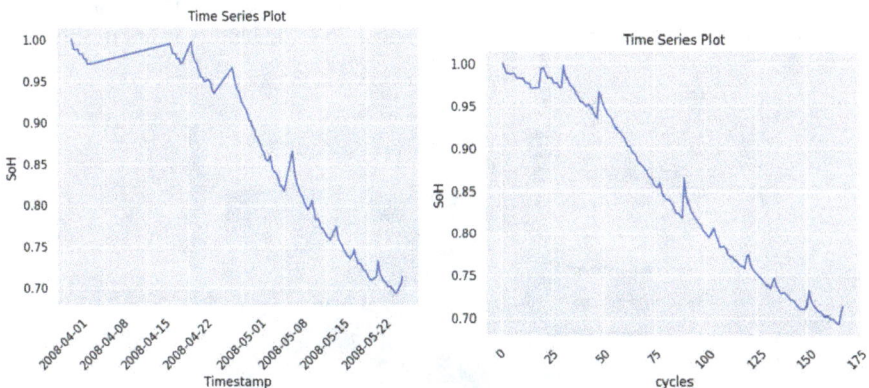

Fig. 4. Timestamp Vs SoH plot, showing the change in State of Health (SoH) over time for one battery.

State of Health (SoH) denotes the prevailing health condition of a battery relative to its initial or nominal capacity. The evolution of SoH over time is depicted through a graphical representation, allowing observation of battery degradation throughout its lifespan. The gradient of the SoH versus time graph serves as an indicator of the battery's degradation rate. Once the SoH of the battery falls below 75%, it signifies that the battery has reached its End of Life. The SoH decreases over the time period of the experiment, as shown in Fig. 4.

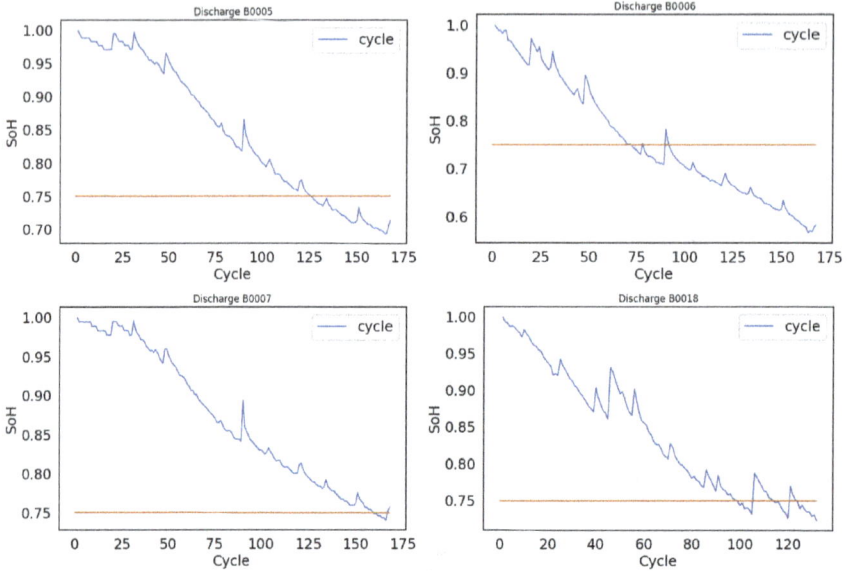

Fig. 5. Batteries: Cycle Vs SOH Plots. Plots showing the capacity of each battery normalized to the initial capacity (State of Health) over cycles.

As shown in Fig. 5, the plot with SoH vs cycle also shows a similar decreasing trend for all four batteries. If the threshold value is below 75% the battery is not usable.

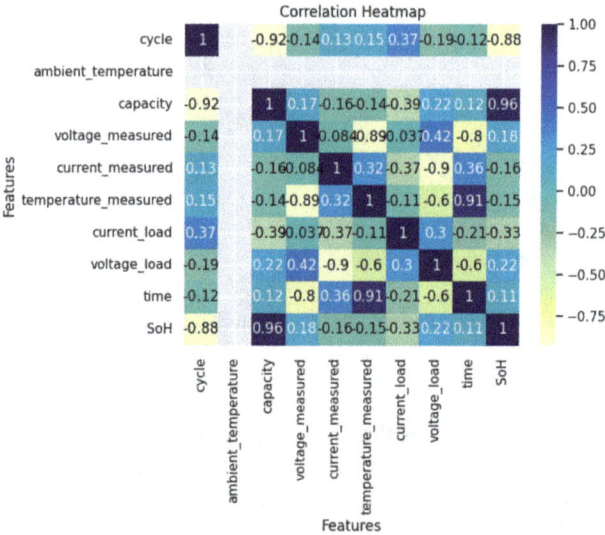

Fig. 6. Correlation Heatmap, showing the correlation between different features in the merged dataframe.

A correlation heatmap is done on the merged data and shows the correlation of each feature in the dataset, as shown in Fig. 6. The features with values closer to 1 have a greater correlation than lower values. These features can be used to create ML models. Capacity is considered as the target variable and the model used in our system for capacity prediction takes all features as input variables. Using the predicted capacity, SoH of the battery can be obtained. The data is split into train data and test data with 0.3 test size.

3.1 ML Models

The code sequentially builds the model for the following with the merged dataset:

Linear Regression: A linear regression model used for predicting continuous values based on linear relationships between the features and the target variable.

Decision Tree: A decision tree model used for regression, which creates a flowchart-like model of decisions and their possible consequences.

Random Forest: A random forest model used for regression, which combines multiple decision trees to make predictions.

Support Vector Regression (SVR): A support vector machine model used for regression, which finds the best fitting hyperplane to predict continuous values.

Machine Learning Model Evaluation: Evaluation metrics (Mean Squared Error, Root Mean Squared Error, and R2 Score) for different ML models (Linear Regression, Decision Tree, Random Forest, Support Vector Regression) used to predict battery capacity based on the given features. A low value for MSE, RMSE and high value for R2_score gives a better model for battery capacity prediction (Table 1).

Table 1. Model, MSE and RMSE with R2 Score.

Model	MSE	RMSE	R2 Score
Linear Regression	0.0164	0.1279	0.5504
Decision Tree	0.0007	0.0256	0.9820
Random Forest	0.0003	0.0174	0.9917
Support Vector Regression	0.0030	0.0548	0.9173

From the Model Evaluation, we can see that random forest has the highest R2_score with 99.1% making it an excellent model for battery capacity prediction. Following Random Forest, decision tree and Support vector regression also have comparatively high accuracy with 98.2% and 91.7% indicating a good fit. Linear regression had a low R2 score. The model is only able to explain 55% for the variance in the battery capacity.

Other models like redox flow batteries (RFBs) are an important type of energy storage systems (ESS) that store electrical energy. The capacity and current output of RFBs are in their design flexibility for (Yanxin Yao, 2021) large-scale energy storage. In RFB, basic cell attributes and performance metrics, such as electromotive force, capacity fade rate, and Coulombic efficiency, are defined considering factors like volumetric capacity, areal capacity, and decay mechanisms in evaluating RFB performance is emphasized [9].

Solid oxide fuel cells (SOFCs) are cleaner renewable energy electrochemical devices that convert chemical energy into electrical energy with high efficiency and low emissions. In a simulation study (Fiammetta Rita Bianchi, 2021), the modeling work has been implemented in simulation software such as COMSOL and Aspen Plus. Based on physical principles, it models the system behavior through the resolution of material, energy, momentum, and charge balances on a single cell plane. The simulations included variations in fuel compositions and temperatures. The results of the simulations were evaluated using I-V and current-power (I-P) curves [10].

Electrochemical Impedance Spectroscopy data of a bio-based photovoltaic device, using a graphical/analytical/numerical procedure was conducted (Eleonora Alfinito, 2020). EIS is a powerful technique used in electrochemistry to measure the electrical and dielectric properties of a system over a range of frequencies. The EIS dataset can provide information on a range of electrochemical processes, including charge transfer resistance, double-layer capacitance, diffusion processes, and adsorption/desorption phenomena. The technique is particularly useful for studying electrochemical systems that involve complex redox processes, such as bioelectrochemical systems and photoelectrochemical cells. These models involve a series of electrical components, such as resistors, capacitors, and inductors, which are arranged in a circuit that mimics the behavior of the electrochemical system. Current does not depend linearly on the applied bias, but it is governed by Nernst's law [11]. The (Haotian Chen, 2023) highlights the potential of machine learning approaches in the analysis of voltammograms and other methods in fundamental electrochemistry. These approaches can provide a powerful tool to extract quantities of interest from complex data sets and overcome the challenges involved in traditional analysis methods. For example, machine learning algorithms can be trained on simulated data to analyze experimental results, including the analysis of stochastic amperometric data and the modeling of complex geometries via physics-informed neural networks (PINNs) [7].

One popular application of stochastic electrochemistry is the amperometric detection and analysis of individual nanoparticles, often referred to as nano impacts. Manual analysis of nano impact data can be time-consuming, limiting the number of spikes that can be analyzed and introducing the possibility of human bias and error. Various automated or semi-automated rule-based approaches have been discussed in literature to increase throughput and prevent bias. Machine learning approaches have the potential to offer greater versatility in processing data without additional parameterisation [7]. The well-known Kinetic modeling can be done using empirical rate constants or by performing kinetic Monte Carlo simulations, which do not treat the reacting mixture as an ideal solution. The modeling of electrocatalytic processes remains a highly interesting and relevant field of research due to its technological importance in electrochemical energy conversion and storage [4].

4   Conclusion

In this study explored NASA Battery Dataset to understand battery health. To understand the battery health, State of charge is considered as a metric for evaluation. SoH can be found based on the initial capacity and current capacity of the battery. From the dataset the discharging cycles were considered for the prediction as it also provided the capacity of the battery during discharge. Machine learning techniques like linear regression, decision tree, random forest, and support vector regression are used to predict the battery capacity based on features such as voltage, current, temperature, and time. From the dataset four battery data were available and three of the battery data were used for training the model and one for testing the model. State of Health (SoH) for the predicted values can also be found based on initial capacity. Correlation heatmaps were to analyze linear relationships between different features. And Capacity/SoH Vs cycle graphs were also plotted to see the relationship of the target variable as the number of cycles increased. The model evaluation shows that the random forest has the highest R2_score making it a very good model for capacity prediction with MSE of 0.003. The decision tree and SVR have relatively good fit models with 98.2% and 91.7% accuracy respectively. Linear regression has the lowest accuracy with only 55%. These models were evaluated with other battery data and were able to make good predictions. The results provide insights into the accuracy and performance of the models, enabling informed decision-making regarding battery maintenance and replacement. Using machine learning approaches increased accuracy, reduced computational time, increased flexibility.

Acknowledgement. We express our deep gratitude to our Chancellor and renowned humanitarian Sri Mata Amritanandamayi Devi (Amma). We are also grateful for the help of faculty members and staff in the Center for Wireless Networks and Applications, Amrita Vishwa Vidyapeetham, Amritapuri Campus, Kerala, India.

References

1. Dubey, R.S., Chaturvedi, K.M.: Investigation of carbon-silica-titanium dioxide composite as an electrochemical sensor 1department of applied chemistry, amity institute of applied science. In: Proceedings International Conference on Recent Trends in Materials and Devices (ICRTMD-2013), pp. 123–125 (2013)
2. Chaturvedi, K.M.: Synthesis and investigation of thermo-electric properties of Skutterudites CoSb3/Graphene particles nanocomposite. Nat. Phys. Lab. New Delhi (2014)
3. Tran, M.-K., et al.: Comparative study of equivalent circuit models performance in four common lithium-ion batteries: LFP, NMC, LMO, NCA. Batteries 7(3), 51 (2021)
4. Groß, A.: Challenges in the modelling of elementary steps in electrocatalysis. Curr. Opin. Electrochem. 37, 101170 (2022)
5. Misra, D.K., et al.: High yield strength bulk Ti based bimodal ultrafine eutectic composites with enhanced plasticity. Mater. Des. 58, 551–556 (2014)
6. Kumar, A., et al.: Enhanced thermoelectric performance of p-type ZrCoSb0. 9Sn0. 1 via tellurium doping. Mater. Chem. Phys. 258, 123915 (2021)
7. Chen, H., Kätelhön, E., Compton, R.G.: Machine learning in fundamental electrochemistry: recent advances and future opportunities. Curr. Opin. Electrochem. 38, 101214 (2023)

8. NASA battery dataset. https://www.kaggle.com/datasets/patrickfleith/nasa-battery-dataset. Accessed 07 June 2023

9. Hu, X., Li, S., Peng, H.: A comparative study of equivalent circuit models for li-ion batteries. J. Power Sources **198**, 359–367 (2012)

10. Padinjarethil, A.K., et al.: Electrochemical characterization and modelling of anode and electrolyte supported solid oxide fuel cells. Front. Energy Res. **9**, 668964 (2021)

11. Alfinito, E., et al.: A biological-based photovoltaic electrochemical cell: modeling the impedance spectra. Chemosensors **8**(1), 20 (2020)

Prediction of Mechanical Properties of Cr-Mn-N Austenitic Stainless Steel Using Machine Learning Approach

F. M. Ayub Khan[1], V. Narsimha Rao[2], Abhijit Ghosh[3], Anish Karmakar[4], and Sudipta Patra[1](✉)

[1] Indian Institute of Technology (BHU), Varanasi, Uttar Pradesh 221005, India
sudipta.met@iitbhu.ac.in
[2] Jindal Stainless (Hisar) Limited, Hisar, Haryana 125005, India
[3] Indian Institute of Technology Indore, Khandwa Road, Simrol, Indore, Madhya Pradesh, India
[4] Indian Institute of Technology Roorkee, Roorkee, Uttarakhand 247667, India

Abstract. In view of designing new alloys as well as optimizing the mechanical properties of annealed Cr-Mn-N alloys, a machine learning model has been developed using large industrial data. In this work, continuous multi-output regression models were developed to predict mechanical properties such as yield strength, tensile strength and elongation of Cr-Mn-N austenitic stainless steel based on the chemical composition, thickness and grain size. In the present study, the performance of several model such as KNN, ETR, XGboost, RF etc. has been compared. The ETR model outperformed other models and thereby it has been selected for further analysis. The relative importance of different parameters affecting the mechanical properties of the steels have been investigated. While the yield strength and ultimate tensile strength could be predicted very well but the percentage elongation showed slight deviation from actual data. The deviation has been explained in the light of metallurgical fundamentals. Further, the developed model was validated using separate data points outside the training data. The optimized model will be useful in designing and optimizing the composition of Cr-Mn-N austenitic stainless steels.

Keywords: Austenitic stainless steel · mechanical properties prediction · machine learning · feature importance

1 Introduction

Austenitic stainless steel (ASS) mainly contains large amount of chromium and nickel, commonly 18Cr-8Ni. Popular austenitic steel of Cr-Ni 300 series can be replaced by low cost Cr-Mn-N in many application where moderate corrosion resistance, high strength, high drawability is required [1 2, 3]. Cr-Mn-N series now used in diverse industries like white goods, engineering, railways, automobiles, nuclear energy due to excellent mechanical properties along with corrosion [4–7]. Well known factors which controls the

© The Author(s), under exclusive license to Springer Nature Singapore Pte Ltd. 2024
S. Patra et al. (Eds.): METCENT 2023, *Proceedings of the International Conference on Metallurgical Engineering and Centenary Celebration*, pp. 119–130, 2024.
https://doi.org/10.1007/978-981-99-6863-3_13

mechanical properties are chemical compositions, grain size, dislocation density, precipitates and impurities [8–11]. In recrystallized austenitic Stainless steels mechanical properties are mostly dependent on chemical composition and grain size as dislocation density in recrystallized matrix is almost constant and there is no as such precipitates in the microstructures [12]. Moreover, comparatively clean steel produced in commercial scale will not have large impurities and inclusions which can impact the mechanical properties. These properties can also be affected by the temperature at which the material is exposed [13]. Investigating how these variables affect the mechanical properties through experiments, one could only get preliminary trends through changing a part of variables for a long time. Conventional linear fitting regression also finds difficulty in capturing the correlation among the variables due to complex nonlinear relationships among them [14]. In such scenario, machine learning (ML) techniques become efficient methods to deal with linear or nonlinear relationships between the variables. The popular machine learning techniques include Random Forest, Xgboost, Extra Trees, Decision Trees, Artificial Neural Network (ANN), etc.

P.L. Narayana et al. [15] designed ANN model to establish the complex relationships among the chemical compositions, temperature and tensile properties of 18Cr-12Ni-Mo ASS. The yield strength, tensile strength, elongation and percentage reduction in area (RA) were the output features. The predicted yield strength and tensile strength were compared with that of results generated using empirical formula given by Ohkubo et al. [16]. Y. Wang et al. [17] analyzed the prediction of tensile properties of austenitic stainless steels (ASS) using artificial neural network (ANN). They tried to predict the tensile properties such as yield strength and ultimate tensile strength based on the chemical compositions, heat treatment and test temperatures. For the implementation of ANN, Bayesian regularization technique was used with three layers of neural network. Hyperbolic tangent sigmoid function (TRANSIG) was used in the hidden layer, that is, in the second layer while linear function (PURELIN) was used in the third or output layer. They implemented the neural network through nntool of MATLAB. Feature analysis was carried out using mean impact value (MIV) method to know the influence trend of the variables on the output. Z Chen et al. [18] tried to implement ANN to predict mechanical properties of 301 ASS. The effect of cold rolling of plates on yield strength and tensile strength was studied through ANN. Original thickness, thickness of plate after rolling and rolling reduced were used as input parameters. The yield and the tensile strength were the output parameters. The optimum ANN architecture was found to be 3–12-2. It was found that when the rolling reduction is from 0 to 50%, the yield and the tensile strength improved with the increase in rolling reduction. D. Shin et al. [19] proposed a data analytic approach to predict creep of high-temperature alloy such as alumina-forming austenitic (AFA) stainless steel. AFA alloys have been designed to use MC, $M_{23}C_6$, Laves and/or L12 strengthening precipitates in an austenite single-phase matrix. In their work, bulk alloy compositions were analyzed by a combination of inductively-coupled plasma (ICP), combustion techniques, and inert gas fusion analysis. The analyzed alloy compositions were used in the computational thermodynamic calculations used to populate scientific features into the AFA dataset. Larson-Miller Parameters (LMP) was used to represent the creep behavior of AFA alloys. Feature selection was carried out through Pearson's correlation coefficient and with the consultation of AFA alloy design expert. Five different representative ML models were used in the study: random forest (RF), linear regression (LR), nearest neighbor (NN), kernel ridge (KR), Bayesian ridge (BR). They were implemented in the Python-based open-source data analytics toolkit, scikit-learn.

Though significant amount of work has been done in Cr-Ni stainless steel and other alloys, ML algorithm has not been used for predicting mechanical properties for Cr-Mn series stainless steel till now. Behaviors of Cr-Mn series stainless steel become little bit more complex because of the addition of Nitrogen and thus it is the interest of many researchers and industries.

The present work aims at developing models for predicting the mechanical properties Cr-Mn-N ASS grade with chemical compositions and sample thickness as the input features. In this work, Extra Tree Regressor (ETR) and Random Forest Regressor (RF) were used for building the models. The performance and evaluation metrics of both the model were compared with other classical ML algorithms and the better model was chosen for reference in further analysis of the present study.

2 Methodology

2.1 Details of the Data and Data Preprocessing

The present work is based on the industrial data of high Nitrogen Grade Austenitic Stainless steel (Cr-Mn-N). The original data consist of 2863 samples and 40 features consisting of chemical compositions, sample thickness, mechanical properties and other processing parameters. The samples were roughly divided into four industrial grades such as JT, JSLU SD, JSLU DD and 204Cu. The details composition of all grades are given in Table 2. The data were processed to remove missing and incorrect data. Out of all features, chemical compositions and sample thickness were selected as input variables and yield strength, ultimate tensile strength and % elongation were as the output parameters. The selected features include chromium (Cr), nickel (Ni), carbon (C), nitrogen, silicon (Si), molybdenum (Mo), etc. The well-known elements which are in significant quantities and which can affect the mechanical property that has been taken as input parameters. Some mechanical property data was removed due to possible testing error as the % Elongation was very low and they were assumed to be broken due to error in sample dimension or sample preparation error. So, a total of 2630 samples were used for developing the model. The detail and statistical description of all the features used in the model are given in Table 1. Some features from the original dataset such as tin (Sn), cobalt (Co), lead (Pb), aluminum (Al) and hydrogen are not used in building the model as they have either lower correlation or seems to be not affecting the mechanical property particularly when they are present at low fractions.

2.2 Machine Learning Models

For developing the model, the data were divided into two groups: training and testing data. The training data constitute 90% of the original data and the remaining constitutes testing data. The Pearson's Correlation matrix in Fig. 1 shows the relationship between the chemical compositions and the mechanical properties. Ensemble-based models such as Random Forest (RF) and ExtraTreesRegression (ETR) were developed using the training set through python based *Scikit-learn* module. To compare the performance of these two models with other ML algorithms, three tree-based ML regression models such as K-Nearest Neighbors (KNN), Gradient Boosting (GBR), and XGBoost were developed. All the models were developed at default parameters given by *scikit-learn*.

Table 1. Statistical description of the chemical data

	Cr%	C%	Mn%	N%	Ni%	P%	Si%	Cu%	Mo%	Sample thk (mm)	YS (MPa)	UTS (MPa)	El% per 50mm
Mean	15.305	0.104	9.458	0.165	0.924	0.0622	0.388	1.536	0.08	0.887	427.41	859.335	53.837
Std	0.738	0.0101	0.799	0.0105	0.468	0.0087	0.0927	0.623	0.121	0.408	29.553	79.682	10.99
Min	13.25	0.084	8.07	0.135	0.4	0.008	0.18	0.41	0.01	0.26	344	625	26
Max	16.55	0.163	10.45	0.1979	1.83	0.095	0.69	2.2	1.2	3.17	607	1243	76

The trained models are evaluated and ranked using agreement criteria such as coefficient of determination (R2) and error criteria such as mean absolute error (MAE). The low performing models were discarded. The remaining models were re-trained by redefining their hyperparameters. Further, feature analysis was carried out to know the influence of the variables on the output parameters. The detail framework for developing the model system is given in Fig. 2.

Table 2. Maximum chemical Compositions of the different grades

Grades	Cr	C	Mn	Ni	Si	P	Mo	Cu	N	Sample thickness
JT	14.65	0.163	10.45	0.98	0.58	0.09	1.2	1.36	0.176	2.7
JSLU DD	15.95	0.126	10.3	1.15	0.69	0.092	0.2	2.2	0.18	3.17
JSLU SD	15.18	0.129	10.3	1.04	0.69	0.095	0.15	1.34	0.188	2.05
204Cu	16.55	0.1	8.82	1.83	0.63	0.06	0.23	2.18	0.198	3.12

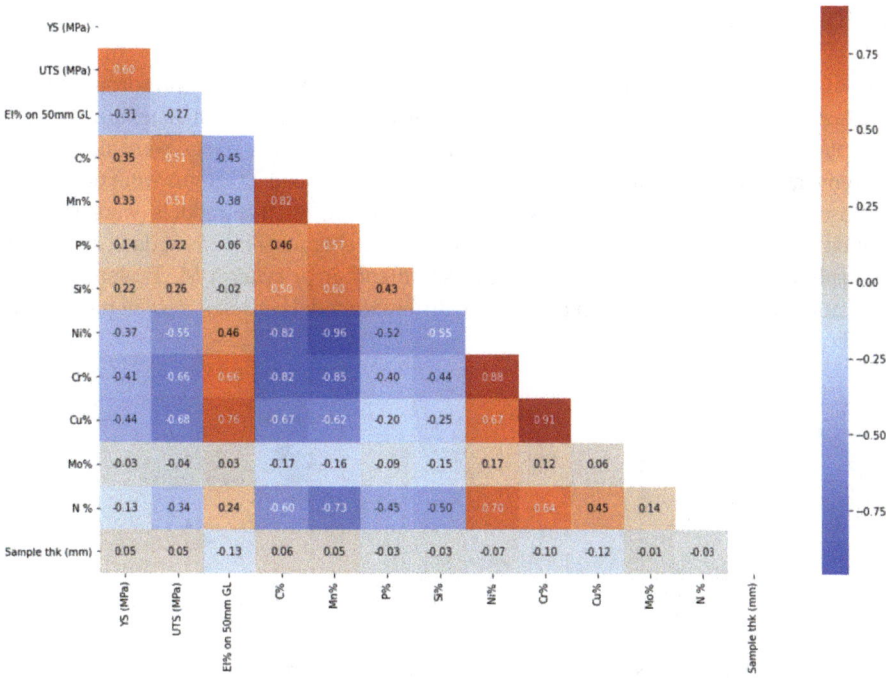

Fig. 1. Pearson's Correlation matrix of input and output features

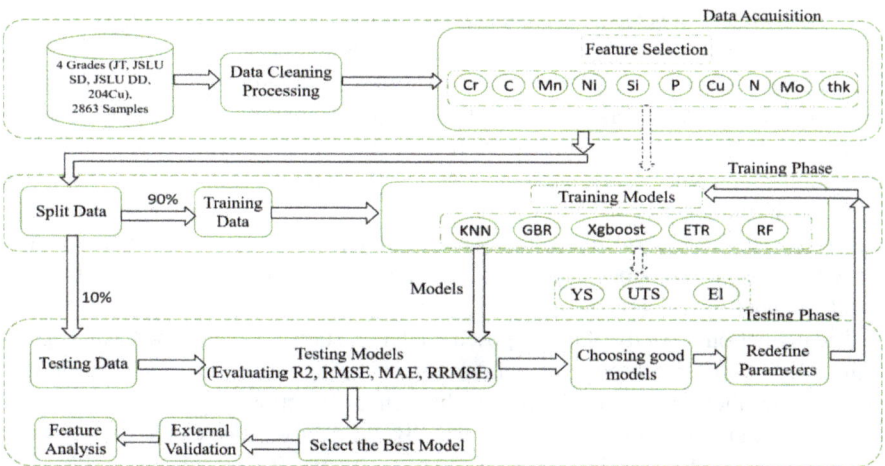

Fig. 2. Framework for developing ML design system

3 Results and Discussion

The key purpose of the current study is to implement the viability of a rigorous method of Machine learning model for the prediction of yield strength, ultimate tensile strength and ductility. Two models such as Random Forest and Extra-Trees were developed. Moreover, three models such as KNN, Gradient Boosting and XGBoost were developed as baseline models. Based on scikit-learn parameters, all five models were initially trained and analyzed to see which algorithms are suitable for the current data. For initial evaluation, agreement criteria such as R2 score and error criteria MAE were used as evaluation metrics. The performance of the different models at *scikit* default parameters are shown in Fig. 3.

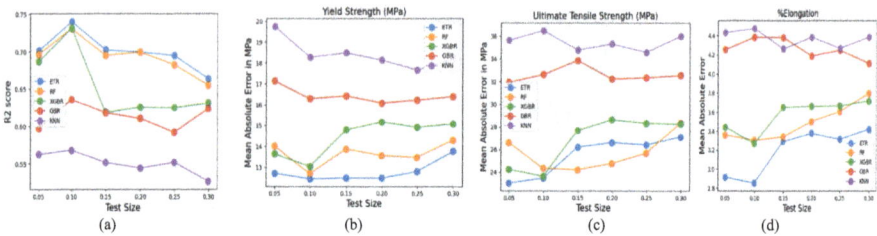

Fig. 3. Plot of test size versus (a) R2-score (b) MAE in Yield strength (c) MAE in UTS and (d) MAE in %Elongation per 50 mm.

From the figure, it is seen that RF and ETR outperforms others at default parameters. The other low performing model were discarded as tuning these would not be able to outperform RF and ETR model. These two models were chosen and re-trained by tuning their hyperparameters. Grid Search techniques were used to search the optimum parameters. Different statistical measure indexes including agreement criteria such as correlation coefficient (R) and coefficient of determination (R2) and error criteria such as root mean square error (RMSE), mean absolute error (MAE) and relative root mean square error (RRMSE), are utilized for assessing the performance of the selected models. In addition, visual assessment tool like scatter plots is used for evaluating the proposed models in order to select the best predictive models.

The prediction capacity of all proposed models is statistically assessed using agreement and error criteria through the calibration or testing phase. Based on Fig. 4, it is observed that there are some significant differences in terms of prediction precision between ETR and RF models. More specifically, the performance of ETR in all target variables is noticed to be slightly better than that of the RF model, where ETR has lower values of prediction errors. The model R2 score of ETR was found to be 0.78 while that of RF was 0.765. The R2 score of yield strength is 0.70 and 0.69 in ETR and RF respectively. Similarly, R2 score of UTS for both model was found to be 0.84 while that of elongation is 0.76 and 0.805 in RF and ETR respectively. Even though there is high R2 score of elongation in both cases, its RRMSE is high in both model which indicates significant relative error in the prediction of elongation. This can be explained from the metallurgical point of view as discussed in the latter section. The

Cr-Mn-N austenitic stainless steels are metastable in nature and convert to martensite during plastic deformation like in tensile loading. Generally, this transformation from austenite to martensite takes place either in single step (direct γ to α' transformation) or in through a intermediate step like (γ -ε- α' transformation) [20]. The austenite stability is related to the stacking fault energy (SFE), which is mainly dependent on composition and deformation temperature [21]. Further, the strain rate, stress state and grain size also plays a very vital role in determining the degree of the martensitic transformation in final microstructure, and thereby the mechanical properties of the material [22]. The % elongation is affected by both test temperature and strain rate. The data which is collected throughout the tensile testing of full year where testing temperature in northern India varies from 10° to 45 °C. As the Ms temperature of the steel is closer to room temperature, slight change in the temperature will lead to significant change in the % elongation. So, apart from the chemical composition and grain size, testing temperature will also lead to significant change in the % elongation. Probably that could have led to the significant error in the prediction.

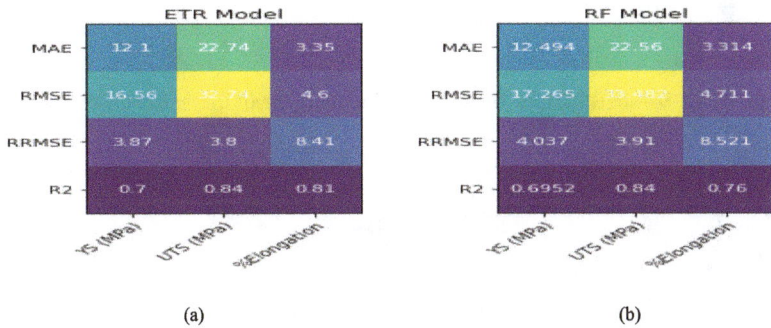

(a)   (b)

Fig. 4. Performance metrics of (a) ETR (b) RF during testing phase

In order to examine the efficiency of the proposed model, the visualization assessment is vital in predicting each sample separately. A scatter plot is one of the most significant figures used frequently to assess the performance efficiency of the predictive models. It provides significant information about the diversion of each point from the actual observation. Figure 5 and 6 show the scatter plots of actual values versus predicted values of all three output variables during training phase and testing phase. From the figures, it is seen that t the yield strength and UTS are predicted within 5% error of the regression line whereas in case of % Elongation, significant number of outliners are there in 5% error line but it is well within 10% error. Based on the evaluation metrics and visualization of the model prediction performance, it is observed that the ETR model performed better in terms of prediction capability. So, ETR model has been chosen for further analysis (Figs. 7 and 8).

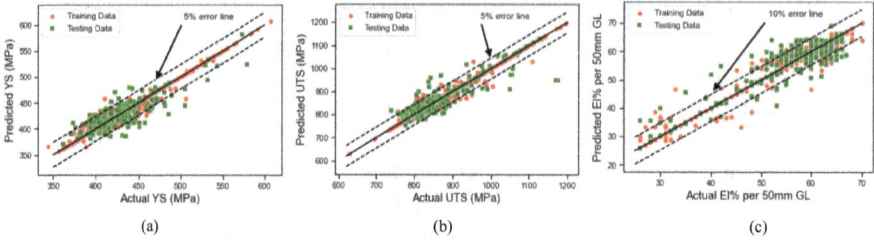

Fig. 5. Predicted vs actual values of (a) yield strength, (b) ultimate tensile strength and (c) Elongation: ETR model

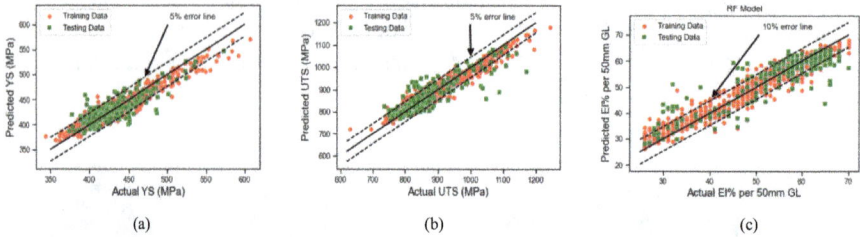

Fig. 6. Predicted vs actual values of (a) yield strength, (b) ultimate tensile strength and (c) Elongation: RF model

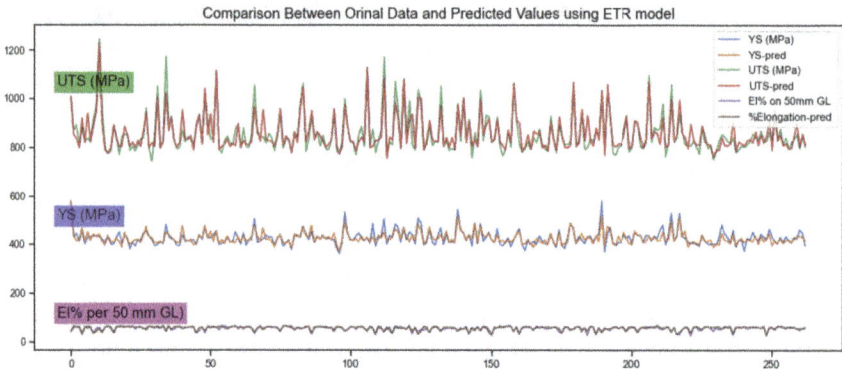

Fig. 7. Comparision between the actual and the predicted values in the testing samples

3.1 External Validation

To analyze the capability of the proposed models to predict the unseen data outside the database, external validation was carried out with some new data of mechanical properties of ASS obtained from the industry. The details of these data are listed in Table 3. The prediction results are shown in Fig. 10. The model have relative better prediction in case of UTS ($R^2 = 0.74$) which is similar to the testing phase. The external validation scores for YS was found to be 0.66 which was close to that of testing score of 0.7. The R2 score of the % elongation was found to be 0.68 which was far lower than

0.81 of testing score. This indicates the model could not capture the trend in case of elongation as compared to YS and UTS.

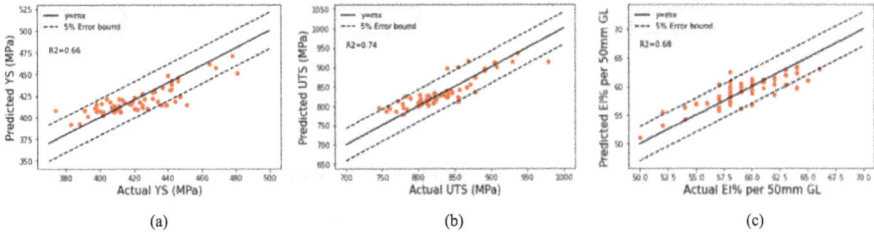

(a)   (b)   (c)

Fig. 8. Comparison between the predicted and original values of the unseen data outside the database for: (a) YS and (b) UTS (c) Elongation

Table 3. Description of new data outside of database

	Cr%	C%	Mn%	N%	Ni%	P%	Si%	Cu%	Mo%	Sample thickness (mm)	YS (MPa)	UTS (MPa)	El% per 50mm
Mean	15.538	0.102	9.461	0.161	0.982	0.0647	0.422	1.819	0.068	0.702	420.8	828.739	59.348
Std	0.418	0.0086	0.748	0.0113	0.365	0.0072	0.0999	0.322	0.0241	0.182	21.823	44.517	3.195
Min	14.7	0.82	8.02	0.141	0.51	0.051	0.18	1.15	0.03	0.34	374	746	50
Max	16.38	0.123	10.2	0.185	1.74	0.08	0.68	2.2	0.15	1.26	481	978	66
Count	65	65	65	65	65	65	65	65	65	65	65	65	65

3.2   Feature Analysis using SHapley Additive Explanation (SHAP)

Feature analysis was carried out using SHAP method which is given below. SHAP value is a smooth tool for machine learning interpretation. It works on both classification and regression problems. In regression problems, SHAP value can correctly assign the feature importance. This method is based on a famous theory called Game Theory. In this method, the impact of a feature is calculated as the difference between the performance of the model with and without the feature [27]. In fact, the SHAP value is the marginal contributions to all possible coalition. Mathematical expression for SHAP value of i_{th} feature is given by

$$\varphi_i = \sum_{S \subseteq M \sim i} \frac{|S|!(|M| - |S|)!}{|M|!} [f(S \cup i) - f(S)]$$

S = subsets of the features without i feature, $S \cup i$ = subsets of the features with i feature, $S \subseteq M \sim i$ = all sets S that are subsets of the full set of features M, excluding feature i.

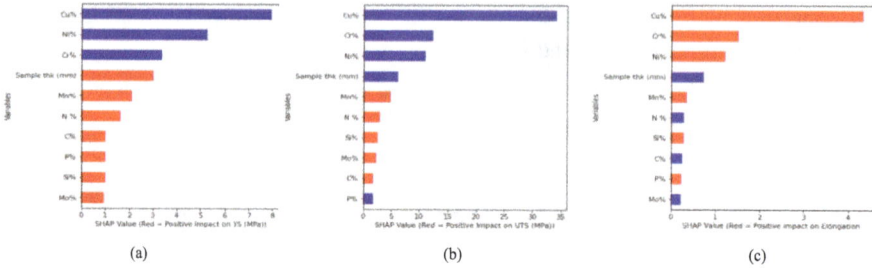

Fig. 9. Feature Importance plot using Shap Method (a) YS (b) UTS and (c) Elongation

From the Fig. 9, it is seen that Cu and Ni causes the solid solution softening of austenitic stainless steel whereas Mn, C, N will provide solid solution hardening which is in agreement with the published literature. Though the extent of hardening provided by C and N is much more than substitutional solid solution strengthener like Mn which is a mild strengthener, the model could not predict the extent of hardening by C and N. It may be due to the low range in the variation of Carbon (0.084% to 0.163%) and Nitrogen (0.135% to 0.1979%) while the variation of Mn (8.07% to 10.45%) is little bit wider in range and thus the model could not pick up the effect of C and N properly. Though Cr should be a mild solid solution strengthener as per earlier study [23] but here it shows softening. It also increases the elongation as it can increase the Md30 temperature which helps in increasing the % elongation by TRIP effect. Further expanding the training set to a wide-ranging Cr may lead to better prediction of the properties. Leffler and Nordberg et al. [24, 25] found hardening effect of Cu but Okhubu and Ellisan et.al. [16, 26] found softening. Ni and Cu is well known for their solid solution softening effect and the model predicts it very well. In case of C and N, some very low Carbon (0.02%) and high C (0.5%) and low Nitrogen (0.02) and high Nitrogen (0.6%) are needed to properly predict the effect of C and N on CrMnN stainless steel. Moreover, there is significant influence of the sample thickness on the strength and elongation.

4   Conclusion

The aim of this study was to implement vigorous method of machine learning techniques to predict the mechanical properties of Cr-Mn-N grade austenitic stainless steel. Two different machine learning approaches, namely, ETR and RF have been used to predict the mechanical properties of low nickel austenitic stainless steel by using chemical composition and sample thickness. Three classical ML algorithms such as KNN, Gradient Boosting, and Xgboost were also trained as baseline models to compare against the proposed models. It was found that proposed models, RF and ETR outperformed the baseline models. The two models were tuned and the outcomes of both are summarized as follows:

(1) ETR model outperformed RF model in both training and testing phases in terms of every evaluation metrics. The problem of over fitting is lesser in ETR

(2) The overall R2 (coefficient of determination) score of ETR model was found to be 0.78 and that of RF model was 0.766. So, ETR model was chosen for further analysis and feature analysis was carried out using this model testing the impact of variables on the model and their relative importance's on the mechanical properties of the stainless steel

(3) The selected model is tested with data outside the data range of the database to check the reliability of the model and R2 score in new data were found to be 0.7.

(4) However, processing parameters and treatment conditions are not included in the analysis and development of the model. These parameters can seriously affect the mechanical properties of SS. This study recommends inclusion of these parameters for future researches.

References

1. Charles, J., et al.: A new European 200 series standard to substitute 304 austenitics, La Revue de Métallurgie - CIT - Février (2009)
2. Rasouli, D., Kermanpur, A., Najafizadeh, A.: Developing high-strength, ductile Ni-free Fe–Cr–Mn–C–N stainless steels by interstitial-alloying and thermomechanical processing. J. Mater. Res. Technol. **8**(3), 2846–2853 (2019)
3. Yang, K., Ren, Y.: Nickel-free austenitic stainless steels for medical applications. Sci. Technol. Adv. Mater. **11**, 014105 (2010)
4. Gardner, L.: Stability and design of stainless steel structures – Review and outlook. Thin-Walled Struct. **141**, 208–216 (2019)
5. 200 Series Stainless Steel CrMn Grades, ASSDA Technical Bulletin
6. Ramesh, R., Gopal, R.D.: Sustainable Stainless Steel for Wagons. National Workshop on Stainless Steel Coach & Wagon Manufacture Jadavpur University, 8–9 September 2011, Kolkata (2011)
7. Structural Materials for Liquid Metal Cooled Fast reactor Fuel Assemblies — Operational Behaviour, IAEA Nuclear Energy Series No. NF-T-4.2
8. Pozuelo, M., Witting, J.E., Jimenez, J.A., Frommeyer, G.: Enhanced mechanical properties of a novel high-nitrogen Cr-Mn-Ni-Si austenitic stainless steel via TWIP/TRIP effects. Metall. and Mater. Trans. A. **40**, 1826–1834 (2009)
9. Fluch, R., Kapp, M., Spiradek-Hahn, K., Brabetz, M., Holzer, H., Pippan, R.: Comparison of the dislocation structure of a CrMnN and a CrNi austenite after cyclic deformation. Metals **9**, 784 (2019)
10. Järvenpää, A., Jaskari, M., Juuti, T., Karjalainen, P.: Demonstrating the effect of precipitation on the mechanical stability of fine-grained austenite in reversion-treated 301LN stainless steel. Metals **7**, 344 (2017)
11. Duan, J., Farrugia, D., Davis, C., Li, Z.: Effect of impurities on the microstructure and mechanical properties of a low carbon steel. Iron Steel Making **49**(2), 140–146 (2021)
12. Tikhonova, M., Kaibyshev, R., Belyakov, A.: Microstructure and mechanical properties of austenitic stainless steels after dynamic and post-dynamic recrystallization treatment. Adv. Eng. Mater. **20**, 1700960 (2018)
13. Krauss, G.: Steels: Processing, Structure, and Performance; ASM International: Russell. OH, USA (2015)
14. Reifsnider, K.L., Tamuzs, V.: On nonlinear behavior in brittle heterogeneous materials. Compos. Sci. Technol. **66**, 2473–2478 (2006)

15. Narayanaa, P.L., et al.: Modeling high-temperature mechanical properties of austenitic stainless steels by neural networks. Comput. Mater. Sci. **179**, 109617 (2020)
16. Ohkubo, N., Miyakusu, K., Uematsu, Y., Kimura, H.: Effect of alloying elements on the mechanical properties of the stable austenitic stainless steel. ISIJ Int. **34**(9), 764–772 (1994)
17. Wang, Y., Xuebang, W., Li, X., Xie, Z., Liu, R.: Prediction and analysis of tensile properties of austenitic stainless steel using artificial neural network. Metals **10**, 234 (2020)
18. Chen, Z., Zoub, D., Yuc, J., Hand, Y.: Artificial neural network approach to predict mechanical properties of 301 austenitic stainless steel. Mater. Sci. Forum **658**, 145–148 (2010)
19. Shin, D., Yamamoto, Y., Brady, M.P., Lee, S., Haynes, J.A.: Modern data analytics approach to predict creep of high-temperature alloys. Acta Mater. **168**, 321–330 (2019)
20. Talonen, J.: Effect of strain-induced α'-martensite transformation on mechanical properties of metastable austenitic stainless steels. Ph.D. Thesis, TKK Dissertations 71, Espoo, Finland (2007)
21. Fussik, R., Egels, G., Theisen, W., Weber, S.: Stacking fault energy in relation to hydrogen environment embrittlement of metastable austenitic stainless CrNi-steels. Metals **11**, 1170 (2021)
22. Kisko, A., Misra, R.D.K., Talonen, J., Karjalainen, L.P.: The influence of grain size on the strain-induced martensite formation in tensile straining of an austenitic 15Cr–9Mn–Ni–Cu stainless steel. Mater. Sci. Eng. A **20**, 408–416 (2013)
23. Pawar, P.B., Utpat, A.A.: Effect of chromium on mechanical properties of a487 stainless steel alloy. Int. J. Adv. Res. Sci. Eng. **5**, 112–118 (2016)
24. Nordberg, H.: La Metallurgia Italiana **85**, 147–154 (1994)
25. Leffler, B.: Proceeding of the Nordic Symposium on Mechanical Properties of Stainless Steels, SIMR, Sigtuna, Sweden, pp. 32–42 (1990)
26. Eliasson, J, Sandstrom, R.: Steel Res. **71**, 249–254 (2000)
27. Nohara, Y., Matsumoto, K., Soejima, H., Nakashima, N.: Explanation of Machine learning models using improved shapley additive explanation. In: ACM-BCB 2019, 7–10 September 2019, Niagara Falls, NY, USA (2019)

On the Deformation Mechanism and Dislocation Density Evolution in A Polycrystalline Nano Copper at 10 K–700 K/10^8 s^{-1}–10^9 s^{-1} Employing Molecular Dynamics Simulations

Prashant Kashyap[1], G. Sainath[2,3], Nilesh Kumar[1], and Surya D. Yadav[1(✉)]

[1] Department of Metallurgical Engineering, Indian Institute of Technology (BHU), Varanasi 221005, Uttar Pradesh, India
`sury.met@iitbhu.ac.in`
[2] Materials Development and Technology Division, Metallurgy and Materials Group, Indira Gandhi Centre for Atomic Research, Kalpakkam 603102, Tamilnadu, India
[3] Homi Bhabha National Institute, Indira Gandhi Centre for Atomic Research, Kalpakkam 603102, Tamilnadu, India

Abstract. In this work, molecular dynamics (MD) simulations have been utilized to explore the effect of temperature and strain rate on the deformation mechanisms that occurs in nano grain polycrystalline Cu having special $\sum 3$ grain boundaries along with other high and low angle grain boundaries. The temperature and strain rate ranges explored in this work are 10 K–700 K and 1×10^8 s^{-1}–1×10^9 s^{-1}, respectively. The results show that the yielding starts by the nucleation of the Shockley partials of $\frac{1}{6} <112>$ character at the $\sum 3$ grain boundaries. Furthermore, the subsequent plastic strain is evidenced to be accommodated by the formation of stacking faults and twins. The formation of stacking faults at the expense of Shockley partials occurs at lower strain while twin formation from Shockley partials is favorable at high strain. Low temperature (10 K) deformation leads to intergranular failure via void nucleation, growth and crack propagation.

Keywords: Molecular dynamics (MD) simulations · Polycrystalline Cu · Dislocations · Twins · Grain boundaries

1 Introduction

In recent years, the development of the nano/micro electromagnetic components (NAMS/MEMS) [1] fabricated employing metallic nanomaterials have attracted immense interest for research. Metallic nanomaterials possess excellent mechanical, electrical, thermal, optical and chemical properties [2–7] useful for such applications. Copper based nanomaterials finds a wide range of applications due to their excellent performance with respect to conductivity, strength coupled with ductility, transmittance, easy and inexpensive synthesis [8, 9]. The components fabricated out of copper nano-materials may experience different temperatures and stresses during the operations.

S. Patra et al. (Eds.): METCENT 2023, *Proceedings of the International Conference on Metallurgical Engineering and Centenary Celebration*, pp. 131–144, 2024.
https://doi.org/10.1007/978-981-99-6863-3_14

Therefore, understanding of the mechanical properties and the deformation mechanism becomes utmost important to get insights in order to design the optimum microstructure with adequate properties. Understanding of the deformation mechanism in case of nano materials is limited due to the complexities of the experimental routes. With the enhancement in the computational capabilities and availabilities of the adequate interatomic potentials, molecular dynamics (MD) simulation technique opens an alternative pathway to study the mechanical properties of the materials. MD simulations can help in studying the real-time deformations occurring in a nanomaterial at varying temperatures.

Several researchers have investigated the various aspects of deformation in nanomaterials employing MD-simulations. Sainath et al. [10, 11] studied the interaction of longitudinal twin boundaries with the stair-rod dislocations during tensile loading of Cu nano pillars. It was demonstrated that single crystal nano pillar deforms by twinning on two independent slip systems while the longitudinally twinned Cu nano pillar deforms on four independent slip systems by full slip with leading and trailing dislocation partials. Rohith et al. [11] carried out MD simulation studies to obtain the insights about the effect of temperature, strain rate and the specimen size on dislocation exhaustion and starvation in a single Cu nano pillar. During the tensile loading and post yielding, nanowire experiences dislocation exhaustion at lower strain and dislocation starvation at higher strain. Zhigao et al. [12] have shown the effect of temperature and strain rate on the mechanical properties of single crystal Al. Zhang et al. [13] studied the effect of the strain rate on the deformation mechanism of the copper nanocrystals using MD simulation. It was revealed that maximum flow stress is independent of the strain rate at the mean grain size range of 10–20 nm. It was further realized that, at lower strain rate the grain boundary deformation mechanisms are easy to operate while at higher strain rate the partial dislocations are difficult to nucleate at the onset of the plastic deformation. In another study of polycrystalline copper using MD simulation, Zhang et al. [14] studied the effect of grain size and working temperature on the mechanical properties of nanocrystal. It was observed that increasing the grain size leads to gradual increase of elastic modulus. Opposite trend was observed with respect to temperature, and elastic modulus as well as yield strength decrease with an increase in temperature. Yazdani et al. [15] studied the deformation mechanisms in polycrystalline Cu, Ni and equimolar Cu-Ni alloy. The study reveals that the accumulation of dislocations near grain boundaries is observed in Cu and Ni polycrystals, while dense accumulation of dislocations occurs inside grains in the equimolar Cu-Ni alloy due to the formation of shearing bands. While the hardening mechanism in Cu and Ni polycrystals follows the Taylor hardening, the deformation behaviour of the Cu-Ni alloy deviates from this theory. Another MD-simulation study conducted by Tian et al. [16] to explore the deformation behaviour of polycrystalline copper under different loading revealed the importance of grain boundaries and free surfaces with respect to plastic deformation. The results show the emergence of full dislocations, pile-up of dislocations and twin boundaries in the grains along with stress-assisted grain growth. Rohith et al. [17] investigated the tensile deformation behavior of polycrystalline Cu nanowires of different grain sizes. The results reveal an inverse Hall-Petch relation between yield stress/flow stress and grain size, and partial dislocation-mediated plasticity along with twinning was observed. The formation of fivefold twins and extensive grain growth during plastic deformation have also been observed. The study also found that the ductility of the nanowires is insensitive to grain size, however the failure mode depends on it.

As per the aforementioned literature it can be deduced that, the presence of different interfaces or grain boundaries in materials greatly affects the deformation mechanism and hence, influences the properties [18]. In line with this fact, Wantable [19] introduced the concept of Grain Boundary Engineering (GBE) that involved introducing a fraction of 'special' boundaries in microstructure. The idea was further used for improving the ductility [20], creep [21, 22] fatigue resistance and corrosion resistance [23] in materials. Similarly, the presence of special boundaries such as $\sum 3$ grain boundaries in the copper nanomaterials are of particular interest due to the unique structural and mechanical properties [24]. The structural stabilities implicate enhanced mechanical, corrosive and electrical properties. Significant efforts have been made by the various researchers for investigating the deformation behavior of randomly oriented polycrystalline Cu at different temperature, strain rate and grain size. However, the dislocation density evolution and deformation behavior of nano grain polycrystalline Cu having special $\sum 3$ grain boundaries along with other high and low angle grain boundaries are yet to be explored. In order to bridge the gap, this study focuses on the deformation mechanism and dislocation density evolution in the presence of the $\sum 3$ boundaries along with other high and low angle grain boundaries. The model/representative microstructure was designed in such a way that it resembles with real copper in terms of grain shape and $\sum 3$ grain boundaries.

MD simulations were performed in order to investigate the dislocation density evolution and deformation behavior of nano polycrystalline Cu that contains $\sum 3$ grain boundaries on (111) planes along with other high and low angle GBs. Uniaxial tensile loading has been applied with the temperature and strain rate ranges of 10 K–700 K and $1 \times 10^8 \, s^{-1}$–$1 \times 10^9 \, s^{-1}$, respectively. The flow response in varying conditions has been discussed thoroughly in light of different mechanisms and microstructural evolution.

2 Methodology

2.1 Description of Microstructure

The copper polycrystals have been generated employing the Atomsk software developed by the Hirel [25]. The detailed procedure of creating polycrystal Cu is as follows. First, a unit cell of copper has been created with the crystallographic orientation of $[1\bar{2}1][111][10\bar{1}]$ as a seed for producing the Cu polycrystal. A box size of 106.43 × 106.24 Å has been produced and seeds are positioned at the fixed nodes as shown in the Fig. 1a. The incoherent $\sum 3$ boundary is created that lies in the (111) plane of fcc crystal having an angle of misorientation of 109.47°. The grains (a, b, c and d) are oriented in the supplements of the misorientation angle around the $[10\bar{1}]$ axis to form $\sum 3$ CSL boundary and wrapped to form 2D polycrystal. The 2D polycrystal is duplicated along the Z-axis with a width of 12.72 Å to obtain the desired polycrystal as shown in Fig. 1b.

2.2 Simulation Details

Large atomic/molecular massively parallel simulator (LAMMPS) package developed by Plimpton at Sandia National Laboratories has been used to perform MD simulations

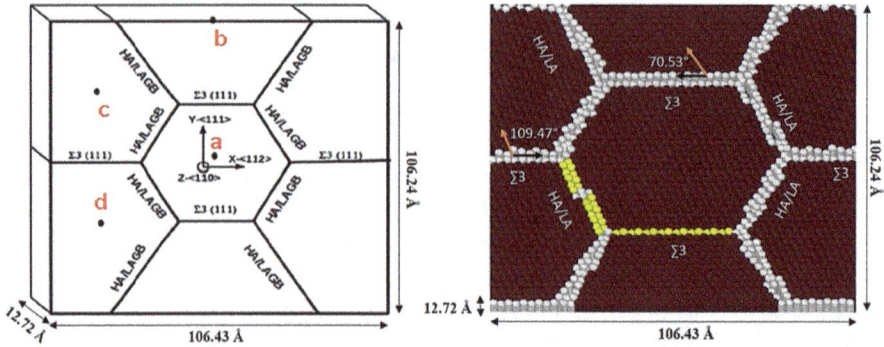

Fig. 1. (a) Proposed nano polycrystalline structure of copper (b) Nano polycrystalline copper structure generated using Atomsk

[26]. In order to mimic the inter-atomic forces, modified embedded atomic method (MEAM) potential for copper, developed by Baskes et al. [27] has been used in this work. This potential is successful in calculating the equation of state, elastic moduli, lattice constants, simple defects and surfaces energies. The potential is advantageous in the sense that it can equilibrate and generate desired stable microstructure. The total energy of the atom using MEAM potential is stated as,

$$E_{total} = \sum_i F_i(\overline{\rho}_i) + \frac{1}{2} \sum_{i \neq j} \phi_{ij}(r_{ij}), \qquad (1)$$

where F_i is the "embedding energy" function (energy required to embed an atom in the background electron density $\overline{\rho}_i$ at site i), ϕ_{ij} is the pair potential interaction which sums up the interaction over all the neighbours of atoms i within the cut-off distance of r_{ij}. Following the initial construction, all the atoms are assigned with random velocities from a finite maxwell distribution [28]. Velocity Verlet algorithm is used to integrate the equations of motion with a time step of two femto seconds for ten pico seconds. Finally, the system is thermally equilibrated at the temperature of 10 K, 300 K and 700 K in an isothermal-isobaric ensemble (NPT) with a Nose-Hoover thermostat to maintain a constant temperature during equilibration.

In order to study the deformation mechanisms occurring at different temperatures, tensile deformation conditions were imposed along [1 $\overline{2}$1] direction at 10 K, 300 K and 700 K employing a stain rate of 1×10^9 s^{-1}. Similarly, at 700 K, tests with strain rate of 1×10^8 s^{-1} and 1×10^9 s^{-1} were carried out to find out the strain rate effect. The average stresses are being calculated using the virial expression [29]. The strain has been calculated as $e = (l\text{-}l_0)/l_0$, where l is the length at each time step and l_0 is the length of the system after equilibration. Furthermore, the results were analysed employing the Dislocation Extraction Algorithm (DXA) developed by A. Stukowski and K. Albe [30] to identify all the dislocation line defects and corresponding Burgers vector. Dislocation density has been obtained as the total dislocation length divided by the simulated cell volume using OVITO [31]. The visualization of atomic configurations is achieved in OVITO using common neighbour analysis (CNA) modifier [32]. The CNA parameter helps in characterizing local structural environment such as FCC (CNA = 1), HCP (CNA

= 2), BCC (CNA = 3) and other crystalline structure (CNA = 5). Stacking fault and twin boundaries has been identified as HCP environment [32]. The distinction between stacking fault and twin boundary has been done through the numbers of HCP layers. The stacking faults contain two layers of HCP atoms whereas twin boundaries contain one layer of HCP atom.

3 Results and Discussion

3.1 Microstructural Evolution at 300 K and Strain Rate of 1×10^9 s^{-1}

The variation of flow stress response and the corresponding dislocation density as a function of strain is depicted in the Fig. 2 at 300 K / 1×10^9 s^{-1}. The figure shows that polycrystal undergoes elastic deformation showing almost linear behaviour up to a peak followed by a drop in the stress (depicted by a vertical black dotted line). The very negligible perturbation in stress response before yielding is due to the micro plasticity, in synergy with grain boundary dislocations. In polycrystalline metal and alloys, the initiation of plastic deformation instigates at low applied stresses that are below the yield stresses [33]. The initiation occurs at very small grains in form of microplastic deformation that further extends with ongoing time. This is also in line with the observation that, before yielding the dislocation density tends to increase slightly, due to the micro plasticity and grain boundary dislocations. At the onset of yielding, the dislocations are nucleated from the some of the $\sum 3$ grain boundaries and move freely across the polycrystal causing deformation. With further straining, the deformed structure provides the scope for new sites for the nucleation of the dislocations, hence increase in the dislocation density can be witnessed after yielding. Once the yielding has occurred, graph follows a trend that whenever dislocation density increases, stress decreases or vice versa. This can be understood from the fact that strain can be accommodated by the movement of dislocations. Initially, when the dislocation density was low it was difficult to accommodate the strain, hence a higher value of stress was required. Once plenty of dislocations are nucleated they help in easy deformation via gliding and lead to drop in stress. During the deformation these dislocations may get absorbed in the grain boundaries or may interact with other microstructural features and eventually ends up on the free surface of specimen. Once the dislocations end up on the free surface, the specimen is starved of dislocations or achieves a state of lower dislocation density, hence a higher stress is required for the further deformation. The higher value of the dislocation density in this study as compared to the commonly reported experimental values shall be attributed the high strain rate used for the simulations [11, 34].

The yielding strength observed herein is higher in comparison to the copper nanocrystals having random GB's [35]. It has been observed that grain boundaries act as a barrier to the dislocation motion, thus delaying the plastic deformation in polycrystals. Figure 3 depicts the substructure evolution at relatively lower strains at 300 K/1×10^9 s^{-1}. The yielding starts by the nucleation of the Shockley partials of $\frac{1}{6} <112>$ character at the $\sum 3$ grain boundaries (see Fig. 3a and b and c and 3e, f and g). These Shockley partials glides away from the parent GBs in the direction towards the free surface leaving a trail of stacking faults behind it as shown in Fig. 3d. This is similar to the mechanism of the formation of stacking faults in a Cu Nano pillars [36, 37]. The stacking faults and

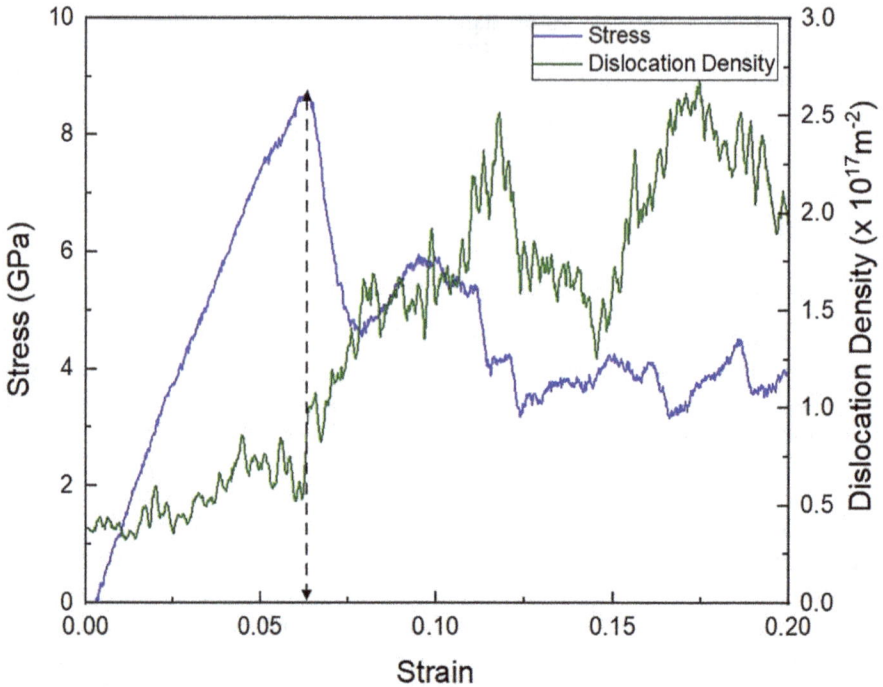

Fig. 2. Variations in flow stress along with dislocation density as a function of strain at 300K at a strain rate of 1×10^9 s^{-1}.

twins are responsible for plastic deformation in such materials. The stacking faults are stable at the lower strain responsible for initiation of yielding whereas formation of both, stacking faults and twins take place at higher strains.

Figure 4 depicts the substructure evolution at relatively higher strains at 300 K/1 $\times$ 10^9 s^{-1}. At higher strain, the formation of twin was more favourable (see Fig. 4a-d). The mechanism for the formation of twin initiates with the nucleation of the Shockley partials at the three-layer non-crystalline structure formed at the grain boundary (see marked rectangular area of Fig. 4a and b). The three-layered structure is unstable in nature resulting in nucleation of the Shockley partials (see marked arrow of Fig. 4 a & e and b & f). The partials glides from the GBs towards the free surface forming a trail of twins behind it as shown in Fig. 4d and g [38].

3.2 Effect of the Temperature and Strain Rate

Influence of temperature and strain rate on yielding stress and dislocation density is depicted in Fig. 5 and 7, respectively. Figure 5a depicts the variation of yielding stress with respect to temperature at constant strain rate. It can be observed that yielding strength decreases with increase in temperature from 10 K to 700 K. This can be attributed to the increased vibrational energy of atoms causing enhanced dislocation mobility. Figure 5b shows the influence of temperature on dislocation density at fixed strain rate. It can

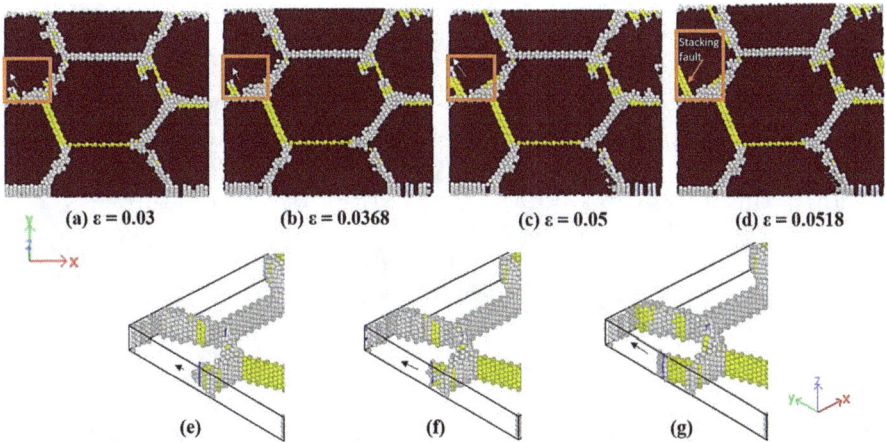

Fig. 3. Atomic snapshots showing the formation of stacking fault at 300 K and strain rate of 1 $\times 10^9$ s^{-1}. Images (a-d) (e-g) shows the atomic snapshot at selected strain obtained from MD-simulation using CNA and DXA methods. The brown atoms represent FCC and other structure are also represented by different colors. The blue lines in the vicinity of arrows represent the Shockley partials. The motion of the Shockley partials is represented by the arrows. Image (a) shows the nucleation of Shockley partials at $\sum$3 grain GBs, (b) shows the movement of Shockley partial away from GB leaving a stacking fault behind, (c) shows the Shockley partial leaving the specimen at the free surface and (d) shows the stacking fault manifested during the deformation. Images (e-g) represents the magnified images of orange rectangular marked sections in images a, b and c, respectively, showing the motion of the Shockley partial.

be observed that, in general the dislocation density is high at higher temperatures. The result is contrary to the results observed in bulk materials. High dislocation density at high temperature is due to fact that increased vibrational energy of atoms at elevated temperature results in more nucleation of dislocations at deformed structures and grain boundaries. Higher vibrational frequency at elevated temperature that leads to enhanced dislocation mobility also explains why yielding occurs at lower strain at higher temperature (first vertical arrow in Fig. 5b). At 300 K the variation of dislocation density follows the similar trend as 700 K. The yielding starts at higher strain about $\varepsilon \sim 0.0634$ (second vertical arrow in Fig. 5b) for 10 K due to the less vibrational energy that leads to reduced dislocation mobility. At 10 K, the dislocation density is almost constant around a mean value with increasing strain (Fig. 5b)) and lower compared to 300 K and 700 K. This is due to the fact that, during the deformation at 10 K, the voids nucleation can be observed at triple junction that acts as a stress concentrator resulting in intergranular fracture. Suppression of dislocation mediated plasticity due to to less dislocation mobility as well as void nucleation and growth resulted in low dislocation density and less strain accumulation at 10 K. It need to be emphasized that at 10 K the sample starts to fails after the strain of 0.175, due to void nucleation and growth, thus, the dislocation density plot is depicted only up to that point.

Figure 6 depicts the substructure evolution observed at 10 K/1 $\times 10^9$ s^{-1}. It can be visiualized that low temperature deformation leads to intergranular failure (see Fig. 6).

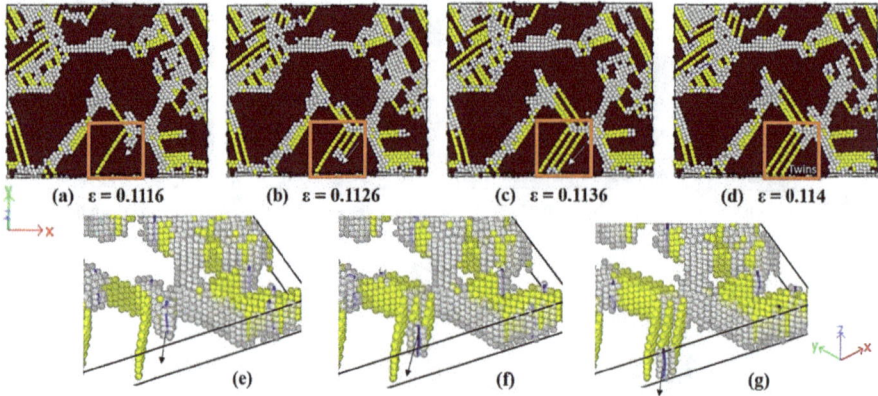

Fig. 4. Atomic at 300 K and strain rate of 1×10^9 s^{-1}. Images (a-d) (e-g) shows the atomic snapshot at selected strain obtained from MD-simulation using CNA and DXA methods. The brown, yellow and white atoms represent FCC, HCP and other structure respectively. The blue lines represent the Shockley partials. The motion of the Shockley partials is represented by the arrows. Image (a) shows the formation of triple layered deformed structure at the GB, (b) instability of the triple layer leads to formation of twins, (c) Shockley partial leaving the specimen and (d) depicts the Twins. Images (e-g) represents the magnified images of orange rectangular marked section in images a, b and c, respectively, showing the motion of the Shockley partial.

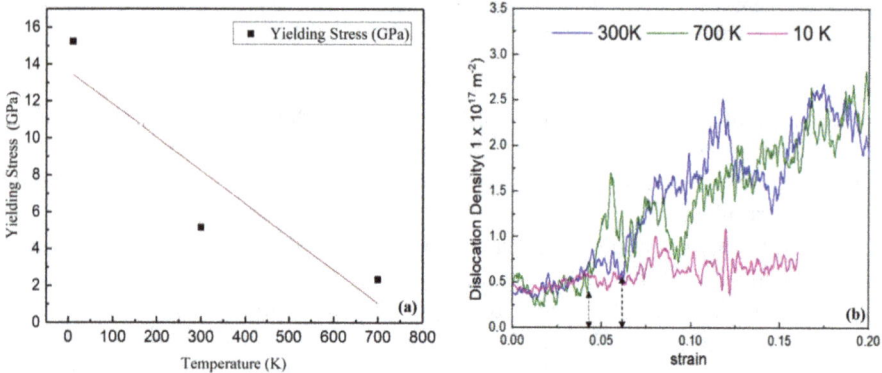

Fig. 5. Variation of (a) Yielding stress (b) Dislocation density of copper nano polycrystal at temperature of 10 K, 300 K and 700 K and strain rate of 1×10^9 s^{-1}.

Figure 6a depicts the nucleation of void at triple junction of grains. The triple nodes are the sites of high energy and the system tries to minimize the energy by formation of void at the triple juntion during the deformation [39]. With ongoing deformation growth of the void and crack opening can be observed in Fig. 6b. It has been reported and discussed in the litrature that the voids act as the cite for the crack opening and propagation. Furthermore, this stable crack depicted in Fig. 6c can be observed to be growing along the grain boundaries resulting in intergranular fracture of the sample during the tensile loading (see Fig. 6d).

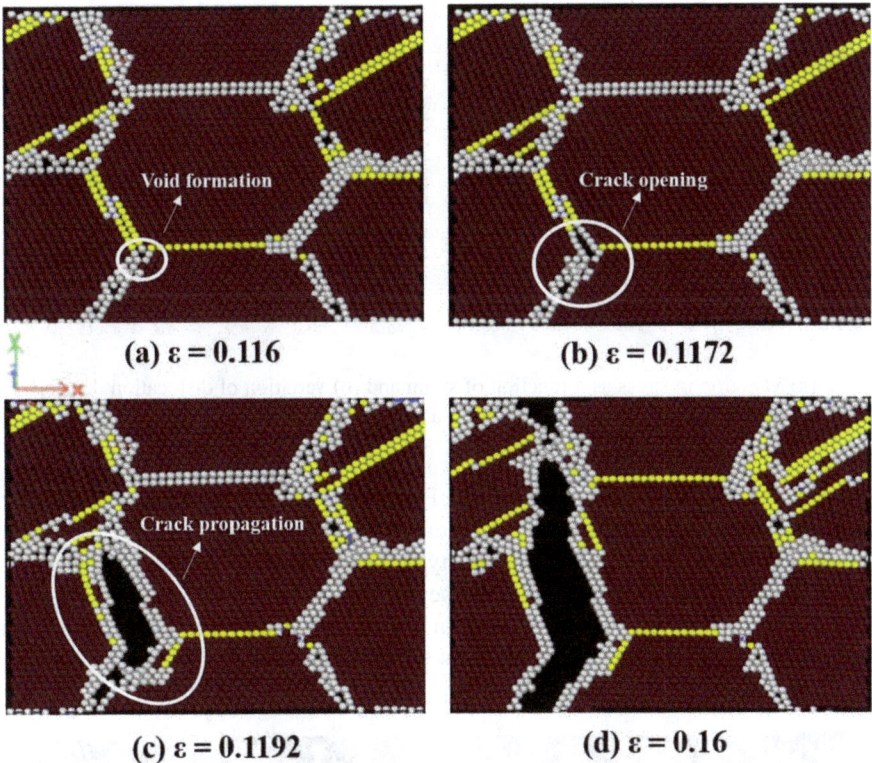

Fig. 6. Atomic snapshots representing the failure of copper polycrystal at 10 K and strain rate of 1 $\times 10^9$ s^{-1}. Image (a) shows the void at triple junction, (b) shows crack opening along HA/LA GBs, (c) shows the crack propagation along the HA/LA GBs and (d) shows failed copper polycrystal.

The effect of stain rate on flow stress response and dislocation density at a constant temperature of 700 K has been shown in Fig. 7. In flow curves, the very negligible perturbation in stress response before yielding is due to the micro plasticity, in synergy with grain boundary dislocations. In polycrystalline metal and alloys, the initiation of plastic deformation instigates at low applied stresses that are below the yield stresses [33]. Figure 7a depicts that, the specimen yields at higher stress when subjected to higher strain-rate of 1 $\times 10^9$ s^{-1} compared to that of 1 $\times 10^8$ s^{-1}. The lower yielding stress at lower strain rate is due to the higher dislocation density observed in this case, that assists the deformation at a lower strain and lower yielding stress [12]. These observations are entirely different from the bulk deformation results, as the mechanisms are different at nanoscale.

The increased dislocation density of polycrystal at lower strain-rate may be accounted to the additional mechanism for the formation of twins at lower strain rate, that produces dislocations. At lower strain rate, apart from regular mechanism observed at higher strain rate, twin formation takes place by another mechanism, resulting in more sites for the dislocation nucleation. Initiation takes place by formation of four layers of non-crystalline structure (See Fig. 8a). The four layered non-crystalline structure expands by

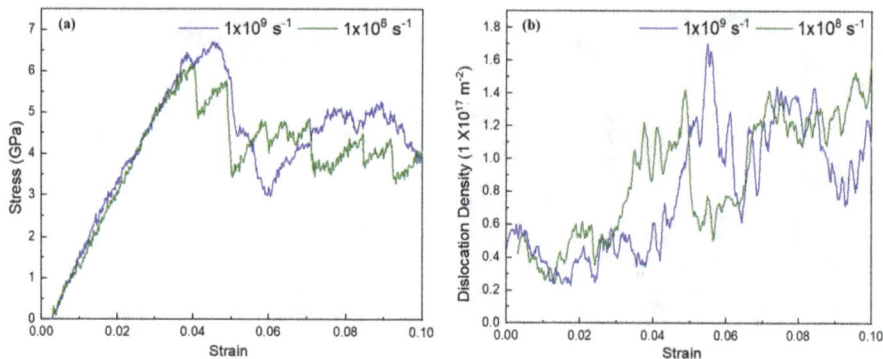

Fig. 7. (a) Variation in stress as a function of strain and (b) variation of dislocation density as a function of strain, at strain rates of 1×10^8 s^{-1} and 1×10^9 s^{-1} at 700 K.

the nucleation of the Shockley partials (see Fig. 8b-c). The four layer atoms expansion requires larger energy therefore the formation of twins is more energetically favourable (see Fig. 8d). Therefore, it leads to instability of the four layered structure by the nucleation of the Shockley partials resulting in the formation of the twins as shown Fig. 8f-i. This mechanism is similar to the mechanism of twin formation from the stacking faults by the nucleation of the Shockley partials at higher imposed stress [8].

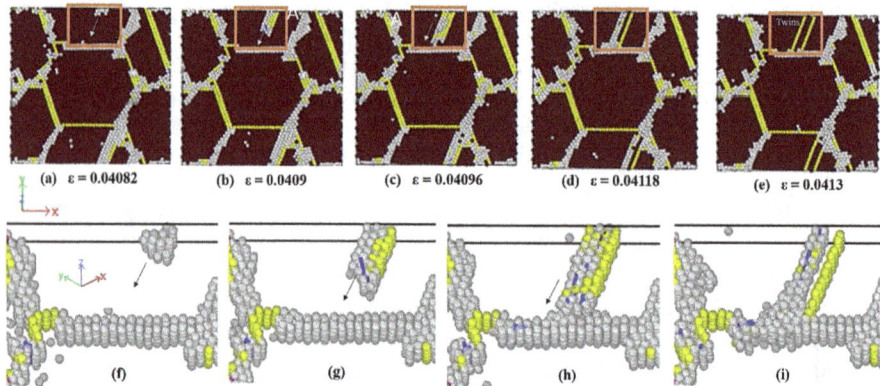

Fig. 8. Atomic snapshots showing the formation of Twins at 300 K and strain rate of 1×10^8 s^{-1}. Images (a-e) (f-i) shows the atomic snapshot at selected strain obtained from MD-simulation using CNA and DXA methods. The brown, yellow and white atoms represent FCC, HCP and other structure, respectively. The blue lines represent the Shockley partials. The motion of the Shockley partial is represented by the arrows. (a) formation of four layers non-crystalline structure (b-c) expansion of the four layered structure by the motion of Shockley partials (d) instability of four layered deformed structure and (e) twinned structure. Images (f-i) are the magnified images of orange rectangular marked section in images a, b, c and d, respectively, showing the motion of Shockley partials leading to formation of twins.

4   Conclusions

Molecular dynamics simulations have been performed at 10 K–700 K/1 $\times$ 10^8 s^{-1}–1 $\times$ 10^9 s^{-1} to understand the effect of temperature and strain rate on the dislocation density evolution and deformation mechanisms occurring in the nano Cu polycrystals with special $\sum 3$ grain boundaries along with other high and low angle grain boundaries. Following conclusions can be drawn from the explored conditions,

1. Yielding begins with the nucleation of Shockley partials of $\frac{1}{6} < 112 >$ character at the $\sum 3$ grain boundaries. It was observed that low temperature (10 K) deformation leads to intergranular failure.
2. The motion of the Shockley partials results in formation of the Stacking faults at lower strain and twins at higher strain. As the stacking faults are stable for relatively smaller deformation, only responsible for accommodating the small strains. Twins are stable at relatively large deformation, thus accommodating the higher strains.
3. Yielding stress decreases with the increase of temperature due to increased thermal vibration of atoms leading to higher dislocation mobility at higher temperature. Yielding occurs at lower strain at higher temperatures and for lower temperature it is shifted to higher strain due to reduced dislocation mobility in later case.
4. At higher strain rate, twin formation takes place by formation of triple layered deformed structure at the GBs. Expansion of unstable triple layered deformed structure is facilitated by the formation of Shockley partials, that eventually leads to the formation of twins. Nucleated Shockley partials were observed to be leaving the specimen at the free surface. In case of lower strain rate, twin formation takes place by the development of four layered non-crystalline structure that expands by the nucleation of the Shockley partials and eventually results in formation of twins. In this process, nucleated Shockley partials are observed to be getting trapped inside the grains. Thus, the dislocation density was observed to be high in case of lower strain rate, compared to the former case.
5. The yielding stress was observed to be decreasing with decrease in strain rate. Contrary to conventional observations, the dislocation density is higher at low strain rate, due to additional mechanism of twin formation that leads to manifestation of more dislocations, thus, resulting in yielding at the lower stress.

Acknowledgments. The authors would like to thank DST India for the financial support and project DST/INSPIRE/04/2018/003390 is acknowledged.

Appendix A

Figure A1 depicts the variation of potential energy with respect to time during the relaxation process. Initially the is high and decreasing over the period of time during the equilibration. Potential energy of the system tends to come to a minimum/equilibrium shown by flat region of the all three curves. Tensile tests were simulated, once the system was equilibrated and potential energy curves reaches minimum.

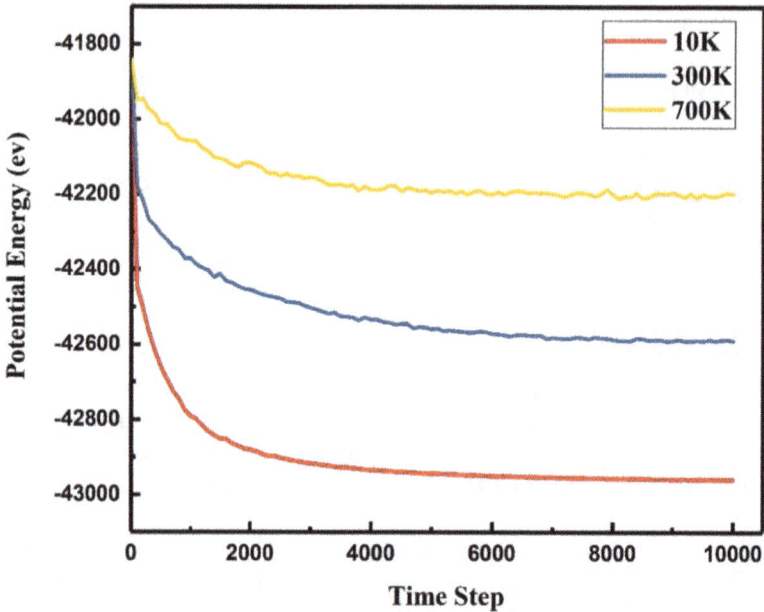

Fig. A1. Evolution of potential energy at the temperature of 10 K, 300 K and 700 K, once the system was thermally equilibrated.

References

1. Crone, W.C.: A brief introduction to MEMS and NEMS. In: Sharpe, W.N. (ed.) Springer Handbook of Experimental Solid Mechanics, pp. 203–228. Springer, New York (2008). https://doi.org/10.1007/978-0-387-30877-7_9
2. Iskandar, F.: Nanoparticle processing for optical applications – a review. Adv. Powder Technol. **20**, 283–292 (2009)
3. Bisoyi, H.K., Li, Q.: Discotic liquid crystals for self-organizing photovoltaics. In: Li, Q. (ed.) Nanomaterials for Sustainable Energy, NanoScience and Technology, pp. 215–252. Springer, Cham (2016). https://doi.org/10.1007/978-3-319-32023-6_6
4. Fuchs, H., Webster, T.J., Tang, Z., Banhart, F.: Functional nanomaterial and their applications: from origins to unanswered questions. Chem. Phys. Chem. **13**, 2423–2425 (2012)
5. Bhullar, S.K., Singh, H.P., Kaur, G., Buttar, H.S.: An overview of the applications of nanomaterials and development of stents in treating cardiovascular disorders. Rev. Adv. Mater. Sci. **44**, 286–296 (2016)
6. Arivalagan, K., Ravichandran, S., Rangasamy, K., Karthikeyan, E.: Nanomaterials and its potential applications. Int. J. Chemtech. Res. **3**, 534–538 (2011)
7. Lieber, C.M.: Nanoscale science and technology: building a big future from small things. MRS Bull. **28**(7), 486–491 (2003). https://doi.org/10.1557/mrs2003.144
8. Wang, J., Huang, H.: Shockley partial dislocations to twin: Another formation mechanism and generic driving force. Appl. Phys. Lett. **85**, 5984 (2004)
9. Karthik, P.S., Singh, S.P.: Copper conductive inks: synthesis and utilization in flexible electronics. RSC Adv. **5**(79), 63985–64030 (2015)
10. Sainath, G., Choudhary, B.K.: Molecular dynamics simulation of twin boundary effect on deformation of Cu nanopillars. Phys. Lett. A **379**, 1902–1905 (2015)

11. Rohith, P., Sainath, G., Srinivasan, V.S.: Effect of size, temperature and strain rate on dislocation density and deformation mechanisms in Cu nanowires. Phys. B: Condens. Matter. **561**, 136–140 (2019)
12. Li, Z., Gao, Y., Zhan, S., Fang, H., Zhang, Z.: Molecular dynamics study on temperature and strain rate dependences of mechanical properties of single crystal Al under uniaxial loading. AIP Adv. **10**, 075321 (2020)
13. Zhang, T., Zhou, K., Chen, Z.Q.: Strain rate effect on plastic deformation of nanocrystalline copper investigated by molecular dynamics. Mater. Sci. Eng. A **648**, 23–30 (2015)
14. Zhang, Z., Chen, P., Qin, F.: Molecular dynamic simulation of grain size and work temperature effect on mechanical properties of polycrystalline copper. In: International Conference on Electronic Packaging Technology (ICEPT), pp. 228–232 (2018)
15. Yazdani, S., Vitry, V.: Using molecular dynamic simulation to understand the deformation mechanism in Cu, Ni, and equimolar Cu-Ni polycrystalline alloys. Alloys **2**, 77–88 (2023)
16. Tian, X., Cui, J., Li, B., Xiang, M.: Investigations on the deformation behavior of polycrystalline Cu nanowires and some factors affecting the modulus and yield strength. Model. Simul. Mater. Sci. Eng. **18**, 055011 (2010)
17. Rohith, P., Sainath, G., Choudhary, B.K.: Molecular dynamics simulations study on the grain size dependence of deformation and failure behavior of polycrystalline Cu. In: Prakash, R., Suresh Kumar, R., Nagesha, A., Sasikala, G., Bhaduri, A. (eds.) Structural Integrity Assessment, pp. 253–262. Springer, Singapore (2020). https://doi.org/10.1007/978-981-13-8767-8_21
18. Sinha, S., Kim, D.I., Fleury, E., Suwas, S.: Effect of grain boundary engineering on the microstructure and mechanical properties of copper containing austenitic stainless steel. Mater. Sci. Eng. A **626**, 175–185 (2015)
19. Watanabe, T.: An approach to grain boundary design for strong and ductile polycrystals. Res. Mech. (Int. J. Struct. Mech. Mater. Sci.) **11**, 47–84 (1984)
20. Watanabe, T., Tsurekawa, S.: Toughening of brittle materials by grain boundary engineering. Mater. Sci. Eng. A **387–389**, 447–455 (2004)
21. Don, J., Majumdar, S.: Creep cavitation and grain boundary structure in type 304 stainless steel. Acta Metall. **34**, 961–967 (1986)
22. Spigarelli, S., Cabibbo, M., Evangelista, E., Palumbo, G.: Analysis of the creep strength of a low-carbon AISI 304 steel with low-Σ grain boundaries. Mater. Sci. Eng. A **352**, 93–99 (2003)
23. Deepak, K., Mandal, S., Athreya, C., Kim, D.-I., De Boer, B.: Implication of grain boundary engineering on high temperature hot corrosion of alloy 617. Corros. Sci. **106**, 293–297 (2016)
24. Wang, J., Li, N., Misra, A.: Structure and stability of Σ3 grain boundaries in face centered cubic metals. Philos. Mag. **93**(2013), 315–327 (2013)
25. Hirel, P.: Atomsk: A tool for manipulating and converting atomic data files. Comput. Phys. Commun. **197**, 212–219 (2015)
26. Plimpton, S.: Fast parallel algorithms for short-range molecular dynamics. J. Comput. Phys. **117**, 1–19 (1995)
27. Baskes, M.I.: Modified embedded-atom potentials for cubic materials and impurities. Phys. Rev. B **46**, 2727 (1992)
28. Mohazzabi, P., Shankar, S.P.: Maxwell-Boltzmann distribution in solids. J. Appl. Math. Phys. **6**, 602 (2018)
29. Zimmerman, J.A., WebbIII, E.B., Hoyt, J., Jones, R.E., Klein, P., Bammann, D.J.: Calculation of stress in atomistic simulation. Model. Simul. Mater. Sci. Eng. **12**, S319 (2004)
30. Stukowski, A., Albe, K.: Dislocation detection algorithm for atomistic simulations. Model. Simul. Mater. Sci. Eng. **18**, 025016 (2010)
31. Stukowski, A.: Visualization and analysis of atomistic simulation data with OVITO–the open visualization tool. Model. Simul. Mater. Sci. Eng. **18**, 015012 (2009)

32. Tsuzuki, H., Branicio, P.S., Rino, J.P.: Structural characterization of deformed crystals by analysis of common atomic neighborhood. Comput. Phys. Commun. **177**, 518–523 (2007)

33. Dudarev, E.F., Pochivalova, G.P., Kolobov, Y., Bakach, G.P., Skosyrskii, A.B., Zhorovkov, M.F.: Microplastic deformation of submicrocrystalline and coarse-grained titanium at room and elevated temperatures. Russ. Phys. J. **55**, 825–834 (2012)

34. Begau, C., Hua, J., Hartmaier, A.: A novel approach to study dislocation density tensors and lattice rotation patterns in atomistic simulations. J. Mech. Phys. Solids **60**, 711–722 (2012)

35. Xiang, M., Cui, J., Tian, X., Chen, J.: Molecular dynamics study of grain size and strain rate dependent tensile properties of nanocrystalline copper. J. Comput. Theor. Nanosci. **10**, 1215–1221 (2013)

36. Rohith, P., Sainath, G., Choudhary, B.: Molecular dynamics simulation studies on the influence of aspect ratio on tensile deformation and failure behaviour of< 1 0 0> copper nanowires. Comput. Mater. Sci. **138**, 34–41 (2017)

37. Veerababu, J., Manzoor, U., Sainath, G., Goyal, S., Sandhya, R.: Deformation behaviour of Cu nanowire with axial stacking fault. Mater. Res. Express **6**, 075056 (2019)

38. Liu, S., Yin, J., Zhao, Y.: Revealing twinning from triple lines in nanocrystalline copper via molecular dynamics simulation and experimental observation. J. Mater. Res. Technol. **11**, 342–350 (2021)

39. Priester, L.: The triple junction. In: Grain Boundaries. Springer Series in Materials Science, vol. 172, pp. 305–336. Springer, Dordrecht (2013). https://doi.org/10.1007/978-94-007-4969-6_10

Thermodynamic Assessment of Tin Based Molten Binary Indium-Tin Solder Alloys

M. R. Kumar[✉]

Department of Metallurgical and Materials Engineering, National Institute of Technology, Raipur, C.G 492010, India
mrkumar.mme@nitrr.ac.in

Abstract. It has been focused to develop the lead free solder alloy by substitute the In in place of Pb in Sn-Pb alloy to reduce the harmful effect of lead in conventional Pb-Sn solder alloy. For development of alloy depend on its thermo dynamical properties. In this study, integral enthalpies of mixing for the molten binary In-Sn system at 757, 809, and 845 K were obtained using a drop calorimeter spanning the whole indium range. Automated motorized dropping equipment was used to drop the element. Then, four pieces of α-Al2O3 were dropped at the falling temperature to calibrate the calorimeter. When the integral mixing enthalpies were compared to the composition, it was discovered that all compositions were exothermic. At $x_{In} \sim 0.55$, there was a discernible minimum enthalpy of mixing. There is a small degree of temperature dependence. The enthalpy of mixing was utilised to derive the binary interaction parameters using the Redlich-Kister polynomial. The experimental results are contrasted with the estimated results obtained using the Redlich-Kister polynomial. Its show that they were substantially in good agreement.

Keywords: Binary alloy · Indium-Tin · Calorimeter · mixing enthalpy

1 Introduction

Pb-Sn solders are used regularly to join electrical circuits inside of electronic equipment, however because they include lead, they are hazardous. The creation of Pb-free solders is urgently required in light of impending laws that would prohibit or restrict the use of Pb due to worries about its toxicity to the environment and human health. The development of a database of lead-free solders, the creation of innovative lead-free solders with exact phase diagrams, and the prediction of lead-free solders' physical and chemical properties, like surface tension and viscosity, all strongly rely on thermodynamic data. Understanding binary and ternary thermodynamic measurements is crucial for all metallurgical processes because they offer a wealth of information on phase equilibria, such as liquidus line and solidus line, isothermal and vertical section diagrams, mole fractions of the phase constitutions, etc., in addition to other thermodynamic properties such as activities of element, heat of mixing of system, surface energy,

© The Author(s), under exclusive license to Springer Nature Singapore Pte Ltd. 2024
S. Patra et al. (Eds.): METCENT 2023, *Proceedings of the International Conference on Metallurgical Engineering and Centenary Celebration*, pp. 145–153, 2024.
https://doi.org/10.1007/978-981-99-6863-3_15

viscosity, and others properties. It is the starting point for all calculations, including those that involve phase diagrams, wetting behaviour, and surface tension. The objective of thermodynamic investigations is to determine partial and integral quantities as well as their dependence on concentration, temperature, and pressure. Depending on the system and required attributes, various strategies are feasible. The binary and ternary data can be used to calculate higher level systems. It is common practise to use a variety of extrapolation and computation models. It is becoming more and more interested in low melting point alloys that can be used to solder lead-free products. Lead in solders is often viewed as being particularly detrimental to the environment due to the enormous volume of printed circuit boards and other electronic devices that must be recycled from landfills. The electronic industries, the US, Japan, and the European Community have all started programmes to find lead-free solders with technological, physical, and chemical attributes that are on par with or better than those of the Sn-Pb alloys currently in use. Even in India, government restrictions are tightening and there are more rules governing how trash is handled. Targeted materials are frequently the focus of a prohibition in accordance with the history of governmental control; lead in paint, lead in plumbing and lead in petrol have all been abolished.

Recently, lead-free solder has received increased attention from researchers. Due to their excellent alloying propensity with copper and other metals, researchers are focusing increasingly on lead-free solder materials these days, particularly In-Sn based alloys [1–3] for soldering copper wire, particularly in electronic equipment. Kleppa [4] studied the liquid In-Sn system's mixing enthalpy for the first time using direct-reaction calorimetry at 450 °C and 6 to 34 at.% In. Wittig and Scheidt [5] used the same instrument for the duration of the composition range at 371 °C after that. The same technique was used by Yazawa et al. [6] from 19 to 90 at.% In, and by Bros and Laffitte [7] from 248 °C to the entire composition. Values for the integral enthalpy published prior to 1971 were collated by Hultgren et al. [8], and there is generally strong agreement among these authors. These experimental findings reveal a relatively negative enthalpy for mixing liquid In-Sn alloys. A variety of thermodynamic studies of the In-Sn system were provided by Lee et al. [9], Korhonen and Kivilahti [10], and more recently David et al. [11] based on these experimental findings. There was good agreement between the estimated and experimental outcomes in each of these situations. The most recent experimental research of the was carried out by Luef et al. The most recent experimental investigation into the enthalpy of mixing was carried out by Luef et al. [12] using direct-reaction calorimetry at 900 C across the whole composition range. Numerous lead-free systems have been studied by the different authors [13–15]. At a temperature of 520–820 K, R. Hultgren et al. and V. Vassiliev et al. determined the enthalpy of formation and the activity of this liquid system. They noticed that the In-Sn phase diagram [11] was revised with the aid of these experimental results. It differs from the ideal mixing in the wrong direction and is marginally superior to its previous versions [16, 18–20]. The earlier phase diagram of the In-Sn system that was accepted was heavily influenced by the work by Heumann and Alpaut [20]. The most significant difference between the phase diagrams from [21] and [22] is in the stability region of two intermediate phases, the Sn-rich and the In-rich phases.

The phase equilibria between γ and bct (Sn) phases, however, are still a significant area of uncertainty. Heumann and Alpaut [20] discovered that the γ + bct (Sn) in this two phase area ranges from 88 to 93 at% Sn at ambient temperature. It is between 86 and 97 at% Sn, according to Predel and Gsdecke [23]. The martensitic change between the and bct(Sn) structures occurs in Sn-(8.0–9.5) at% In alloys, according to a later report by Koyama et al. [24]. In the temperature range of 80 to 1300C, Wojtaszek and Kuzyk [25] have discovered that In is soluble in Sn using measurements of resistances and the single phase region. This suggests that the bct(Sn) two phase zone is located between 89 and 92.5 at% Sn at 800 C and between 90 and 94 at% Sn at 1200 C. The statistical thermodynamic method developed by Bhatia and Thornton [26] demonstrates the absence of these experimental results. This experiment will be feasible for determining the exact phase diagram.

2 Experimental Details

2.1 Materials

Pure metals (In, Sn) were used as raw materials and Al_2O_3 needles from as the calibration standard in the experiment. The oxide layer from the surface of pure metals of indium and tin were removed using fine sandpaper. Table 1. included a list of the purity and supplier information.

Table 1. Sources and purity of materials used in the present study

Raw materials	Sources	Initial purity (wt. %)
Tin (Shots)	J Matthey, U. K	99.99
Indium (Ingot)	J Matthey, U. K	99.99
Argon gas	Indian oxygen limited, India	>99

2.2 Procedure

High temperature MHTC 96 For the tests, drop calorimeters from Setaram, France, were employed. It produces heat up to 1723 K using a graphite heating element and a thermopile of 20 thermocouples. Using a motorised (Multisample Introducer) device, samples are successively dropped at predetermined intervals of time. It sits on top of the calorimeter and has a 23-sample capacity. Through an alumina tube, the furnace is connected to this. The experiment is carried out at 757,809, 845 K in liquid state for getting equilibrium of this alloy. The crucible used is a 10-mm-diameter, 60-mm-height of alumina material. Before the experiment began, the necessary amount of Sn (nSn = 0.003619 mol) was stored in an alumina crucible. By furnace heating after attaining the required temperature kept it for 1 h and follow dropping of the sample at a fixed interval of time 40 min for base line for each drop is maintained. Four drops of α-Al2O3

(National Institute of Standards and Technology) were dropped after the sample had been dropped to calibrate the apparatus. All experiments were carried out while argon gas was circulating to prevent sample oxidation. The manufacturer-provided Calisto software was utilised for data evolution and control. By conducting the trials again at the same temperatures, the reproducibility and accuracy of the data were verified.

3 Results and Discussion

The data on heat flow as a function of temperature was gathered by CALISTO data collection software after the experiment was finished. The data from the CALISTO data processing programme were then combined to establish the peak's area. The measured enthalpy of mixing is computed with integrations of the heat flow curves from experiment, and shown as below,

$$\Delta H_{In,i}^{Reaction} = \left(\Delta H_{In,i}^{Signal}.K\right) - \left(\Delta H_{In,i}^{T_D \to T_M}.n_{In,i}\right) \tag{1}$$

where $n_{In,i}$ is indium moles of added at i - th drop. $\Delta H_{In,i}^{Signal}$ is the heat effect brought on by the i-th drop of indium sample added to the bath, K is the calorimeter constant t; $\Delta H_{In,i}^{T_D \to T_M}$ is the enthalpy change caused by a drop in temperature for one mole of indium, T_D (in Kelvin's) is bath temperature T_M, in the i-th measurement. The molar enthalpy difference $\Delta H_{In,i}^{T_D \to T_M}$ utilising the polynomial of pure elements from Dinsdale, for indium were calculated [27].

$$\Delta H_{Signal} = n_i\left(H_{Sample,T_C} - H_{Sample,T_D}\right) + \Delta H_{Reaction} \tag{2}$$

where n_i is the quantity of moles added to the sample , H_{Sample,T_C} molar enthalpies at calorimetric temperature and H_{Sample,T_D} denotes molar enthalpies at drop temperature. $\Delta H_{Reaction}$ is the reaction's enthalpy as a result of the addition of ni moles.

Because of the relatively small mass added the Partial enthalpy can be directly calculated as:

$$\Delta \overline{H_i} = \frac{\Delta H_{Reaction}}{n_i} \tag{3}$$

By adding up the corresponding reaction enthalpies and dividing by the total molar amount of the substance, the integral enthalpy of mixing was determined.

$$\Delta H_{mix} = \frac{\Sigma \Delta H_{Reaction}}{(n_{Crucible} + \Sigma n_i)} \tag{4}$$

The molar enthalpy difference $\left(H_{Sample,T_C} - H_{Sample,T_D}\right)$ for Indium was calculated using the following relation. $\left(H_{Sample,T_C} - H_{Sample,T_D}\right) = \int_{T_D}^{T_C} C_P dT$

$$\left(H_{Sample,T_C} - H_{Sample,T_D}\right) = \int_{T_D}^{T_{Tr}} C_P dT + \Delta H_{T_{Tr}} + \int_{T_{Tr}}^{T_m} C_P dT + \Delta H_{T_f} + \int_{T_m}^{T_c} C_P dT \tag{5}$$

where T_D = Dropping (Room) Temperature, T_{Tr} = Transformation Temperature, $\Delta H_{T_{Tr}}$ = Enthalpy change during transformation*, T_m = Melting Point, ΔH_{T_f} = Enthalpy change during melting, T_C = Temperature of the Calorimeter at which experiment is carried out. Cp = heat capacity [28].

There are six common reasons of error in calorimetric measurements: i)Impediments are present. ii) the calorimeter's construction iii) calibration iv) signal integration v) insufficient dissolution. The instrument MHTC 96 line evo made from Setaram, France, has an overall inaccuracy of about 230 J. mol -1. A NIST standard sapphire was used to assess the calibration error, which was found to be less than 1.5%. The materials dropped during the experiment and the predicted integral molar enthalpy of mixing at 757,809 and 845 K are displayed in Table 2 below. It provides details on the liquid alloys' initial and additional quantities as well as their integral mixing enthalpies. Using the composition of indium (xIn) as the abscissa and the integral enthalpy of mixing as the ordinate in a graph has been plotted. It has been observed that, it demonstrates that the nature of mixing enthalpy is exothermic. The plot in Fig. 1 shows how this enthalpy varies when indium molar fractions fluctuate (the refractive condition is a pure liquid component). Enthalpies of mixing fall as indium content rises, approach the minimum value at $x_{In} \sim 0.55$, and eventually reach zero when indium content rises even higher in the alloy. So, it was determined that there was no segregation or grouping of atoms in this system. It demonstrates that no compound can form during alloy production because no abrupt increase or decrease in mixing enthalpies is visible.

Theoretical Modelling

It is frequently used to determine higher order systems' thermodynamic properties. As Redlich-Kister polynomials, the thermodynamic data of binary and ternary systems are increasingly expressed. Therefore, a least squares fit of Eq. (6) was performed on the binary calorimetric data obtained by experiment at 757K using the Redlich-Kister polynomial shown below. For substitutional solutions according to the CALPHAD approach suggested by Ansara & Dupin [29].

$$\Delta H_{mix} = \Sigma_i \Sigma_{j>i}\left[x_i x_j \Sigma_v L_{i:j}^{(v)}(x_i - x_j)^v\right] \tag{6}$$

Here i and j are equal to 1 for Bi and 2 for In and $L_{i:j}^{(v)}$ ($v = 0, 1, 2..$) are the so called binary interaction parameters of the order v. In this case the polynomial has the form:

$$\Delta H_{mix} = L_{1:2}^{(0)}x_1 x_2 + L_{1:2}^{(1)}x_1 x_2 (x_1 - x_2) + L_{1:2}^{(2)}x_1 x_2 (x_1 - x_2)^2 \tag{7}$$

Table 2 Integral mixing enthalpy as determined by calorimetric tests on the In-Sn System with an indium drop at 757, 809, and 845 K, Standard states: pure liquid metals.

No. of moles of Indum (n_{In}) [mole],	mole fraction x_{In}	Standard uncertainties $u\,(x_{In})$	Heat effect $\Delta H_{Signal}\cdot K$ [J]	Standard uncertainties $\Delta H_{Signal}\cdot K$ [J]	Drop enthalpy $H_{Diss\text{-}x}$ [J]	Integral enthalpy ΔH^m [Jmole-1]
Series 1: In-Sn alloys; Atmosphere: Argon; pressure p = 0.1 MPa, Initial amount n_{Sn} = 0.003619 mol, K = 0.003400 J μV, T^D = 298 K, T^M = 757 K						
0.0003917	0.1020	0.0001	6.58	0.01	−0.33	−84
0.0004044	0.1930	0.0001	7.30	0.01	−0.35	−156
0.0007211	0.3060	0.0001	11.78	0.02	−0.54	−238
0.0008039	0.3994	0.0002	13.18	0.03	−0.53	−298
0.0010020	0.4993	0.0002	19.54	0.04	−0.72	−349
0.0019550	0.6101	0.0001	33.60	0.08	−1.10	−389
0.0021446	0.6938	0.0002	41.61	0.10	−1.35	−419
0.0051366	0.7887	0.0002	88.77	0.21	−1.40	−371
0.0040443	0.8300	0.0002	69.50	0.16	−0.46	−322
0.0046765	0.8609	0.0001	80.40	0.19	−0.30	−274
0.0058151	0.8868	0.0001	101.03	0.24	−0.59	−241
0.0030920	0.8974	0.0001	53.85	0.13	−0.06	−220
Series 2: In-Sn alloys; Atmosphere: Argon; pressure p = 0.1 MPa, Initial amount n_{Sn} = 0.003619 mol, K = 0.003460J μV, T^D = 298 K, T^M = 809 K						
0.0003705	0.0949	0.0001	6.61	0.018	−0.29	−72
0.0005209	0.2009	0.0001	9.31	0.025	−0.33	−138
0.0006286	0.2995	0.0001	11.12	0.030	−0.49	−219
0.0008746	0.4018	0.0001	15.53	0.04	−0.57	−281
0.001029	0.4960	0.0001	20.01	0.054	−0.63	−329
0.0022656	0.6219	0.0001	41.89	0.113	−1.24	−378
0.0024001	0.6990	0.0001	44.21	0.11	−1.33	−411
0.0058538	0.7985	0.0002	107.53	0.29	−1.04	−331
0.0054544	0.8461	0.0002	101.10	0.27	−0.16	−261
0.0055911	0.8760	0.0002	103.71	0.28	−0.10	−211
0.0059800	0.8971	0.0002	111.10	0.30	0.18	−170
Series 3: In-Snalloys; Atmosphere: Argon; pressure p = 0.1 MPa, Initial amount n_{Sn} = 0.003619 mol, K = 0.003580 J μV T^D = 298 K, T^M = 845 K						
0.0004150	0.1050	0.0001	7.92	0.02	−0.28	−69
0.0004681	0.1996	0.0001	9.00	0.02	−0.28	−130
0.0007331	0.3122	0.0002	14.01	0.04	−0.42	−190
0.0007846	0.4023	0.0002	14.92	0.04	−0.47	−251
0.0011875	0.5010	0.0001	22.58	0.07	−0.69	−309
0.0017080	0.5980	0.0002	33.20	0.10	−0.94	−350

(continued)

Table 2 (*continued*)

No. of moles of Indum (n_{In}) [mole],	mole fraction x_{In}	Standard uncertainties u (x_{In})	Heat effect $\Delta H_{Signal} \cdot K$ [J]	Standard uncertainties $\Delta H_{Signal} \cdot K$ [J]	Drop enthalpy $H_{Diss\text{-}x}$[J]	Integral enthalpy ΔH^m [Jmole-1]
0.0031232	0.7021	0.0002	59.51	0.18	−1.46	−386
0.0050920	0.7910	0.0001	100.50	0.31	−0.54	−303
0.0061185	0.8462	0.0001	121.06	0.37	−0.01	−221
0.0063677	0.8792	0.0001	125.92	0.39	0.01	−171
0.0044278	0.8950	0.0001	88.73	0.27	0.56	−134

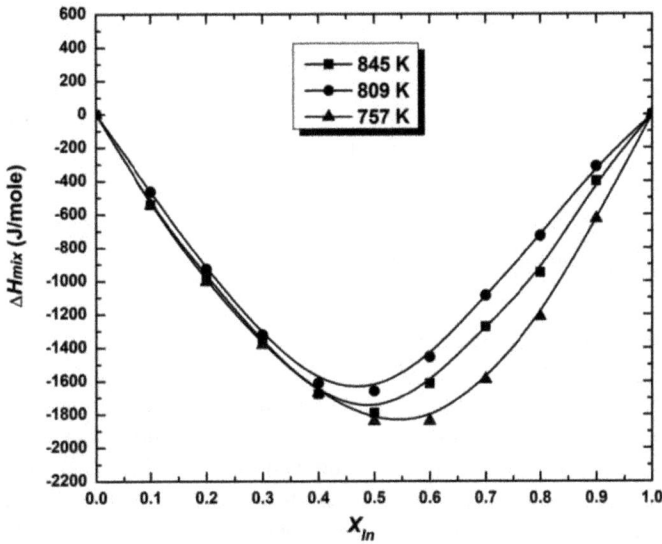

Fig. 1. Integral molar mixing enthalpies of liquid In-Sn alloys at three different Temp 757, 809 and 845 K; standard states: pure liquid metals.

Table 3. Intraction parameters for Tin based molten binary In-Sn system

Interaction parameter	υ	(J/mol)
$L_{In\text{-}Sn}(\upsilon)$	0	−4410 + 3T
	1	−4139 + 4T

According to the Redlich-Kister polynomial, Eq. (7) is used to fit the experimental data. The binary interaction parameters are the ones that were fitted. Table 3. displays the binary-calculated interaction parameters. Binary interaction parameters have been seen to vary linearly with temperature. In Fig. 2, a relative of the excess molar free energy obtained experimentally and using the Redlich-Kister polynomial extrapolation

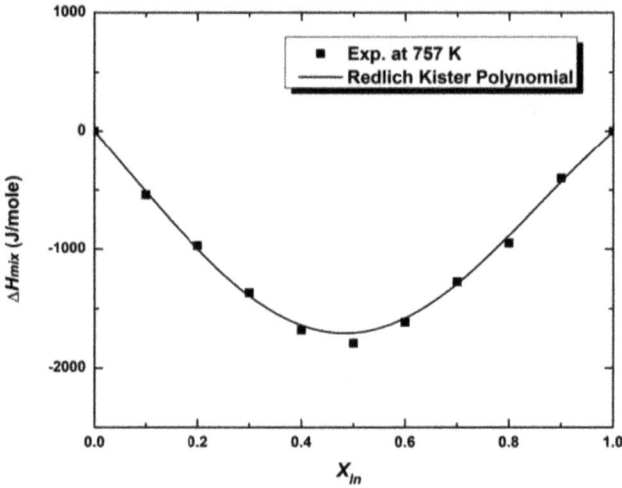

Fig. 2. Comparing the integral mixing enthalpies for In-Sn alloys using this experimental investigation at 757 K and the theoretical Redlich-Kister polynomial model.

shows that the experimental values for 757 K. It shows that experimental data are in good agreement with polynomial fitting.

4 Summary and Conclusions

The integral enthalpy of mixing of the molten binary indium-tin solder alloys over the complete composition range was measured using a drop calorimeter at 757,809 and 845 K. It has been shown that the In-Sn system's mixing enthalpy is exothermic at these temperatures. The enthalpy curve minima was discovered to be around at $x_{In} \sim 0.55$. It has been found that the enthalpy of mixing depends only marginally on temperature. It presents the outcomes of calculating the binary interaction parameters in the In-Sn system using the Redlich-Kister polynomial and values of empirically measured enthalpy of mixing, and it illustrates linearly how temperature influences the properties of the binary interaction.

References

1. Morris, J.W., Jr., Freer Goldstein, J.L., Mei, Z.: Microstructure and mechanical properties of tin-indium and tin-bismuth solders. JOM **45**, 25–27 (1993)
2. Moelans, N., Hari Kumar, K.C., Wollants, P.: Thermodynamic optimization of the lead-free solder system Bi-In-Sn-Zn. J. Alloys Compd. **360**, 98–106 (2003)
3. Gnecco, F., et al.: Wetting behaviour and reactivity of lead free Au-In-Sn and Bi-In-Sn alloys on copper substrates. J. Adhes. Adhesiv **27**, 409–416 (2007)
4. Kleppa, O.J.: The thermodynamic properties of the moderately dilute liquid solutions of copper, silver, and gold in thallium, lead, and bismuth. J. Phys. Chem. **60**, 446–452 (1956)
5. Wittig, F.E., Scheidt, P.: Energetics of metallic systems. XIV. Heats of mixing in the binary systems of indium and thallium with tin and lead. Z. Phys. Chem. **28**, 120–142 (1961)

6. Yazawa, A., Kawashima, T., Itagaki, K.: Thermodynamics of the In-Sn system. J. Jpn. Inst. Met. **32**, 1281–1287 (1968)
7. Bros, J.P., Laffitte, M.: Enthalpies of formation in indium+tin alloys in the liquid state. J. Chem. Thermodyn. **2**, 151–152 (1970)
8. Hultgren, R., Desai, P.D., Hawkins, D.T., Gleiser, M., Kelley, K.K.: Selected values of the thermodynamic properties of binary alloys. ASM International, Metals Park, Ohio (1973)
9. Lee, B.J., Oh, C.S., Shim, J.H.: Thermodynamic assessments of the Sn-In and Sn-Bi binary systems. J. Electron. Mater. **25**, 983–991 (1996)
10. Korhonen, T.M., Kivilahti, J.K.: Thermodynamics of the Sn-In-Ag solder system. J. Electron. Mater. **27**, 149–158 (1998)
11. David, N., Aissaoui, K.E., Fiorani, J.M., Hertz, J., Vilasi, M.: Thermodynamic optimization of the In-Pb-Sn system based on new evaluations of the binary borders In-Pb and In-Sn. Thermochim. Acta **413**, 127–137 (2004)
12. Luef, C., Flandorfer, H., Ipser, H.: Enthalpies of mixing of liquid alloys in the In-Pd-Sn system and the limiting binary systems. Thermochim. Acta **417**, 47–57 (2004)
13. Zivkovic, D., Mitovski, A., Balanovic, L., Manasijevic, D., Zivkovi, Z.: Thermodynamic analysis of liquid In–Sn alloys using Oelsen calorimetry. J. Therm. Anal. Calorimeter **102**, 827–830 (2010)
14. Janke, D., Madara, J., Lukacs, S., Pokol, G.: Dsc investigations of amalgam formation in bi–sn–hg system doped with indium. J. Therm. Anal. Calorimetr. **96**, 443–447 (2009)
15. Dragana Zivkovi, A. Milosavljevi, A. Mitovskiand B. Marjanovi, Comparative thermodynamic study and characterization of ternary Ag–In–Sn alloys, J. Therm. Anal. Calorimetr. **89**,137–142 (2007)
16. Hultgren, R., Desai, P.D., Hawkins, D.T., Gleiser, M., Kelly, K.K.: Selected values of thermodynamics properties of binary alloys. ASM International, Metals Park, OH, p. 1031 (1973)
17. Vassiliev, V., Feutelais, Y., Sghaier, M., Legendre, B.: Liquid-state electrochemical study of the system indium-tin. Thermochim. Acta **315**, 129–134 (1998)
18. Massalski, T.B. (ed.), Binary Alloy Phase Diagrams, A. Society for Metals, vol. 1, p. 1401 (1986)
19. David, N., El Aissaoui, K., Fiorani, J.M., Hertz, J., Vilasi, M.: Thermodynamic optimization of the In-Pb-Sn system based on new evaluations of the binary borders In-Pb and In-Sn Thermochim. Acta **413**, 127–137 (2004)
20. Heumann, T., Alpaut, O.: The phase diagram of indium-tin. J. Less-Common Met. **6**, 108–117 (1964)
21. Hansen, M., Anderko, K.: Constitution of Binary Alloys, pp. 59–59. McGraw-Hill, New York (1958)
22. Massalski, T.B., et al.: Binary alloy phase diagrams. Massalski, T.B., et al. (eds.) (Metals Park, OH: ASM, vol. 38, p. 19 (1986)
23. Predel, B., GSdecke, T.: Ternary system tin-indium-thallium. Z. Metallkd **66**, 654–659 (1975)
24. Koyama, Y., Suzuki, H., Nittono, O.: Phase transformation in tin-(8.0-9.5)at.%indium alloys Scripta Met. **18**, 715–717 (1984)
25. Wojtaszek, Z., Kuzyk, H.: Phase diagram of the indium-tin system in the range 60–100 atomic % tin Zesz. Nauk. Univ. Jagiellon Pr. Chem. **19**, 281–288 (1974)
26. Bhatia, A.B., Thornton, D.E.: Structural aspects of the electrical resistivity of binary alloys. Phys. Rev. B **82**, 3004–3012 (1970)
27. Dinsdale, A.T.: SGTE data for pure elements. Calphad **15**, 317–425 (1991)
28. Kubaschewski, O., Alcock, C.B.: Metallurgical thermochemistry, Maxwell Macmillan International Editions, Pergamon Press, p. 337 (1989)
29. Ansara, I., Dupin, N.: Cost 507 thermo chemical database for light metal alloys. In: European Commission DG X11, European Commission, Luxembourg (1998)

Advanced Material

Nanosized Hybrid Polymer Modifiers (HPM) for Improved Mechanical and Thermal Behavior of Carbon Fiber Reinforced Composites

D. Dhakal[1], P. Lamichhane[2], L. Baxter[2], B. Sedai[2], and R. Vaidyanathan[1(✉)]

[1] School of Materials Science and Engineering, Oklahoma State University, Tulsa, OK 74106, USA
vaidyan@okstate.edu

[2] MITO Material Solutions, 8902 Vincennes Cir B, Indianapolis, IN 46268, USA

Abstract. Lightweight carbon fiber-reinforced polymer composites are replacing metallic components due to improved strength-to- weight ratios and fatigue resistance. However, they crack and delaminate due to low-velocity impact reducing their mechanical properties. Some of the approaches include surface modification of carbon fibers and addition of graphene or graphene oxide nanoparticles, improving resistance to crack propagation and reduces delamination. Graphene oxide (GO) can achieve excellent dispersion with polymer matrices due to functional groups compatible with composite matrix systems. Prior research from our group demonstrated successful grafting of GO with polyhedral oligomeric silsesquioxane (POSS) to optimize the thermal stability of GO. Due to cage-like structure of POSS, dispersion of these hybrid nanoparticles within polymer matrices could result in an enhancement in mechanical and thermal behavior of composite materials. Methacryl polyhedral oligomeric silsesquioxanes (MAPOSS) molecules were hybridized to GO to MEGO and characterized. The dispersion of HPM was studied in thermoset resin system Epon 862 Epoxy. In addition, the effect of HPM on polymer and carbon fiber reinforced polymer composite were studied. Results confirmed that the addition of HPM at very low wt.% can enhance the viscoelastic, mechanical, and thermal properties of composites. The presence of MEGO enhanced the interlaminar fracture toughness by ~ 70% at 0.5% MEGO loading without drastically affecting the strength and modulus, as confirmed by SEM images.

Keywords: Carbon fiber reinforced-epoxy composites · nano-sized hybrid polymer modifiers · Graphene oxide (GO) · polyhedral oligomeric silsesquioxanes (POSS) · mechanical properties · thermal properties

1 Introduction

1.1 Polymer Matrix Composites

Lightweight carbon fiber-reinforced polymer composites are replacing metallic components in the aerospace and automotive industries due to their improved strength-to-weight ratios and fatigue resistance. Despite their attractive properties, these composites, however, crack and delaminate due to low-velocity impact causing a drastic drop

S. Patra et al. (Eds.): METCENT 2023, *Proceedings of the International Conference on Metallurgical Engineering and Centenary Celebration*, pp. 157–165, 2024.
https://doi.org/10.1007/978-981-99-6863-3_16

in their mechanical properties. Some of the approaches evaluated to overcome these issues include surface modification of the carbon fibers and the addition of nanoparticles such as carbon-based nanomaterials (graphene, graphene oxide, carbon nanotubes, etc.) that can improve the interlaminar toughness of the composites. The addition of these nanoparticles shows improved resistance to crack propagation and reduced delamination in these composites. One carbon-based nano-additive additive, graphene oxide (GO), can achieve excellent dispersion with organic solvents in polymer matrices due to the presence of specific functional groups compatible with most composite matrix systems.

Prior research from our group has demonstrated successful grafting of GO with other molecules such as ç (POSS) to optimize the thermal stability of GO [1, 2]. Due to the robust cage-like silica structure of POSS, dispersion of these hybrid nanoparticles within polymer matrices could result in an overall enhancement in the mechanical as well as thermal behavior of the composite materials.

1.2 Polyhedral Oligomeric Silsesquioxanes (POSS)

Polyhedral oligomeric silsesquioxane or POSS has a unique silicon cage with eight organic moieties that have both organic and inorganic features. It has a nanosized cage structure with empirical formula $Rn(SiO1.5)n$, where R represents organic functional groups like alkyl, alkylene, acrylate, or epoxide, etc. [3, 4]. This is shown in Fig. 1. Due to its organic and inorganic structure, it is compatible with several composite resin systems. The siloxane (Si-O-Si) moiety imparts thermal stability, chemical resistance, rigidity, and flame-retardant properties to POSS and, therefore, to the matrix resin it is added to. The organic groups can be either inert or reactive. POSS can be blended into a polymer matrix by either chemical or physical mixing. Additionally, the organic groups attached to the Si cage can be modified with different substituents leading to better solubility in common solvents [3, 5].

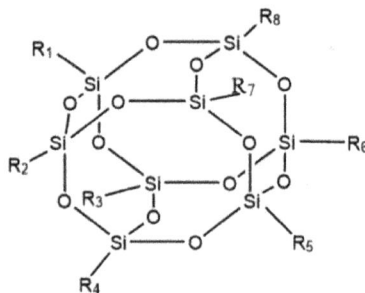

Fig. 1. POSS structure

1.3 Graphene Oxide and Graphene

The natural state of graphene is in the form of a stack of multiple sheets, held together by van der Waals attraction. Graphene-based products are available in different forms: graphene, graphene oxide (GO), reduced graphene oxide (rGO), and monolayer graphene. The structure of graphene and GO is shown in Fig. 2. The properties of graphene are influenced by the number of layers and their functionality. However, the applications of graphene are limited due to the difficulty of dispersing graphene in solvents. This issue is directly related to the van der Waals interaction and π- π stacking between graphene sheets [6, 7].

Graphene oxide (GO), obtained from the oxidation and exfoliation of graphite, contains oxygen-containing carbonyl, hydroxyl, and epoxy functional groups. The presence of a functional group imparts several advantages to GO over pristine graphenes, such as 1) flexibility of further modification, 2) excellent dispersion in polymer matrices and solvents, and 3) excellent interfacial interaction between the functional groups and the polymer matrix.

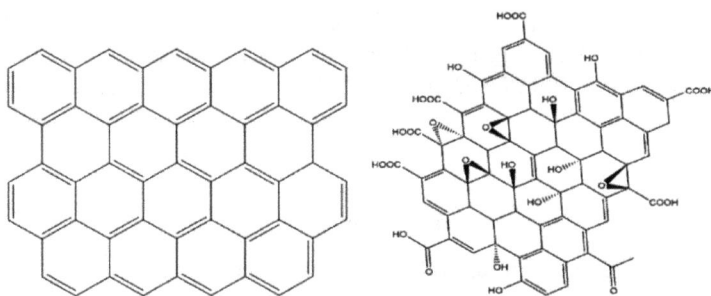

Fig. 2. Structure of graphene and graphene oxide

1.4 Hybrid Polymer Modifiers

The dispersion of nanoparticles such as POSS and GO nanosheets can be difficult due to the aggregation and forces between the particles and the sheets. One of the potential approaches to overcome particle agglomeration is to utilize the presence of various functional groups on the surface of GO that opens the platform for chemical modification and produce new hybrid materials with compatible properties and a lower probability of agglomeration [7–11]. This method is more desirable because the species of the grafting monomer can be selected and tailored to the grafting conditions and a specific composite resin system [12].

Muthu and Vaidyanathan studied the effect of octa ammonium POSS (OAM POSS) grafted GO in epoxy composites. They reported a 3 °C improvement in glass transition temperature (Tg) and a 260% increment of onset degradation temperature of GO at 5% weight loss after grafting with OAM POSS [13].

The work presented here describes the effect of a nano-sized hybrid polymer modifier termed MEGO (Methacryl POSS (MAPOSS) grafted on the GO surface) on carbon fiber reinforced Epon 862. The effect of the polymer modifier is studied using different mechanical and thermal testing methods such as 4-point flexural testing and double cantilever test (DCB), Dynamic mechanical analysis (DMA), and Differential Scanning

Calorimetry (DSC). The grafting reaction is carried out using a redox reaction mechanism for grafting methacryl POSS (MAPOSS) on the GO surface in the presence of aqueous Ce(IV)/HNO3 [14]. The MEGO nanoparticles were characterized by Raman spectroscopy to analyze the distortion of the GO due to the incorporation of MAPOSS while XRF was utilized to evaluate the composition of MEGO.

2 Materials and Methods

2.1 Materials for Hybrid Polymer Modifiers and Composites

A 2.5% graphene oxide dispersion in water was purchased from Graphenea Inc. (Cambridge, MA). Nitric acid, acetonitrile, and cerium ammonium nitrate (CAN) were purchased from Alpha Aesar (Haverhill, MA). Methacryl POSS (MAPOSS) cage mixture was purchased from Hybrid Plastics (Hattiesburg, MS). Epoxy resin EPON 862 was purchased from Hexion LLC (Columbus, OH). Acetone was purchased from BDH VWR Analytical (Radnor, PA). All the reagents and chemicals were used without further modifications. Epoxy resin EPON 862 and EPIKURE 3370 were purchased from Hexion LLC (Columbus, OH). T-300 6K, 2 × 2 Twill Wave Carbon Fabric Fiber was purchased from Fiberglast (Brookville, OH).

2.2 Preparation of Hybrid Polymer Modifiers

Methacryl POSS (MAPOSS) was grafted on the surface of graphene oxide (GO) layers using a redox polymerization method. More details on the preparation method are given in Dhakal's Ph. D thesis [14].

2.3 Preparation of Composites with Hybrid Polymer Modifiers

MAPOSS hybridized GO, i.e., MEGO, was prepared in a cake form [14]. Cake MEGO was dried under a vacuum oven at room temperature such that the MEGO: Acetone equals 1:3 (wet MEGO). A 3% MEGO master batch was prepared by mixing 300 g of Epon 862 resin with 36 g of wet MEGO (equivalent to 9 g Dry MEGO), and then the hand-mixed MEGO with resin was poured into the feed roller and collected at the apron of a three-roll mill. The sample preparation process is summarized in reference [14].

Composite samples were prepared according to ASTM standards for flexural testing and Double Cantilever Beam (DCB) testing (ASTM D6272 and ASTM 5528). For the flexural test, four sheets of carbon fiber (CF) sheets (~200 mm x 200 mm or 8″ x 8″) were used, whereas 12 CF sheets (~200 mm x 200 mm) were used for DCB samples. Additionally, in DCB samples, a thin Teflon sheet (9μm thick, 63mm width) was inserted in the middle layer, which serves as a crack initiation during the test. Epoxy containing MEGO dispersion with different loading wt.% (Neat, 0.05, 0.1, 0.5) of wet MEGO in Epon 862 plus Epikure 3370 was prepared [15]. The epoxy/MEGO dispersion was applied by hand on the CF layers and kept under vacuum overnight. The CF to epoxy ratio was maintained 1:1 prior to curing. After 24 h room temperature cure process, composites were post cured at 100 °C for 2 h. For all composite samples, the ratio of Epon 862 to Epikure 3372 was maintained at 100: 44, as suggested by the supplier. The fabrication process and the samples are shown in Fig. 3.

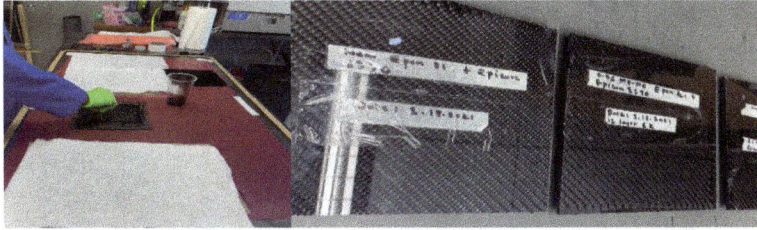

Fig. 3. CFRP sample preparation process and samples

2.4 Characterization of Composites with Hybrid Polymer Modifiers

XRD spectrum for the MEGO powders was collected using Bruker AXS D 8 Discover X-ray Diffractometer with general area detector diffraction system (GADDS) Vantec 500 2D detector. The wavelength of the X-ray used is 1.54056 Å. The scanning time was set to 90 s with the range of diffraction angle (2θ) from 5 to 40 degrees. The CFRP fracture surface morphology was investigated using a Hitachi S-4800 field emission scanning electron microscope (FE-SEM). CFRP samples were metal (Iridium) sputter coated for 90 s using a Leica EM ACE600. The scanning was carried out with 10 or 15 kV for magnification 400X, 1,000X, 2,000X, and 4,000X.

The flexural strength and modulus of polymer composite samples were determined by a four-point bend test on an Instron 5582 according to ASTM D6272. The mode I interlaminar fracture toughness (G_{IC}) of continuous carbon fiber-reinforced composite samples were determined using the ASTM 5528 test method [15].

3 Results and Discussion

3.1 XRD Results

The exfoliation of carbon-based nanomaterial in polymer matrix directly correlates with the interlaminar spacing and its randomness in orientation. One major objective was to increase the interlaminar distance as well as enhance the randomness of the MEGO particles so that the van der Waal's force of attraction can be reduced between each GO layer. This could help to reduce the agglomeration of the GO additive. Normalized powder X-ray diffraction of GO and MEGO is shown in Fig. 4. The peak of graphene oxide at 11.01° represents an interlayer spacing of 8.2 Å. The modified GO, i.e., MEGO (GO-g-MAPOSS), shows the XRD peak shift by 1.16° towards the left representing the interlayer spacing of 9.0 Å with an increase of 0.8 Å. The shift in peak position of MEGO indicates the chemical modification of GO [15] that corresponds with an increase in interlaminar spacing and randomness in orientation. The modification of graphene oxide (GO) to MEGO has thus been confirmed by the XRD spectra.

3.2 Flexural Test Results

The flexural tests were performed with a span of ~ 45 mm by maintaining a span-to-thickness ratio of 16:1 and a speed of ~ 0.1 mm/min. At least five specimens were tested

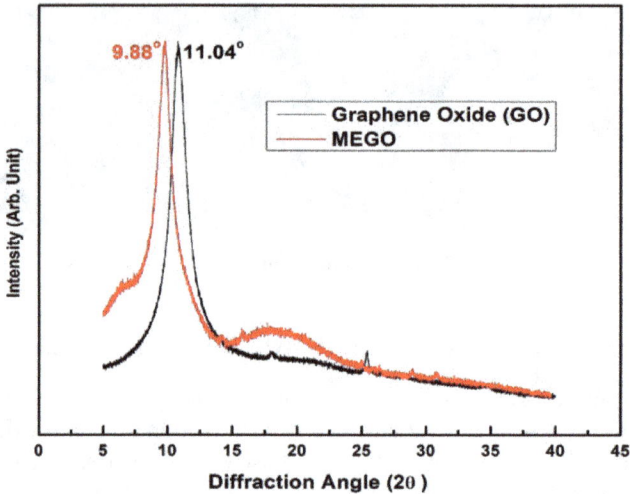

Fig. 4. XRD Spectroscopy of Graphene Oxide and MEGO

under each condition. The average flexural strength and flexural modulus vs MEGO loading are presented in Table 1. The experimental data showed that the maximum improvement in flexural strength and modulus was achieved at 0.1wt% MEGO loading by 11.5% and 27.4%, respectively. It was observed that there was a slight decrease in the flexural strength and modulus beyond 0.1% MEGO loading, which is expected to because of potential agglomeration of the nanofillers, although the difference was minimal.

Table 1. Flexural properties of MEGO toughened CFRP

Sample	Flexural Modulus (GPa)	Flexural Strength (MPa)
Neat CF + Epon 862	106 ± 3	705 ± 7
0.05 MEGO + CF + Epon 862	101 ± 12	697 ± 35
0.1 MEGO + CF + Epon 862	95 ± 7	663 ± 27
0.5 MEGO + CF + Epon 862	99 ± 6	654 ± 28

3.3   Mode I Interlaminar Fracture Toughness Test Results

Mode -I fracture toughness provides an insight into the energy absorption capability of multi-layer fiber reinforced polymer composites. This was measured using the Double Cantilever Beam test. Figure 5 summarizes the mode I interlaminar fracture toughness values obtained from the double cantilever beam tests. The load versus extension curves displays a saw tooth pattern inferring that the crack growth is not continuous but is a sequence of growth and arrest. Compared to the neat sample, sawtooth patterns are more dominant in the samples with MEGO loading. It has been reported that the residual oxygen functional groups on carbon fiber surface will enhance the interfacial bonding between fiber, GO, and matrix through hydrogen bonding, $\pi - \pi$ interaction, and van der Walls interactions [15]. The results showed that the mode I interlaminar fracture toughness increases by ~ 27%, ~ 48%, and ~ 70% for 0.05, 0.1, and 0.5 wt.% loading of MEGO respectively.

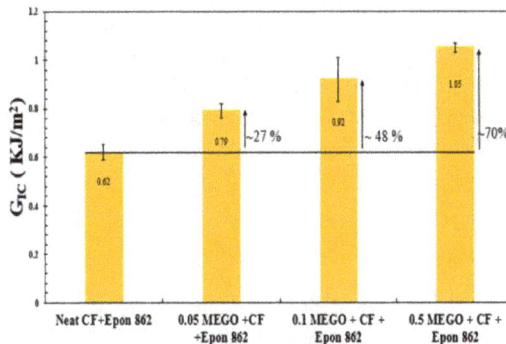

Fig. 5. Mode I Interlaminar fracture toughness for MEGO toughened CFRP

3.4 SEM of MEGO Toughened CFRP Composites

Figure 6 (a) and (b) show SEM images of the fracture surface of the carbon fiber reinforced Epon 862 with different percentages (neat and 0.5 wt%) of MEGO respectively. It was observed that the carbon fibers debonded cleanly from the matrix shown in (a) while the presence of MEGO (b) improved the toughness of the CFRP considerably.

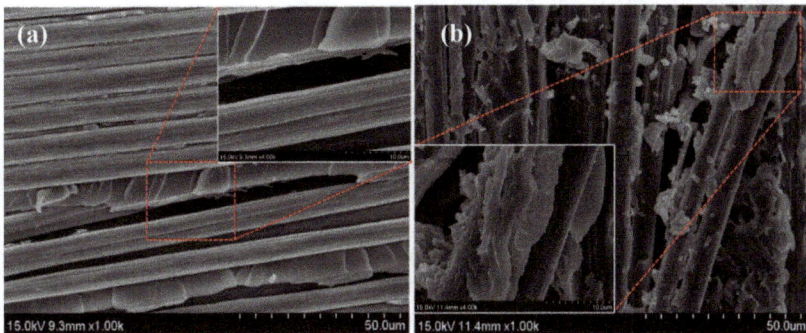

Fig. 6. SEM images of (a) Neat CFRP and (b) 0.5% MEGO + CFRP

4 Conclusions

A simple and scalable process was developed to toughen CFRP composites with hybrid polymer modifiers based on hybridized GO and POSS. It was observed that the addition of MEGO to CFRP enhanced the mode I interlaminar fracture toughness by up to 70% at a loading of 0.5% MEGO without significantly affecting the strength and stiffness of the composites. The toughening behavior was also confirmed in the SEM images.

Acknowledgements. The authors gratefully acknowledge the funding support through a subcontract to Oklahoma State University from MITO Material Solutions under a Phase II Small Business Innovative Research project # 1926906 from the National Science Foundation. Additional support from the Helmerich family endowment funds through the Varnadow Chair funds to Dr. Ranji Vaidyanathan is also acknowledged.

References

1. Baxter, L., Sedhuraman, M., Dhakal, D., Nowak, N., Sedai, B., Vaidyanathan, R.: Study of highly dispersible functionalized graphene oxide/polyhedral oligomeric silsesquioxane additives and the effect of polyvinyl pyrrolidone on dispersion quality and composite properties. In: Proceedings of Composites and Materials Expo (CAMX), 17–20 October, Anaheim, CA (2022)
2. Subramanian, M., Mishra, K., Vaidyanathan, R.: System and method for synthesis of POSS-graphene oxide derivatives as effective fillers for developing high performance composites, US Patent No. 10,011,706, July 3 (2018)
3. Król-Morkisz, K., Pielichowska, K.: Thermal decomposition of polymer nanocomposites with functionalized nanoparticles. In: Polymer Composites with Functionalized Nanoparticles, Elsevier. pp. 405–435 (2019)
4. Kuo, S.-W., Chang, F.-C.: POSS related polymer nanocomposites. Prog. Polym. Sci. **36**(12), 1649–1696 (2011)
5. Raftopoulos, K.N., Pielichowski, K.: Segmental dynamics in hybrid polymer/POSS nanomaterials. Prog. Polym. Sci. **52**, 136–187 (2016)
6. Huang, X., et al.: Graphene-based materials: synthesis, characterization, properties, and applications. Small **7**(14), 1876–1902 (2011)
7. Norhayati, H., et al.: A brief review on recent graphene oxide-based material nano composites: synthesis and applications. J. Mater. Environ. Sci. **7**(9), 3225–3243 (2016)
8. Liu, J., Liu, G., Liu, W.: Preparation of water-soluble β-cyclodextrin/poly (acrylic acid)/graphene oxide nanocomposites as new adsorbents to remove cationic dyes from aqueous solutions. Chem. Eng. J. **257**, 299–308 (2014)
9. Chauke, V.P., Maity, A., Chetty, A.: High-performance towards removal of toxic hexavalent chromium from aqueous solution using graphene oxide-alpha cyclodextrin-polypyrrole nanocomposites. J. Mol. Liq. **211**, 71–77 (2015)
10. Mkhoyan, K., et al.: Atomic and electronic structure of graphene-oxide. Nano Lett. **9**(3), 1058–1063 (2009)
11. Lin, Z., Liu, Y., Li, Z., Wong, C.P.: Novel preparation of functionalized graphene oxide for large scale, low cost, and self-cleaning coatings of electronic devices. In: 2011 IEEE 61st Electronic Components and Technology Conference (ECTC), pp. 358–362. IEEE, (2011)
12. Li, Z., et al.: Control of the functionality of graphene oxide for its application in epoxy nanocomposites. Polymer **54**(23), 6437–6446 (2013)

13. Mohan, M.: Synthesis of POSS grafted graphene oxide and its effect on mechanical and thermal properties of epoxy composites. Masters Thesis, Oklahoma State University (2017)
14. Dhakal, D.R.: Synthesis and characterization of nanosized hybrid polymer modifier (HPM) for improved mechanical and thermal behavior of composites. PhD diss., Oklahoma State University (2022)
15. Mishra, K., Bastola, K., Singh, R.P., Vaidyanathan, R.K.: Effect of graphene oxide on the interlaminar fracture toughness of carbon fiber/Epoxy composites. Polym. Eng. Sci. **59**(6), 1199–1208 (2019)

Efficient Degradation of Pendimethalin via Photo-Catalytic Ozonation Over Ni/Mg@TiO$_2$ Nanocomposites

Immandhi Sai Sonali Anantha[1], Maddila Suresh[1,2(✉)] (iD),
and Sreekantha B. Jonnalagadda[2] (iD)

[1] Department of Chemistry, GITAM School of Sciences, GITAM University, Visakhapatnam, Andhra Pradesh, India
sureshmskt@gmail.com

[2] School of Chemistry and Physics, University of KwaZulu-Natal, Westville Campus, Chiltern Hills, Durban 4000, South Africa

Abstract. The amplifying use of chemical herbicides have led to an accumulation of toxic layers on the domestic crops. These chemical through irrigational practices wash these chemicals into water bodies making it difficult to preserve an acceptable water quality. The most commonly used herbicide which protects the crops against a broad range of unwanted herbs is pendimethalin. Advanced oxidative processes are comprehended to be most efficient techniques compared to the classic waste water treatment methods. In this work we have separated pendimethalin from one of the paddy fields around Vizianagaram, Andhra Pradesh, using liquid-liquid extraction process over ionic liquids and then mineralised using photo-catalytic ozonation over Ni/Mg@TiO$_2$. The catalytic nano composites have been prepared using simple sol-gel method and characterised using various analytical techniques such as BET, SEM, EDX, TEM and PXRD. We achieved complete mineralisation of the target herbicide within a span of 2.5 h with major value added products identified during pendimethalin degradation which are, 5,6-dimethyl-2-(pentan-3-ylamino)cyclohex-4-ene-1,3-dione (DCHO), 3-((4,5-dimethyl-2,6-dioxocyclohex-3-en-1-yl)amino)pentanedioic acid (DOAP), 3-((1-carboxy-4-methyl-1,3-dioxohexan-2-yl)amino)pentanedioic acid (COAP), 2-methylbutanoic acid (MBA), 3-nitropentanedioic acid (NPA), and oxalic acid (OA). The by products were separated using GC-MS at given time. The catalytic nanoparticles showed high recyclability and efficiency for 7-consecutive cycles at a wavelength of 254 nm.

Keywords: Pendimethalin · AOPs · Heterogenous catalyst · Photocatalyst · Ozonation

1 Introduction

Pendimethalin is a pre-emergent that has gained an enduring reputation owing to its stability, persistence and its potential to harm non-targeted species [1]. Pendimethalin can impact the central nervous system, producing convulsions and coma in extreme

S. Patra et al. (Eds.): METCENT 2023, *Proceedings of the International Conference on Metallurgical Engineering and Centenary Celebration*, pp. 166–176, 2024.
https://doi.org/10.1007/978-981-99-6863-3_17

cases, as well as skin irritation, nausea, diarrhoea, and weariness [2, 3]. The U.S. Environmental Protection Agency (EPA, 2020) has also classed it as a potential human carcinogen in light of research revealing thyroid and adrenal cancers in animals [4]. Due to pendimethalin's poor water solubility and strong adsorption to soil particles, it has a propensity to remain in the environment, especially in soil [5]. An extensive application of pendimethalin drastically decreased the versatility of the soil microbial population [6]. This might have a detrimental effect on soil fertility and nutrient cycling [7–9]. It requires creative and practical solutions because conventional remediation techniques have proven insufficient for its total degeneration. Recent research has demonstrated that the use of advanced oxidation processes (AOPs), particularly ozone-based catalytic degradation, offers promising outcomes for successfully removing pendimethalin from water systems [10]. The generation of hydroxyl radicals, which are strong oxidants and the first step in the degradative process, is made possible by the combination of catalytic materials and ozone, a powerful oxidizing agent [11]. This two-pronged approach has been found to dramatically increase the pace and efficiency of deterioration. Titanium based mixed metal oxides serve as promising photocatalysts in the degradation of pendimethalin [12].

2 Materials and Method

2.1 Photocatalytic Ozonalysis Experiment

To guarantee that UV rays were kept at a wavelength less than 420 nm, a 500 W Xenon lamp was employed as the light source coupled with a UV filter. A long sintered glass tube was used to inject ozone into the reaction solution at a steady rate of flow of 10 mL/min using a Fisher ozone generator 500. The temperature of the reactor was maintained at $(25 + 1)$ °C by the movement of water through a network of double-walled pipes. A magnetic stirrer was used to achieve equitable distribution of ozone and catalytic surface throughout the reaction system. In each experiment, a 50 mL water specimen containing pendimethalin was examined in conjunction with various concentrations (from 10 to 50 mg/L) of the produced photocatalyst. The oxygen stream was kept at a steady concentration of 0.05 M while the ozone concentration was measured using a KI test kit. For each batch, several runs were made in order to provide exact and accurate findings (Fig. 1).

2.2 Synthesis of Ni/Mg@TiO$_2$ Nanoparticles

The sol-gel process was employed to create Ni/Mg@TiO$_2$ nanoparticles using, an organic titanium precursor form labelled tetrabutyl orthotitanate which was dispersed in an aqueous solution with a pH of 11. To produce TiO$_2$ particles that were the size of a nanometer, the precursor was cautiously introduced to the medium while being continuously stirred. The nitrates of nickel (Ni(NO$_3$)$_2$·6H$_2$O) and magnesium (Mg(NO$_3$)$_2$·6H$_2$O) were carefully added to the resultant solution in a predetermined ratio, while continually stirring the mixture, to guarantee homogenous dispersion of the dopants. After being stirred for 5–6 h, the resultant solution was heated to 80 °C in order

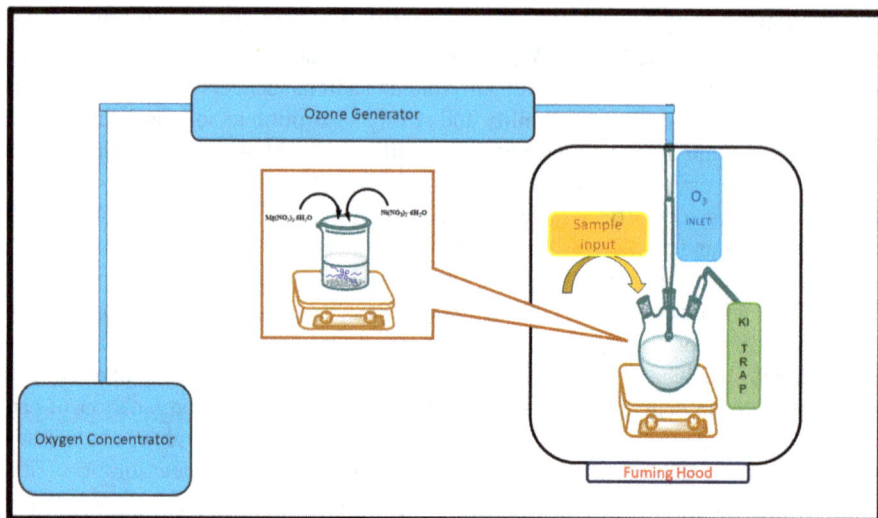

Fig. 1. Experimental Setup

to evaporate excess water to generate a paste. The produced paste was then placed in an oven and dried at a temperature between 150 °C and 200 °C in order to thoroughly eliminate any residual water. Using a mortar and pestle, the dried substance was manually crumbled into a fine powder. The resultant catalytic powder was then put through calcination, where any remaining nitrates in the compound were removed by progressively raising the temperature up to 500 °C and keeping it there for 5 h.

2.3 Instrumentation

PXRD (powder X-ray diffraction), BET (Brunauer-Emmett-Teller) analysis, ICP (inductively coupled plasma), PL (photoluminescence) analysis, SEM-EDX (scanning electron microscopy-energy dispersive X-ray spectroscopy), and TEM (transmission electron microscopy) were used to evaluate the morphology of the catalyst. The identification of intermediates using 1H and 13C NMR spectroscopy, as well as GC-MS (gas chromatography-mass spectrometry) and FTIR (fourier transform infrared) spectral studies, led to the proposal of the reaction mechanism involved in the degradation of pendimethalin.

3 Results and Discussion

3.1 BET Surface and Elemental Analysis

Through nitrogen adsorption investigations, the surface area and structural porosity of Ni/Mg@TiO$_2$ hybrids at different concentrations (1%, 2.5%, and 5%) were measured, and the outcomes were compared with pure TiO$_2$. The nitrogen adsorption isotherms displayed a type IV isotherm's H1 hysteresis loop. Pure TiO$_2$ was discovered to have a surface area of 25.37 m^2/g. The surface area dramatically increased with the addition of Ni/Mg dopants, reaching a value of 62.27 m^2/g at 5% dopant loading. The surface area was reduced to 60.77 m^2/g as a result of the pore-clogging caused by additional increases in the dopant concentration, it should be highlighted (Fig. 2).

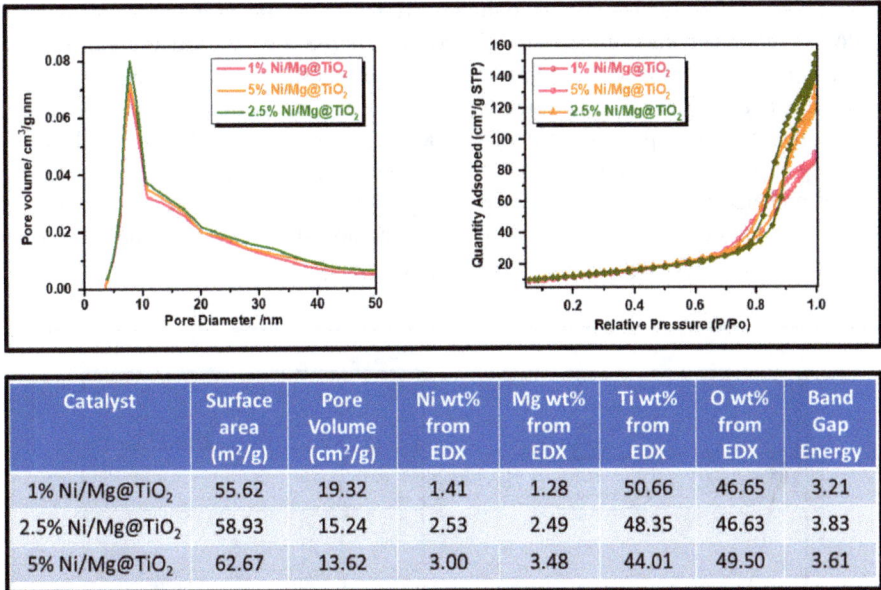

Catalyst	Surface area (m²/g)	Pore Volume (cm³/g)	Ni wt% from EDX	Mg wt% from EDX	Ti wt% from EDX	O wt% from EDX	Band Gap Energy
1% Ni/Mg@TiO₂	55.62	19.32	1.41	1.28	50.66	46.65	3.21
2.5% Ni/Mg@TiO₂	58.93	15.24	2.53	2.49	48.35	46.63	3.83
5% Ni/Mg@TiO₂	62.67	13.62	3.00	3.48	44.01	49.50	3.61

Fig. 2. BET Surface area, Elemental analysis and Band gap of Ni/Mg@TiO2

3.2 UV-DRS

According to the adsorption spectra shown in Fig. 3(a), the band gap values of pure TiO_2 were calculated and compared to those of the produced hybrid composite. Pure TiO was found to have a band gap energy (Eg) of 3.18 eV, or a wavelength range of 360–380 nm. 1% Ni/Mg@TiO_2, 2.5% Ni/Mg@TiO_2, and 5% Ni/Mg@TiO_2 were found to have Eg values for the hybrid composite that were 3.21 eV, 3.83 eV, and 3.61 eV, respectively. The band gap values for these data were computed using the method below for the wavelength range of 300–600 nm.

$$\alpha h\upsilon = B(h\upsilon - Eg)^{1/2}$$

where Eg = Energy gap; B = band gap energy; α = adsorption co-efficient; h = Plank's constant; and υ = Frequency of light.

3.3 Photoluminescence Spectra (PL)

To evaluate the materials' photogenerated electron-hole junction, photoluminescence (PL) examination was conducted. According to the PL analysis, there are peaks at 370 nm (Fig. 3(b)). As the dopant percentage grew, it was found that the strength of these peaks reduced. The intensity of the peak may be a clue for the inverse relation with the combination of holes and electrons in the material. The decline in the peak intensity as the dopant increment may suggest a good separation of the holes and electrons allowing a good conductivity for the nanocomposites. These results suggest that 5% Ni/Mg@TiO$_2$ could serve as a more efficient photocatalyst than lower concentrations. This is due to the fact that the reduced dopant concentration leads to lesser number of holes and electrons favouring easy recombination, whereas by increasing the dopant percentage the number of holes and electrons are also increased giving additional spam before recombination. Therefore, number of charge carriers (electrons and holes) increase with the increase in the dopant percentage, suggesting increased photocatalytic activity.

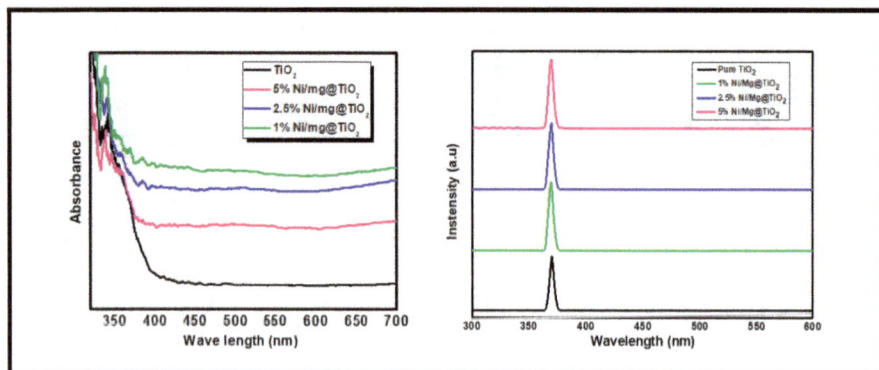

Fig. 3. (a) UV DRS spectra (b) PL spectra

3.4 SEM and TEM

The SEM and TEM images supporting the morphology of the prepared Ni/Mg@TiO$_2$ (1%, 2.5% and 5%) nanocomposites were displayed in Fig. 3. The SEM and TEM images show that the Ni and Mg on TiO$_2$ surface are evenly dispersed. The increase in the surface particles i.e. Ni and Mg is found to be more in Fig. 4(a) and (c) as compared to Fig. 4(b), suggesting that when 1% each of the dopant was used the surface was dispersed with the dopant particles in a monolayer while in 5% doping agglomeration is witnessed. In 2.5% doping of each dopant the surface is dispersed with the dopant particles leading to active sites. The crystalline structures of TiO$_2$ core is clearly wrapped with clusters of Ni/Mg. The elemental analysis of the prepared composites was perceived by EDX studies. The TEM images give a cross-sectional micrograph of the prepared catalysts adding evidence to their structure with a grain size ranging from 29.17 nm to 51.26 nm. The TEM images suggest a spherical to ovoidal shape of the catalytic nanoparticles. The grain size of the catalytic particles increased with the increase in the dopant.

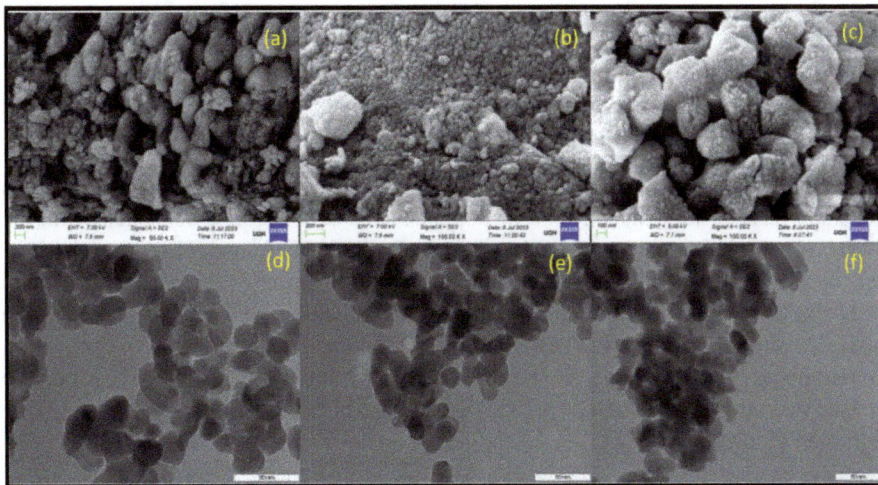

Fig. 4. SEM of (a) 1% Ni/Mg@TiO$_2$ (b) 2.5% Ni/Mg@TiO$_2$ (c) 5% Ni/Mg@TiO$_2$. TEM of (d) 1% Ni/Mg@TiO$_2$ (e) 2.5% Ni/Mg@TiO$_2$ (f) 5% Ni/Mg@TiO$_2$

3.5 XRD Analysis

The X-ray Diffraction patterns of Pure TiO$_2$, and Ni/Mg@TiO$_2$ (1%, 2.5% and 5%) were compared in Fig. 4. The peaks of Anatase TiO$_2$ at $2\theta = 24.7°$, 37.35°, 47.67°, 55.12° and 62.24° represents (JCPDS Card no. 21-1272) while 37.34°, 43.37°, 62.94°, 75.48°, and 79.48° peaks represent NiO (JCPDS Card no. 04-0835) and 36.96°, 38.02°, 42.98°, 62.36°, 74.71° and 78.66° represent MgO (JCPDS No. 87-0653). The hybrid

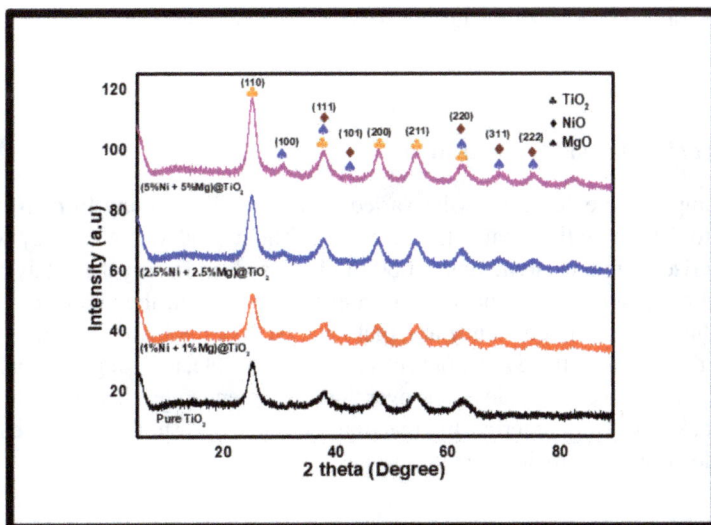

Fig. 5. PXRD of (1%, 2.5% and 5%) Ni/Mg@TiO2

composites of Ni/Mg@TiO$_2$ show a combination of pure crystalline phases of the parent compounds (Fig. 5).

3.6 Effect of Ni/Mg Doping Concentrations

The number of active sites that reduce the activation energy of the degradation process grows along with the concentration of dopants. This conclusion is consistent with the results of morphological investigations, such as BET, SEM, and TEM, which show that the creation of a porous surface and the agglomeration of dopants worsen with increasing dopant percentage. Up to a dopant concentration of 5% Ni/Mg@TiO2, the catalyst's degrading efficiency keeps getting better, but greater dopant concentrations do not appear to significantly increase the efficiency any more. As a result, it may be inferred that a dopant concentration of 5% is the ideal operational setting for maximizing degrading efficiency (Fig. 6).

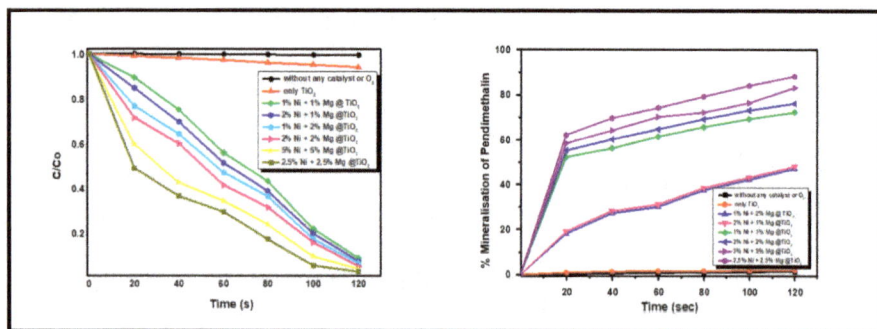

Fig. 6. Degradation of Pendimethalin at varying percentage doping of the catalyst. *Ozone flowrate 100 mL/min; Under Visible light; 10 mg/L of each catalytic hybrid.

3.7 Effect of Catalyst Concentration

Various samples were generated with varied catalyst doses, ranging from 10 mg/L to 80 mg/L, to determine the ideal catalytic dosage. The degradation efficiency was seen to steadily rise up to a catalytic dose of 50 mg/L. However, raising the catalyst dosage further did not result in an improvement in efficiency; rather, the efficiency began to decline. This decrease in efficiency at larger catalyst doses is likely caused by the fact that as the dosage rises, there are more active sites accessible, allowing for more light to flow through. However, if the dose was over the recommended level, extra particles may start to block light from entering the reaction medium. In turn, this has a detrimental effect on deterioration efficiency (Fig. 7).

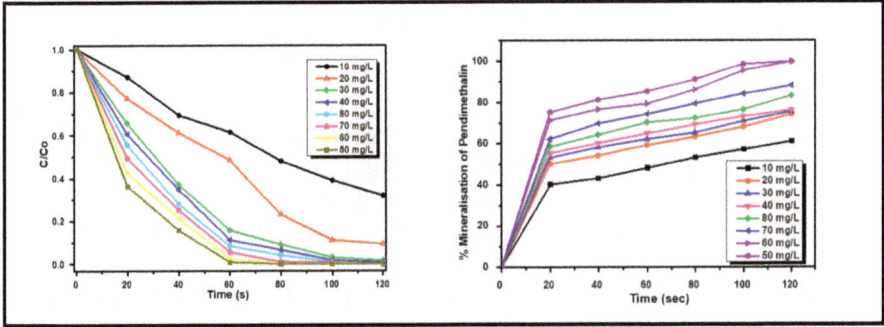

Fig. 7. Degradation of Pendimethalin at varying catalytic dosage. *Ozone flowrate 100 mL/min; Under Visible light; (2.5% Ni + 2.5% Mg @TiO$_2$)

3.8   Effect of pH

Under various acidic to basic pH circumstances (pH = 3, 7, and 11), the degradation of pendimethalin was examined. The degradation efficiency was initially low at an acidic pH and gradually rose with a rise in pH. The sole reactive species that exists at an acidic pH is ozone, which is known to be relatively unstable and prone to deterioration. As a result, the breakdown process was only moderately effective at acidic pH. However, secondary oxidizing agents such hydroxyl radicals were produced more quickly as the pH moved towards the basic range. Hydroxyl radicals are very reactive species that have the capacity to accelerate and improve pendimethalin breakdown. As a result, the presence of hydroxyl free radicals at a basic pH of 11 significantly improved pendimethalin's ability to degrade (Fig. 8).

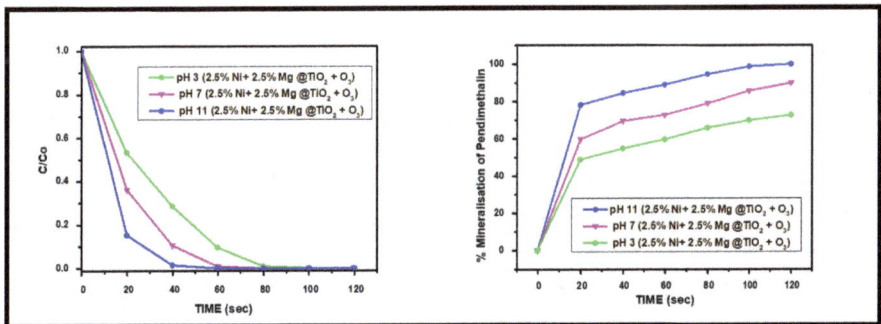

Fig. 8. Degradation of Pendimethalin at varying pH. *Ozone flowrate 100 mL/min; Under Visible light; 50 mg/L (2.5% Ni + 2.5% Mg @TiO$_2$).

3.9   Catalyst Testing and Product Identification

Pendimethalin's photocatalytic degradation was carried out under visible light, together with ozonation. To separate the organic phase, samples of the reaction mixture were taken every 20 min. The intermediates created during the degradation process were then examined in the isolated organic phase. The discovery of these intermediates shed important light on the reaction mixture's pendimethalin degradation mechanism. Several intermediates were found using GC-MS analysis, and their existence was indicated by peaks with retention times between 8 and 18 corresponding to different carboxylic acid products. The discovered compounds were submitted to TLC (thin-layer chromatography), and the samples were eluted from the TLC plates, in order to get pure intermediates. The separated intermediates were then subjected to further examination utilizing a variety of spectrum methods, including mass spectrometry, FT-IR spectroscopy, 1H-NMR, and 13C-NMR. The idea that pendimethalin mineralization occurs during ozone-mediated photocatalytic degradation was further supported by a qualitative test that looked at the release of CO_2 using lime water.

3.10   Reaction Mechanism

The photocatalyzed degradation pathway for pendimethalin under ozonation was proposed upon careful observation and investigation of the obtained intermediates. Highly active oxidizing species are generated during the ozonation process such as hydroxyl free radical (OH˙) and superoxide free radical (·O_2). These free radicals play an important role in the generation and separation of electrons and holes. The key oxidizing species

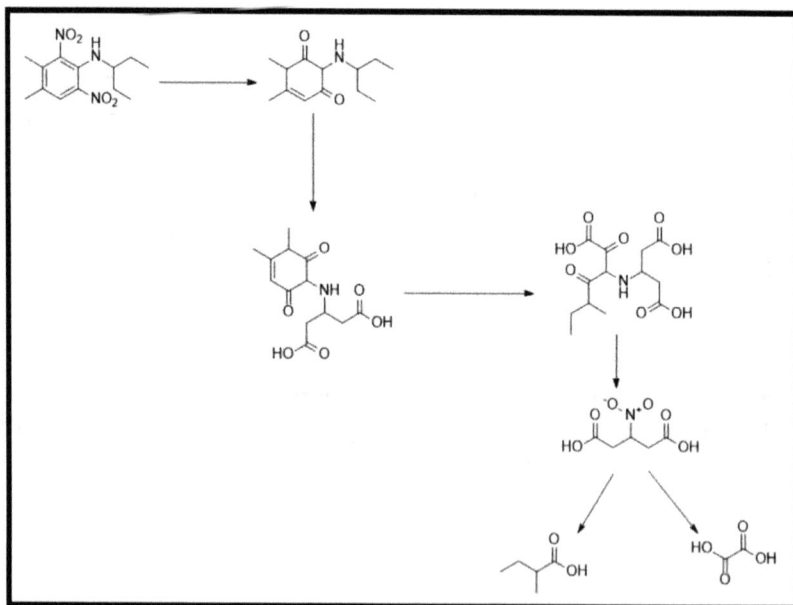

Scheme 1. Degradation pathway of pendimethalin

in this degradation process is assumed to be OH˙, which attributes to the dichlorination of the intermediates and ozonation process. The most important aspect in the ozonated photo-catalyzed degradation is the stability of the generated oxidizing agents, and the surface active sites of the heterogenous photocatalyst (Scheme 1).

4 Conclusions

The goal of the current work is to create an effective synthetic technique for the sol-gel manufacture of $Ni/Mg@TiO_2$ nanoparticles. The methodically planned processes guarantee an even distribution of the necessary dopants throughout the catalyst structure. BET (Brunauer-Emmett-Teller), SEM (scanning electron microscopy), and TEM (transmission electron microscopy) were used to investigate the exterior morphology of the nano-catalyst in order to describe the manufactured catalyst. Additionally, ICP (inductively coupled plasma) and PL (photoluminescence) experiments were used to assess the catalyst's photoconduction efficiency. It was looked into how well the catalyst degraded pendimethalin, a dangerous chemical frequently found in bodies of water close to agricultural areas. High efficiency in the degrading process was shown when ozonation and the photocatalyst $Ni/Mg@TiO_2$ were used. By segregating and identifying the intermediates produced throughout the process, the pendimethalin degradation route was clarified. Notably, the mineralization of all detected intermediates finally led to the formation of CO_2 and H_2O. Importantly, the total destruction of pendimethalin without the creation of any secondary pollutants was achieved using the $Ni/Mg@TiO_2$ photocatalyst in combination with ozone under visible light.

Acknowledgements. The authors are thankful to the National Research Foundation of South Africa, University of KwaZulu-Natal, Durban, South Africa and Department of Chemistry, School of Sciences, Visakhapatnam, India for providing facilities.

References

1. Noor, M., Iqbal, K.M., Kanwal, T., Rehman, K., Ali, M., Shah, M.R.: Synthesis and characterization of modified silica nanoparticles based molecularly imprinted polymers as a sensor for selective detection of pendimethalin from drinking water. J. Mol. Struct. **1287**, 135635 (2023)
2. Faramarzi, M., Avarseji, Z., Alamdari, G.E., Taliei, F.: Biodegradation of the trifluralin herbicide by pseudomonas fluorescens. Int. J. Environ. Sci. Technol. **20**, 3591–3598 (2023)
3. World Health Organisation. Chemical Fact Sheets: Pendimethalin (2022)
4. U.S. Environmental Protection Agency. Pendimethalin Human Health Risk Assessment (2023)
5. Mohamed, I.A., Abdalla, R.M.: Weed control, growth, and yield of tomato after application of metribuzin and different pendimethalin products in upper Egypt. J. Soil Sci. Plant Nutr. **23**(1), 924–937 (2023)
6. Onwuchekwa-Henry, C.B., Ogtrop, F.V., Roche, R., Daniel, K.Y.T.: Evaluation of pre-emergence herbicides for weed management and rice yield in direct-seeded rice in Cambodian lowland ecosystems. Farm. Syst. **1**(2), 100018 (2023)

7. Ratneswar, P., Rajib, K., Soumen, B., Dibaka, G.: Complex weed flora managing efficacy of herbicides in soyabean and their effect on soil properties, microorganisms and productivity of succeeding mustard. Ind. J. Weed Sci. **55**(2), 181–186 (2023)

8. Dennis, P.G., Kukulies, T., Forstner, C., Plisson, F., Eaglesham, G., Pattison, A.B.: The effects of atrazine, diuron, fluazifop-p-butyl, haloxyfop-p-methyl, and pendimethalin on soil microbial activity and diversity. Appl. Microbiol. **3**, 79–89 (2023)

9. European Food Safety Authority: Renewal assessment report: pendimethalin. EFSA J. **19**(1), 6348 (2021)

10. Ergun, C.: A current review on conducting polymer-based catalysts: advanced oxidation processes for the removal of aquatic pollutants. Water Air Soil Pollut. **234**, 524 (2023)

11. Deng, A., et al.: Unlocking the potential of MOF-derived carbon-based nanomaterials for water purification through advanced oxidation processes: a comprehensive review on the impact of process parameter modulation. Sep. Purif. Technol. **318**, 123998 (2023)

12. Sabri, M., Habibi-Yangjeh, A., Pouran, S.R., Wang, C.: Titania-activated persulfate for environmental remediation: the-state-of-the-art. Catalysis Rev. **65**(1), 118–173 (2023)

Effect of A-Site Pr-Doping on Dielectric and Thermoelectric Properties of Lanthanum Based La$_2$FeNbO$_6$ Double Perovskite Oxide Materials

Sanjay Srivastava$^{(\boxtimes)}$ and Lav Kush$^{(\boxtimes)}$

Department of Materials and Metallurgical Engineering (MME), Maulana Azad National
Institute of Technology, Bhopal 462003, M.P., India
s.srivastava.msme@gmail.com, sanjay@manit.ac.in,
verma.lavkushh1991@gmail.com

Abstract. Present study shows the structural, dielectric, and thermoelectric behaviors of Pr-doped La$_2$FeNbO$_6$ double perovskite oxide. The compound was synthesized via sol-gel technique followed by the sintering at 1473 K for 6 h in open air. XRD analysis reveals the observed pattern of La$_{2-x}$Pr$_x$FeNbO$_6$ [LPFNO] shows single phase cubic structure with Fm-3m space group and crystallite sizes were in similar fashion with morphological observation. Analysis of SEM images signifies the uniform distribution of grains in entire surface of the samples. The Maxwell-Wagner model has been used to explain the nature of the frequency-dependence dielectric constant of Pr-doped La$_2$FeNbO$_6$ Double Perovskite. The material undergoes a ferroelectric-para-electric phase transition, represented by a dielectric anomaly in its temperature-dependent dielectric permittivity investigation. At higher sintering temperatures, the investigation of Nyquist plots reveals the presence of nano-grains and grain boundary effects resulting from a conglomerate of nano-particles. The material's conductivity follows Jonscher's Power law and aligns with the large overlapping polaron tunneling model. The activation energy, akin to the energy needed for electron hopping, is determined using the Arrhenius equation. The sample exhibits negative temperature coefficient of resistance (NTCR) characteristics, indicating semiconductor behavior with temperature variation, according to impedance and conductivity analyses. Increasing Pr-content in LPFNO enhances dc-conductivity, power factor, and ZT up to certain high temperatures, indicating nearest neighbor hopping charge carrier conduction, followed by a rapid decline, signaling a transition to metallic-like behavior. Positive S throughout the examined temperature range suggests the presence of p-type charge carriers, while intrinsic behavior remains consistent with increasing Pr-content. Thermal conductivity decreases with rising temperature, suggesting dominant phonon-phonon (anharmonic) scattering. Such double perovskites are valuable for designing modern electronic devices, especially energy storage devices.

Keywords: double perovskite · Sol-gel · Dielectric · thermoelectric · Nyquist plots · ZT

S. Patra et al. (Eds.): METCENT 2023, *Proceedings of the International Conference
on Metallurgical Engineering and Centenary Celebration*, pp. 177–185, 2024.
https://doi.org/10.1007/978-981-99-6863-3_18

1   Introduction

One of the most extensively studied families of oxides is the double perovskite with the chemical formula of $A_2B'B''O_6$ type, where site A refers to alkaline earth or rare earth ions with larger ionic radii and site B' and B'' refer to transition metal cations or lanthanides with smaller ionic radii [1]. Numerous research have discovered that $A_2B'B''O_6$ exhibits a wide variety of behaviors, including semi conductivity, metallic and half-metallicity, dielectricity, ferroelectricity, thermoelectricity, and perhaps even superconductivity [2–4]. As a result, these materials can offer numerous simultaneous answers to issues with energy generation. Indeed, due to its unique features, La-based perovskite-structured has undergone extensive research. These typical applications include magnetic refrigerants, fuel cells, chemical sensors, catalytic devices, gas sensors, memory devices, and fuel cells [5, 6]. This typical multiferroic material displayed a high ferroelectric transition temperature of around 715 K [7, 8]. In fact, adding a different cation to the A or B site can alter the perovskite's neutrality, which has a significant impact on its structural and physical characteristics. The distribution of the B and B' cations over the octahedral sites by the cation ordering, and the octahedral distortion all have an impact on the characteristics of $A_2BB'O_6$.

In light of the aforementioned, novel double perovskite oxides $La_{2-x}Pr_xFeNbO_6$ were synthesized using sol-gel route. The investigations showed that the La/Pr ion distribution on the A-site was disordered, and the material crystallizes with a Pm3m cubic symmetry. The purpose of this work is to investigate the dielectric and thermoelectric response of the multiferroic $La_{2-x}Pr_xFeNbO_6$ [LPFNO] compound as a function of temperature and frequency. The purity and crystal structure were analyzed using X-ray diffraction technique. To investigate the temperature and frequency dependence of the thermoelectric and dielectric properties, dielectric tests were conducted at temperatures ranging from 25 °C to 600 °C and frequencies between 10 Hz and 10^6 Hz.

2   Experimental Procedure

The sol-gel method was opted to prepare the A-site Pr doped La_2FeNbO_6 ($La_{2-x}Pr_xFeNbO_6$) double perovskite. The starting precursors were $La_2(NO_3)_3$ (0.1M) (99.99%), $Pr(NO_3)_3$ $6H_2O$ (0.1M) (99.9%), $Fe_2(NO_3)_3$ $6H_2O$ (0.1M) (99%) niobium (IV) oxide NbO_2. These precursors were mixed with required stoichiometry of deionized water and the citric acid monohydrate (10% by weight) was added to the mixture as a solvent. The mixtures were stirred for 8 h at 353 K until the formation of homogeneous solution by maintaining the pH nearby 3. The resulting material became a highly viscous residual gel, which was then dried overnight at 373 K in oven. After drying, a high energy planetary ball mill (Make: Fretsch) with a maintained BPR (ball-to-powder ratio) of 10:1 was used for wet milling (using toluene) in order to avoid agglomeration of the powder for 24 h at 500 rpm, followed by conventional calcination at 1073 K for 6 h. To achieve the desired nano-size of the calcined powders, the milling process was repeated at 500 rpm for 24 h in a planetary micro mill (Fritsch, PULVERISETTE 7 premium line, Rhineland Palatinate, Germany). The milled calcined powders were bind by using PVA (poly vinyl alcohol) and then uniaxially pressed for 60 s at 600 MPa to form rectangular pellets. The pellets were then visually examined, and no surface cracks were observed. Ultimately, dense pellets of PSFCO ceramic samples were formed by sintering at 1573 K in open air for 6 h.

With a Cu Kα radiation source (= 1.5418) and a 2θ angle range of 10° to 90°, the Rigaku MiniFlex 600 X-Ray Diffractometer was used to characterize the calcined and sintered pellet. The pellet's shape and average grain size were examined using a Phenom ProX scanning electron microscope (SEM) operating at 15 kV. Impedance Spectroscopy (IS) technology was used to observe the electrical properties. The rectangular-shaped sample was used to measure the temperature dependent See-beck coefficient, electrical conductivity, Figure of Merit (ZT) and Power Factor with the help of Turnkey Concept 50, Novo control Technologies, Germany.

3 Results and Discussion

3.1 Analyses of the Morphology and Phase Identification

The powder XRD pattern was refined by using EXPO2014 software, and the structural characteristics of LPFNO ceramic were analyzed, as shown in Fig. 1. La_2MnCoO_6 (ICSD - 4514503) crystallographic data (of La_2FeMnO_6:) were used to refine the $La_{2-x}Pr_xFeNbO_6$ [LPFNO] double perovskite crystallographic data in order to define the atom positions in the unit cell, which already has been characterized as a double perovskite in elsewhere report.

The LPFNO structure with the substitution of Pr^{3+} has two intensity peaks characteristic of the planes (112) and (200), with a bellow $2\theta = 80°$, according to the Rietveld refinement result, as shown in Fig. 1. The monoclinic, with $P2_1/n$ space group, two formulae per unit cell (Z = 2) structure of the LPFNO double perovskite was determined. Our results from the Rietveld refinement approach, obtained using the EXPO software, indicate that the average crystallite size of LFMO is roughly $D_{LPFNO} = 1.05$ μm for x = 0.0, and 1.73 μm for x = 1.0 respectively.

The micrographs of the synthesized chemical compounds were observed on 5 μm scale as displayed in Fig. 2. The SEM micrographs reveal that the sample consists of particles with a quasi-spherical morphology, arranged randomly in clusters. The average crystallite size obtained through Rietveld refinement varies depending on the synthesis technique used. For x = 0.0, the estimated average crystallite size is approximately $DLPFNO = 1.05$ μm, which is in good agreement with the average grain size observed in the micrograph pictures using SEM, approximately 1.32 μm. Similarly, for x = 1.0, the estimated average crystallite size is around 1.73 μm, which aligns well with the observed average grain size of approximately 1.78 μm in the SEM micrographs.

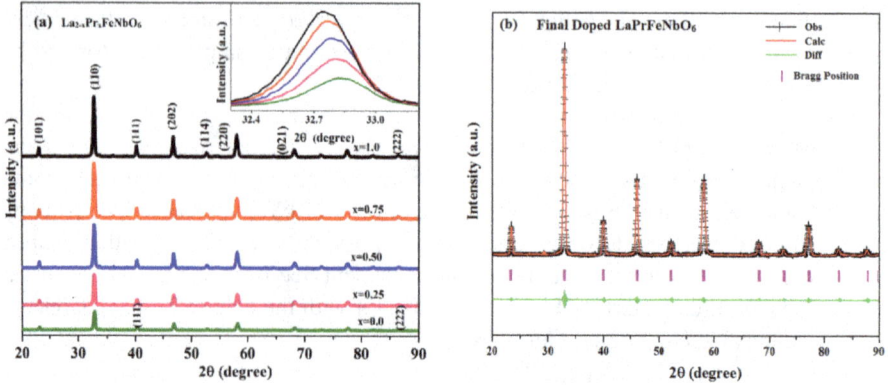

Fig. 1. (a) XRD of $La_{2-x}Pr_xFeNbO_6$ (x = 0 to x = 1), (b) Rietveld refinement of the XRD data for Pr^{3+} substituted $La_{2-x}Pr_xFeNbO_6$ [LPFNO] double perovskite ceramic at x = 1, and number given in parenthesis indicate the Miller indices of the plane

Fig. 2. (a–c) FESEM images of x = 0, 0.5 and 1.0 samples

3.2 Frequency Dependence Dielectric Study at Different Temperature

Figure 3(a & b) shows the variation of dielectric constant (ε') as a function of frequency at various temperatures. For all temperatures, ε' exhibit a constant reduction with rising frequency until becoming essentially frequency independent in the high frequency range. This type of behavior is known as dielectric relaxation, and it is based on the idea that dominating polarization mechanisms progressively changing with frequency due to variations in dipole moment with their characteristic time periods. At the low frequencies, all the polarization mechanism is responsible for the schematic variation of dielectric constant. The contributions of these distinct polarizations decreases with increase the frequency, and as a result, the dielectric constant diminishes along with frequency.

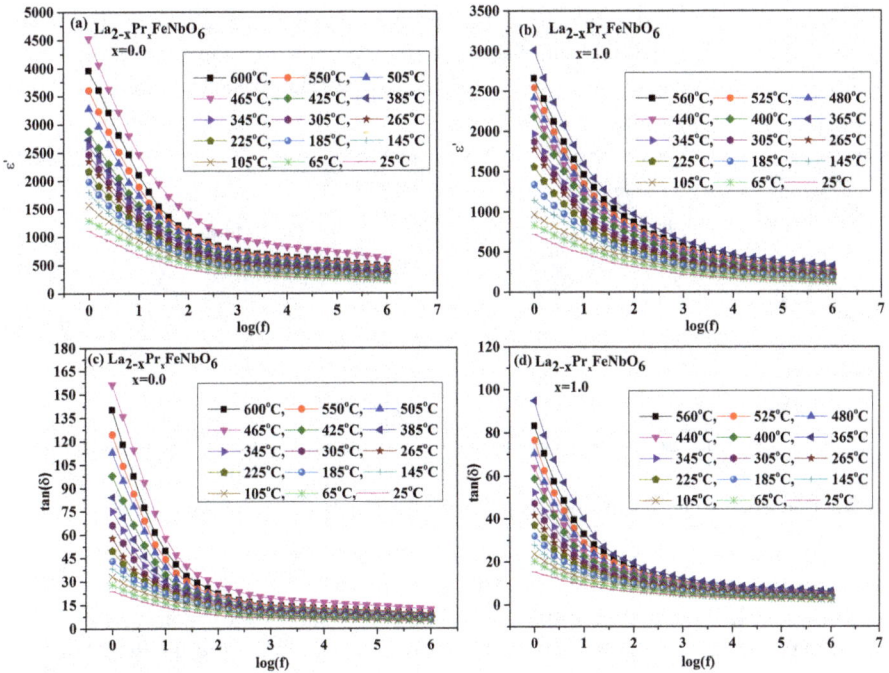

Fig. 3. Frequency dependent (a & b) ε, and (c & d) tanδ of $La_{2-x}Pr_xFeNbO_6$ double perovskite for x = 0.0 and 1.0 composition.

From room temperature to 600 °C for x = 0.0, and to the 560 °C for x = 1.0, the dielectric of the LPFNO increases as expected at their respected phase transition temperature. The dipole rotation mechanism explains the inverse frequency dependency of the tan δ values, as depicted in Fig. 3(c & d). A large value of the loss factor results from a large number of dipoles aligning themselves along the applied field and thoroughly contributing to the polarization at lower frequencies. As the frequency increases, the applied field experiences rapid variations, leading to the cessation of orientational polarization in dipoles. Consequently, there is no need to expend energy to rotate the dipoles. The material exhibits low dielectric loss (tan δ) values in the high-frequency range, indicating its excellent crystalline nature. This characteristic presents promising opportunities for potential applications in the electrical and electronic industries, facilitating the design of various devices.

3.3 Temperature Dependent Dielectric Study

In Fig. 4(a & b), the dielectric characteristics of LFNO ceramics are presented, with a focus on their temperature dependency at different operating frequencies. As the frequency increases, there is a noticeable decrease in the magnitude of the εr value. Additionally, the temperature at which the maximum εr value occurs shifts towards higher temperatures, as depicted in Fig. 4(a & b). This behavior indicates a relaxor-like dielectric relaxation in the material, reminiscent of observations found in various perovskite oxides that have been associated with oxygen vacancy-related phenomena [9].

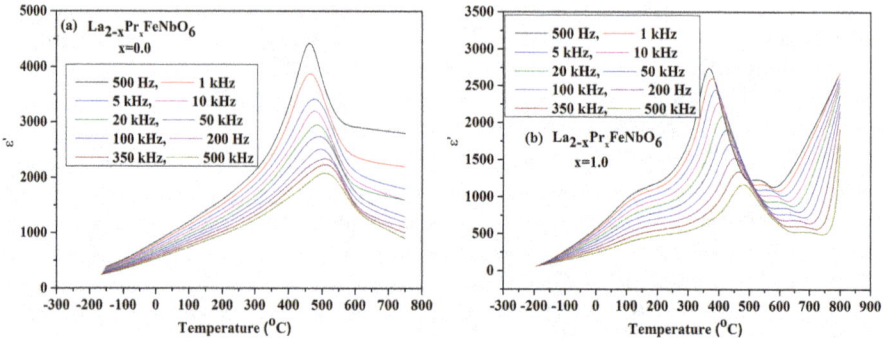

Fig. 4. Temperature dependent ε of $La_{2-x}Pr_xFeNbO_6$ double perovskite for (a) x = 0.0 and (b) 1.0 composition.

During the high temperature sintering of the compounds, the oxygen vacancies ($V_{\ddot{o}}$) are difficult to move in the lattice because of the strong bonding to the defect sites of LPFNO sample. From the study of the reaction, the free electrons are able to release from the neutralization of oxygen vacancies in Eq. (1) and then captured by Fe^{3+} as given in Eq. (4). But Fe^{3+} becomes unstable in oxidizing environment, further changes into Fe^{2+} and released the electron. This electron is further captured by Nb^{4+} and changed into Nb^{3+}.

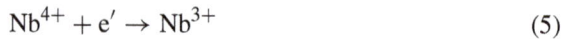

$$O_o^X \rightarrow V_o + 1/2\ O_2 \tag{1}$$

$$V_o \rightarrow V_{\dot{o}} + e' \tag{2}$$

$$V_{\dot{o}} \rightarrow V_{\ddot{o}} + e' \tag{3}$$

$$Fe^{3+} + e' \rightarrow Fe^{2+} \tag{4}$$

$$Nb^{4+} + e' \rightarrow Nb^{3+} \tag{5}$$

It is important to note that domain wall motion is primarily responsible for the high-temperature dielectric behavior. As a result, a greater dielectric constant can be seen from improved domain wall motion observed in x = 0.0.

3.4 Impedance Study

The frequency dependence of the real (Z') and imaginary (Z'') components of impedance at various temperatures are shown in Fig. 5(a–d). It has been observed that in the low-frequency range, the magnitude of Z' decreases as temperature increases, indicating the presence of a compound with a negative temperature coefficient of resistance (NTCR) signifying semiconducting characteristics of the sample. The release of space charge is indicated by the merging of curves in the high-frequency zone. The nature of the curves illustrates a single relaxation phenomenon linked to the increase in conductivity with increasing temperature [10]. Figure 5(c & d) shows the loss spectrum (Z″) with frequency graphs. The loss of the spectrum is characterized by the shift of the peak to higher frequency with reduced intensity. All of the Figures show a clear peak, and as the temperature raises, the peak value declines and moves to a higher frequency. The peak spectrum appears at a specific frequency that meets the requirement of $2\pi\,f\,RC = 1$. The both conditions mentioned above can be used to estimate the values of R and C.

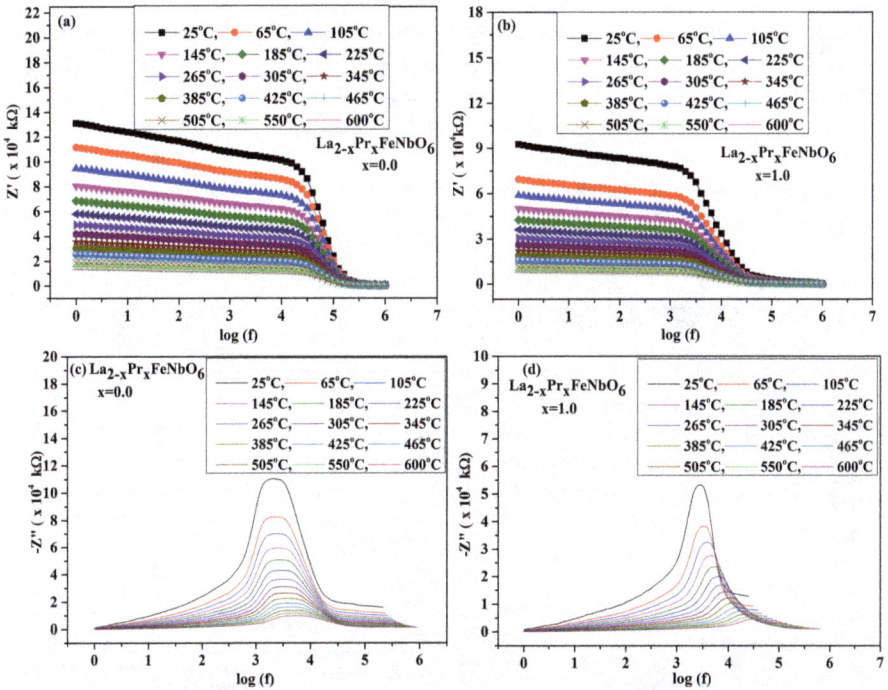

Fig. 5. Frequency dependent plot of (a & b) Z′ and (c & d) Z″ of $La_{2-x}Pr_xFeNbO_6$ double perovskite for x = 0.0 and 1.0 composition respectively.

The asymmetric broadening of the peaks suggests that spreading of relaxation time is related with the electrical processes of the domain wall. The relationship between peak height and resistance is linear. The criterion $\omega_m\tau_m = 1$, met at the highest point [11].

3.5 Thermoelectric Behavior of La$_{2-x}$Pr$_x$FeCrO$_6$ Double Perovskite

The temperature dependent electrical conductivity (σ) of La$_{2-x}$Pr$_x$FeCrO$_6$ (x = 0.0, and 1.0) double perovskite is shown in Fig. 6(a). It can be shown that the σ increases steadily at a slower rate up to 400 K for all compositions, then rapidly increasing. Subsequently, all the compositions attain a peak value of σ at certain higher temperatures, depending upon the compositions, and thereafter decrease with further increase in temperature. Notably, with increasing Pr-content, σ increases drastically in the entire temperature range. In a metallic system, an increase in the frequency of collisions between charge carriers and phonons has been reported to cause a decrease in conductivity as a function of temperature [12–14].

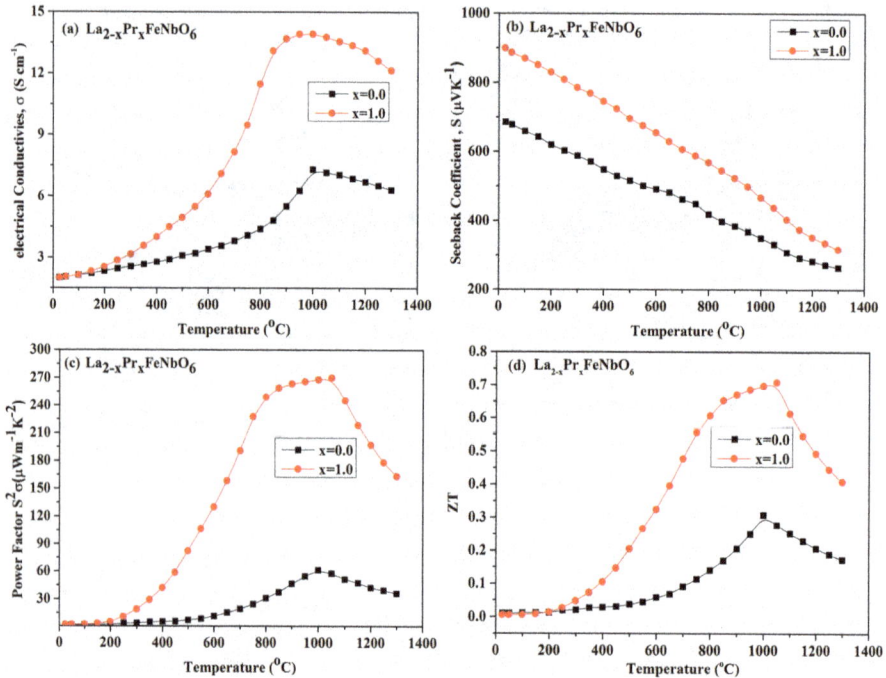

Fig. 6. Temperature dependent (a) electrical conductivity, (b) Seebeck coefficient, (c) power factor, and (d) figure of merit La$_{2-x}$Pr$_x$FeNbO$_6$ double perovskite for x = 0.0 and 1.0 composition respectively.

Figure 6(b) illustrates the Seebeck coefficient (S) of La$_{2-x}$Pr$_x$FeCrO$_6$ material. S values were positive for all compositions over the temperature range studied, indicates the presence of p-type semiconducting behavior [12]. A monotonous decrease in S was observed along with increasing temperature for entire composition (x), signifying the dominant behavior of La$_{2-x}$Pr$_x$FeCrO$_6$ material as a function of temperature. With increasing temperature, this monotonous decrease in S-value was attributed towards the enhanced Screening effect (i.e. fusion of electrons and holes with increasing temperature).

Figure 6(c) represents the calculated temperature dependent power factor (PF = $S^2\sigma$) of $La_{2-x}Pr_xFeCrO_6$ double perovskite. Initially, the PF of all the compositions increases up to a certain high temperature, and thereafter a little drop was noticed up to the maximum examined temperature, which correlates well with our above findings.

Finally, the ZT was calculated and presented in Fig. 6(d) based on the measured data of electrical properties at various temperatures. The combination of a lower S results in relatively high ZT, which further has been enhanced by careful doping with Pr dopant from x = 0.0 to 1.0 in $La_{2-x}Pr_xFeCrO_6$ double perovskite. For all x values, the ZT increases rapidly up to the certain high temperatures, thereafter, decreases with further increase in temperature. This behavior states that at very high temperatures, these compounds might be starts to behave as unstable semiconductors.

4 Conclusion

In conclusion, the LFNO ceramic was extensively studied, revealing its structural and electrical properties. The crystallographic analysis using XRD and Rietveld refinement confirmed the monoclinic structure of the LPFNO double perovskite with Pr^{3+} substitution. The SEM micrographs corroborated the estimated crystallite sizes, showing quasi-spherical particles in clusters. Dielectric studies indicated relaxor-like behavior and low dielectric loss in the high-frequency range due to a dipole rotation mechanism. The impedance study pointed to a compound with NTCR and a single relaxation phenomenon linked to increased conductivity at higher temperatures. Thermoelectric behavior demonstrated positive S-values, indicating p-type semiconducting behavior, with a gradual decrease in S as temperature increased. The PF showed promising trends, and the ZT increased with Pr doping, making these ceramics potential candidates for various applications, particularly in thermoelectric devices.

References

1. Masta, N., Triyono, D., Laysandra, H.: AIP Conference Proceedings, vol. 1862, p. 030036 (2017)
2. Vasala, S., Karppinen, M.: Prog. Solid State Chem. **43**, 1–36 (2015)
3. Lekshmi, P.N., Pillai, S.S., Suresh, K.G., Santhosh, P.N., Varma, M.R.: J. Alloys Compd. **522**, 90–95 (2012)
4. Shein, R., Kozhevnikov, V.L., Ivanovskii, A.L.: J. Phys. Chem. Solids **67**, 1436–1439 (2006)
5. Amira Bougoffa, E.M., et al.: RSC Adv. **12**, 6907–6917 (2022)
6. Kolat, V.S., Gencer, H., Gunes, M., Atalay, S.: Mater. Sci. Eng. B **140**, 217 (2007)
7. Mahapatra, A.S., Mitra, A., Mallick, A., Shaw, A., Greneche, J.M., Chakrabarti, P.K.: J. Alloys Compd. **743**, 274–282 (2018)
8. Jucá, R.F., et al.: Vacuum **202**, 111140 (2022)
9. Yao, L., Pei, Z., Heng, W., Leng, K., Xia, W., Zhu, X.: J. Mater. Sci. **55**, 4179–4192 (2020)
10. Mahato, D.K., Dutta, A., Sinha, T.P.: J. Mater. Sci. **45**, 6757 (2010)
11. Das, R., Choudhary, R.N.P.: J. Adv. Ceram. **8**, 174 (2019)
12. Wang, H.C., et al.: Mater. Res. Bull. **45**(7), 809–812 (2010)
13. Hona, R.K.: Electronic Theses and Dissertations. Paper 3336 (2019)
14. Yu, H.-C., Fung, K.-Z.: J. Mater. Res. **19**(3), 943–949 (2004)

Investigation of Solidification Behavior and Processing Defects in the Continuous Cast Al-Based Composite Sheet

Dheeraj Kumar Saini and Pradeep Kumar Jha[✉]

Process Modelling and Simulation Lab, Department of Mechanical and Industrial Engineering, Indian Institute of Technology Roorkee, Roorkee, India
pradeep.jha@me.iitr.ac.in

Abstract. Owing to high productivity and being highly energy-efficient, continuous casting is among the most widely used processes in manufacturing metallic products like slabs, bools, billet and even sheets. Recently, there has been large attention on manufacturing continuous cast metal matrix composite sheets. However, one of the critical challenges in making these sheets is the formation of porosity in the final product, which significantly affects the product's mechanical properties. In present study, the effect of melt pool height on porosity formation in the continuous casting of composite sheets has been studied using experimental and numerical investigation. Numerical investigation was performed using fluid flow and solidification models to analyze the solidification profile of the melt. The experimental work involved casting of composite sheets under varying melt pool heights (4, 6 and 8 cm) using a vertical continuous casting machine. Microscopic observation was made to analyse the microstructure, distribution of reinforcement particles and shrinkage porosities. Porosity was quantified using the relative density principle. The results showed that increase in the melt pool height led to a significant reduction in porosity formation. The numerical simulations revealed that a broader range of solidification profiles contribute to solidification shrinkage at later stages. Specifically, a melt pool height of 4 cm exhibited a wider range of solid fractions at the roller exit, resulting in a porosity level of approximately 10.9%. For continuous operation at 836 K, however, melt pool heights of 6 cm and 8 cm are preferred over 4 cm, as they lead to reduced porosity levels of 7.7% and 2.7%, respectively.

Nomenclature

ρ	Density
u	Velocity field
P	Static pressure
g	Acceleration due to gravity
θ_l	Liquid fraction
A_{mush}	Mushy zone constant
$T_{solidus}$	Solidus temperature

© The Author(s), under exclusive license to Springer Nature Singapore Pte Ltd. 2024
S. Patra et al. (Eds.): METCENT 2023, *Proceedings of the International Conference on Metallurgical Engineering and Centenary Celebration*, pp. 186–198, 2024.
https://doi.org/10.1007/978-981-99-6863-3_19

k	Kinetic energy
σ_k	Kinetic energy turbulent Prandtl number
P	Static pressure
$C_{\varepsilon 2}$	Turbulent model constant
I_T	Turbulent intensity
l_w	Distance to the closest wall
G	Reciprocal wall distance
C_P	Specific heat capacity
Q_{vd}	Viscous dissipation in fluid
q	Heat flux by conduction
$C_L(T)$	Latent heat distribution
α_m	Mass fraction of liquid phase
T_{pc}	Phase change temperature
s	Solid
Δ	Represent change in variable
μ	Dynamic viscosity
μ_t	Turbulent viscosity
F	Source term
ζ	Constant
T	Absolute local temperature
$T_{liquidus}$	Liquidus temperature
ε	Turbulent kinetic energy dissipation
σ_ϵ	Dissipation energy turbulent Prandtl number
$C_{\varepsilon 1}$	Turbulent model constant
C_μ	Damping function
L_T	Turbulent length scale
$l*$	Dimensionless wall distance
l_{ref}	Reference length scale
ΔT	Temperature difference
k	Kinetic energy
τ	Viscous stress tensor
θ_s	Solid fraction
L	Latent heat of solidification
l	Liquid

1 Introduction

The conventional method of developing a composite sheet involves an energy-intensive process and necessitates secondary metalworking operations [1]. However, continuous production of composite sheets can enhance the ability to manufacture high-quality products in large volumes, all while being cost-effective. This continuous production method eliminates the need for a reheating furnace to convert billets into sheets before hot or cold rolling operations. Continuous casting is widely employed in alloy production, particularly steel, aluminum, and copper [2–4].

Fabricating composite sheets using the continuous casting process is a relatively new area of research that lacks comprehensive information about its processing parameters and potential defects. Saini et al. [5] highlighted that producing Al-Mg$_2$Si composite sheets resulted in macrosegregation and porosity, limiting its suitability for industrial and aerospace applications. Additionally, achieving the uniform particle distribution and dispersion is critical to consider when fabricating composite sheets. A significant challenge in fabricating composite sheets through continuous casting is solidification shrinkage, which leads to the formation of porosity in the cast components. Shrinkage occurs when the feeding material fails to compensate for the volume change during solidification. Saini et al. [5] reported approximately 10% shrinkage porosity in Al-Mg$_2$Si cast sheets, which had a detrimental effect on their mechanical properties. Furthermore, Patil et al. [6] examined the influence of processing parameters, such as superheat temperature, casting speed, and melt pool height, on the successful production of Al-Cu alloy sheets. Overall, the continuous casting process offers advantages for composite sheet production but necessitates further investigation into processing parameters, defects, and particle distribution to optimize the quality and applicability of the resulting sheets.

The present study investigates the effect of melt pool height in the fabrication of Al-Mg$_2$Si composite sheets. A numerical investigation used fluid flow and solidification models to analyze the melts' solidification profile. Experimental studies were also conducted using optimized parameters for various melt pool heights. The microstructure, particle distribution, and shrinkage porosity were observed using an FESEM (Field Emission Scanning Electron Microscope), optical microscope, and the porosity was quantified using the relative density principle.

2 Methodology

The present study primarily comprised of two sections. The first section focused on exploring the impact of melt pool height on the solidification behavior of composite sheets. This investigation used a three-dimensional model that accounted for heat transfer and fluid flow. The second section of the study involved experimental analyses that examined the effects of varying melt pool height. Specifically, this section aimed to understand the behavior of reinforcement particle distribution, porosity, and processing defects within the composite sheet.

2.1 Numerical Methodology

The following assumption has been made for the modeling of the melt pool in the twin-roll continuous casting process [5]:

a. The molten metal is a Newtonian incompressible fluid.
b. The flow of liquid metal is turbulent.
c. Properties of material such as thermal conductivity, specific heat capacity, and density is independent of temperature.
d. Solid fractions vary linearly with temperature.
e. Heat transfer due to radiation is neglected.

2.1.1 Governing Equations

The modified governing equations for the steady-state fluid flow including continuity equation, momentum equations and turbulent flow equations along with energy equations are written in the Appendix.

2.1.2 Process Description

A twin-roll continuous casting setup is used to cast near eutectic Al-Mg$_2$Si composite sheet shown in Fig. 1 [5]. Initially, the melt was poured into the tundish and from the tundish outlet, it came in between the roller and formed a melt pool. The rolls drag the composite sheet from the melt pool as it solidifies. The casting parameters used to cast composite sheets are listed in Table 1.

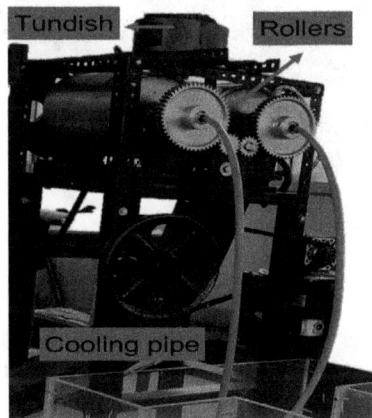

Fig. 1. Twin roll continuous casting setup.

Table 1. Processing parameters used for the fabrication of composite sheet.

Parameter	Value
Melt pool height	4 cm I 6 cm I 8 cm
Sheet thickness	3 mm
Melt inlet temperature	836 K–876 K
Casting speed	0.072 m/s
Sheet width	10 cm

2.1.3 Computational Domain

Metal comes out from the tundish and forms a melt pool between the rolls. After pertaining a specific melt pool height, the flow of molten metal is assumed to be steady. The discretized computation domain used for the simulation study is shown in Fig. 2. The thermophysical properties used to solve the computational domain are tabulated in Tables 2 and 3.

Fig. 2. Computational domain

Table 2. Thermophysical properties for Al-15Mg$_2$Si composite.

Property	Value	Properties	Value
Melting temperature	831.3 K	Melt inlet temperature	876 K
Temperature transition zone width	90 K	Casting speed	0.072 m/s
Latent heat of solidification	402 kJ/kg	Heat transfer coefficient, roller	1000 W/(m^2.K)
Heat capacity at constant pressure, solid phase	897.6 J/(kg.K)	Heat transfer coefficient, air	15 W/(m^2.K)
Heat capacity at constant pressure, liquid phase	1160 J/(kg.K)	Roller radius	115 mm
Ambient temperature	300 K	Density of solid and liquid	2411.3 kg/m^3
		Thermal conductivity	200.1 W/(m.K)

Table 3. Viscosity variation for Al-15Mg$_2$Si composite [7].

Temperature (K)	848	853	858	863	868	873	878	883
Viscosity (mPa-s)	7.43	4.64	1.55	0.76	0.17	0.077	0.077	0.077

2.1.4 Boundary Condition

The boundary conditions are used to solve the fluid flow and energy equations. The inlet, outlet and wall boundary condition for the numerical study is shown in Fig. 3. The melt inlet velocity was selected based on the roller speed.

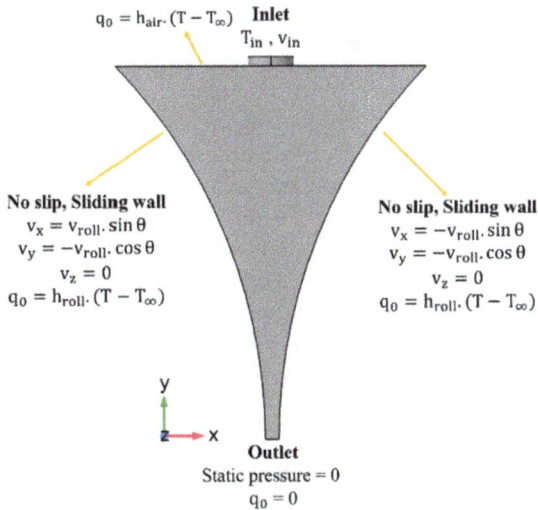

$q_0 = h_{air} \cdot (T - T_\infty)$ **Inlet**
T_{in} , v_{in}

No slip, Sliding wall
$v_x = v_{roll} \cdot \sin\theta$
$v_y = -v_{roll} \cdot \cos\theta$
$v_z = 0$
$q_0 = h_{roll} \cdot (T - T_\infty)$

No slip, Sliding wall
$v_x = -v_{roll} \cdot \sin\theta$
$v_y = -v_{roll} \cdot \cos\theta$
$v_z = 0$
$q_0 = h_{roll} \cdot (T - T_\infty)$

Outlet
Static pressure = 0
$q_0 = 0$

Fig. 3. Boundary condition for melt pool formed between the rollers.

2.2 Experimental Study

2.2.1 Material and Process Description

Commercially pure Al, Mg and Al-12Si were taken in a stoichiometric ratio to maintain the Al-15Mg$_2$Si chemical composition. These master alloys were kept in a muffle furnace held at 650 °C. An extra 20wt.% of the total Mg was added lastly in the melt to avoid oxidation losses. After melting, C$_2$Cl$_6$ was added as a degasser and held for 5 min in the furnace to remove the gaseous porosity from the melt. After cooling the melt near 600 °C with optimum temperature ranges, the melt was poured into the tundish. The composite sheet that comes out from the rollers was further sectioned, and metallography practices were performed.

2.2.2 Material Testing and Characterization

ASTM E8 standard was adopted for the metallography practices. A sequential polishing step was followed on the SiC grit paper. The final polishing was done using colloidal silica. FESEM and Optical microscopy was to capture the reinforcement distribution and porosity inside the material. The porosity was quantified using the relative density measurement. To maintain statistical reliability, the ten-sample dimension of 2 x 2 cm was sectioned from each composite sheet.

$$\% \textbf{ Porosity } = ((\text{Theoretical density} - \text{Experimental density})/\text{Theoretical density}) \times 100$$

3 Result and Discussion

3.1 Effect of Superheat Temperature and Melt Pool Height on Solidification

Increasing the height of the melt pool widens the contact area between the melt and the rolls, resulting in a larger quantity of melt between the rolls. An increase in melt pool height also affects the metallostatic pressure, which impacts the rollers' heat flux and the solidification behaviour of melt. Therefore, studying the melt pool height is an important criterion when performing continuous casting operations [6].

Figure 4 shows the effect of melt pool height and superheat temperature on the solid fraction profile for the Al-Mg$_2$Si composite sheet at the roller exit. Figure 4(a)–(e) shows that as the melt pool height decreased, the solid fraction range at the exit of the sheet became broader. To prevent the late solidification, the temperature profile within the melt pool and the exit of the sheet is required to be uniform. Preventing delayed solidification could reduce the subsequent solidification shrinkage. The inlet temperature between 876 K to 836 K is considered for the study to check the solidification criteria [5]. If the solidified fraction at the exit of the sheet is near 0.7, then the sheet will get enough strength to withstand metallostatic pressure and a successful continuous operation could be possible. In Fig. 4e, at an inlet temperature of 836 K, the solid fraction is near 0.7 for the melt pool height of 6 cm and 8 cm. It shows that a melt pool height of 4 cm still does not have sufficient solid fraction at the exit of the sheet. Hence, a temperature lesser than 836 K is required to ensure the composite sheet fabrication. Figure 4(a)–(e) shows that a solid fraction profile at the roller exit for 4 cm melt pool height varies differently than the other (6 cm and 8 cm). It allows the formation of a solid shell at the extreme sides, and the middle of the melt pool shows a solidified fraction below 0.5 at all the inlet temperatures. A large variation in the solid fraction at the roller exit along the width of the sheet delayed the solidification at the middle portion of the sheet and caused center-line segregation and porosity formation. The range of solid profile variation in the case of 6 and 8-cm melt pool height does not vary significantly and shows a shorter range of solidification variation than the 4-cm melt pool height. Hence, a 6 cm and 8 cm melt pool height are desired for the continuous operation at a melt inlet temperature of 836 K.

Use of 6 cm and 8 cm melt pool heights do not ensure the prevention of solidification shrinkage. It could be possible only if the solidification profile is almost linear. Therefore, by considering the 4, 6 and 8 cm melt pool heights, an experimental study was performed, and the results of solidification shrinkage are reported in terms of porosity in the following section.

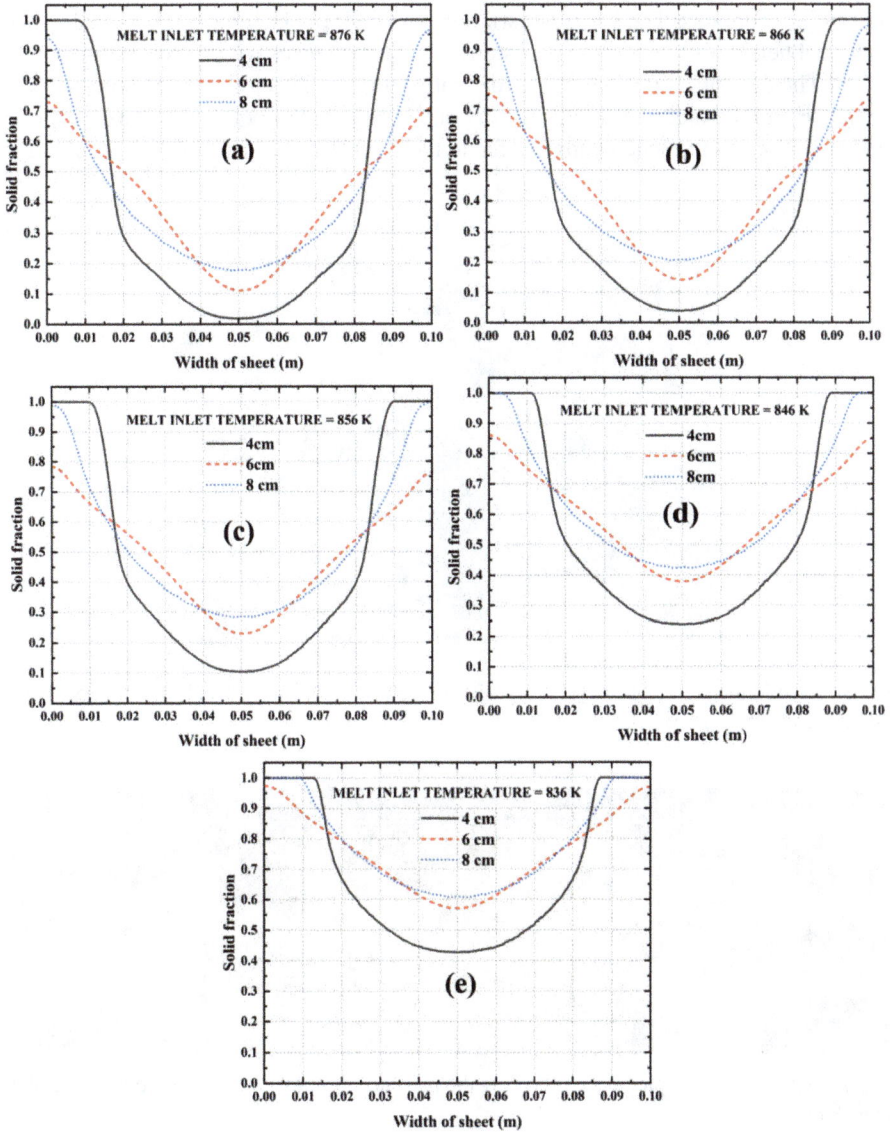

Fig. 4. Variation of solid fraction profile along the width of the sheet with different melt pool heights and inlet temperatures of (**a**) 876 K, (**b**) 866 K, (**c**) 856 K, (**d**) 846 K and (**e**) 836 K.

3.2 Experimental Study

Figure 5 shows the FESEM micrograph of the Al-Mg$_2$Si composite sheet. It consists of α-Al, Primary Mg$_2$Si particles and a eutectic mixture of α-Al and Mg$_2$Si [5]. Microstructure shows the uniform distribution of the primary Mg$_2$Si particles and eutectic mixture. Figure 6 shows the optical micrograph of the Al-Mg$_2$Si composite sheet at a melt pool height of 4 cm, 6 cm, and 8 cm. Microstructure shows a decrease in porosity as the melt pool height increases. The porosity variation with melt pool height is shown in Table 4. Melt pool height also affects the porosity size, which reduces with increasing melt pool height. The dark black portion inside the matrix (Fig. 6) shows shrinkage porosity. The reason behind the increase in the porosity at a shorter melt pool height is explained in the simulation result, which happens due to the extensive variation in the solidification profile.

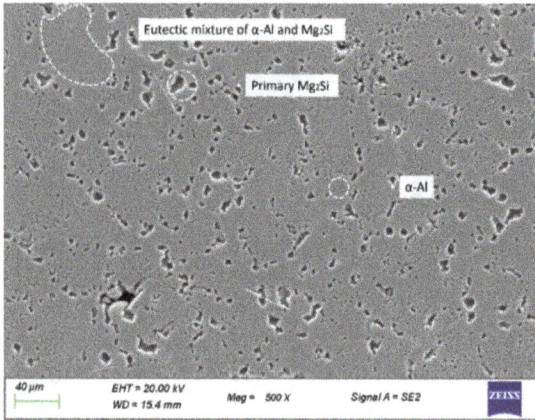

Fig. 5. FESEM images of Al-Mg$_2$Si composite sheet show the presence of alfa Al, primary Mg$_2$Si particles and eutectic mixture.

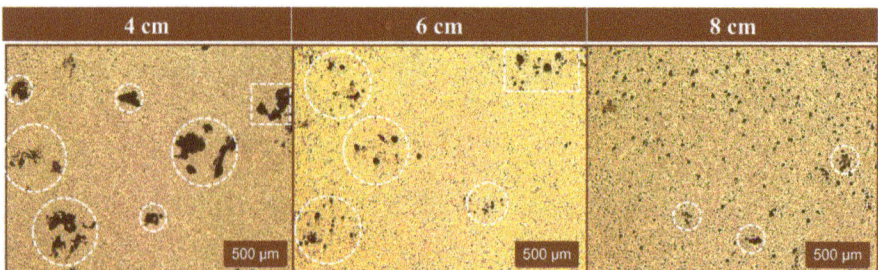

Fig. 6. Optical microscopic images showing the presence of porosity at a melt pool height of 4, 6 and 8 cm.

Table 4. Porosity quantification in an AlMg$_2$Si composite sheet.

Melt pool height	4 cm	6 cm	8 cm
% Porosity	10.93 ± 1.13	7.74 ± 1.76	2.79 ± 2.43

Model Validation

The developed model was validated by Saini et al. [5] in the study of numerical modeling and experimental behavior of Al-Mg$_2$Si composite sheets using continuous casting. The experimental lamellar spacing (λ) of Mg$_2$Si was compared with the numerical results. Equation 1 was used to calculate lamellar spacing. Combining this with the energy equation indicates a relation between lamellar spacing and temperature gradient at the roll surface.

$$\lambda^{1.93}\nu = 5.78 \times 10^{-16}\,\mathrm{m^3/s} \tag{1}$$

$$L \times \nu = h \times (T_m - T_{mi}) \tag{2}$$

$$\lambda^{1.93}\dot{T} = 30.4 \times 10^{-12}\left(\mathrm{m^2 K}\right)/\mathrm{s} \tag{3}$$

Where ν is the solidification front speed, $\dot{T}$ is the average cooling rate along the roll surface, h is the roller heat transfer coefficient, T_m is the melting temperature and T_{mi} is the mold's initial temperature. The temperature difference was calculated using Fig. 7 at the extreme ends of the rollers. Results tabulated in Table 5 show a good agreement with the experimental results.

Table 5. Model validation using lamellar spacing.

Properties	Experimental value	Predicted value	% Error
Lamellar spacing	2.61 ± 1.02 μm	2.86 μm	9.3

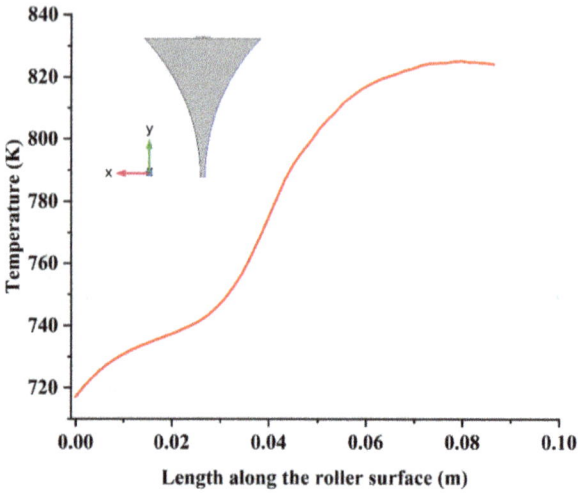

Fig. 7. Temperature variation along the roll surface for the melt pool height of 8 cm and inlet temperature of 836 K.

Conclusion

The present study investigates the effect of melt pool height on porosity formation in the continuous casting of composite sheets using experimental and numerical simulation. Numerical investigation used fluid flow and solidification models to analyze the solidification profile of the melt. The experimental work involved casting samples with varying melt pool heights (4, 6 and 8 cm) using a vertical continuous casting machine. Based on the above study, the following conclusion may be drawn:

(a) Increase in the melt pool height led to a significant reduction in porosity formation. Increasing the melt pool height from 4 cm to 8 cm reduced porosity by four times.
(b) The numerical simulations revealed that a broader range of solidification profiles contribute to solidification shrinkage at later stages. Specifically, a melt pool height of 4 cm exhibited a wider range of solid fractions at the roller exit, resulting in a porosity level of approximately 10.9%.
(c) For continuous operation at 836 K, 6 cm and 8 cm melt pool heights are preferred over 4 cm, reducing porosity levels of 7.7% and 2.7%, respectively.

Appendix

Continuity equation	
$\rho \nabla u = 0$	(A1)
Momentum equation	
$\rho((u.\nabla)u) = \nabla\left(-PI + (\mu + \mu_t)(\nabla u + (\nabla u)^T)\right) + \rho g + F$	(A2)

(*continued*)

(continued)

$F = \frac{(1-\theta_l)^2}{\theta_l^3+\zeta}A_{mush}(u-v_{cast})$	**(A3)**
$\theta_l = \frac{T-T_{solidus}}{T_{liquidus}-T_{solidus}}$	**(A4)**

Turbulent flow equations for low Reynolds k-ε model

$\rho u.\nabla k = \nabla\left[\left(\mu+\frac{\mu_t}{\sigma_k}\right)\nabla k\right]+P_k-\rho\varepsilon$	**(A5)**
$\rho u.\nabla\varepsilon = \nabla\left[\left(\mu+\frac{\mu_t}{\sigma_\epsilon}\right)\nabla\varepsilon\right]+C_{\varepsilon 1}\frac{\varepsilon}{k}P_k-\frac{C_{\varepsilon_2}\rho\varepsilon^2}{k}f_\varepsilon$	**(A6)**
$P_k = \mu_t\left(\nabla u:\left(\nabla u+(\nabla u)^T\right)-\frac{2}{3}(\nabla.u)^2\right)-\frac{2}{3}\rho k\nabla.u$	**(A7)**
$\mu_t = \rho C_\mu f_\mu\frac{k^2}{\varepsilon}$	**(A8)**
$k = \frac{3}{2}(uI_T)^2$ $\varepsilon = C_\mu^{3/4}\frac{k^{3/2}}{L_T}$	**(A9)**
At wall $k=0$ $\varepsilon = \lim_{l_w\to 0}\frac{2\vartheta k}{l_w^2}$	**(A10)**
$f_\mu = \left(1-e^{-l^*/14}\right)^2.\left(1+\frac{5}{R_t^{3/4}}e^{-(R_t/200)^2}\right)$	**(A11)**
$f_\varepsilon = \left(1-e^{-l^*/3.1}\right)^2.\left(1-0.3e^{-(R_t/6.5)^2}\right)$	**(A12)**
$l^* = (\rho u_\varepsilon l_w)/\mu$ $R_t = \rho k^2/(\mu\varepsilon)$ $u_\varepsilon = (\mu\varepsilon/\rho)^{1/4}$ $l_w = \frac{1}{G}-\frac{l_{ref}}{2}$	**(A13)**
$G.\nabla G + \sigma_w G(\nabla.\nabla G) = (1+2\sigma_w)G^4$	**(A14)**

Energy equations

$\rho C_p(u.\nabla T)+\nabla.q = Q_{vd}$	**(A15)**
$q = -k\nabla T$	**(A16)**
$Q_{vd} = \tau.\nabla u$	**(A17)**
$\alpha_p = -\frac{1}{\rho}\frac{\partial\rho}{\partial T}$	**(A18)**

Thermophysical properties variation

$\theta_l = \begin{cases} 0 & if T < T_{pc}-\frac{\Delta T}{2} \\ \frac{T-T_{solidus}}{T_{liquidus}-T_{solidus}} & if T_{pc}-\frac{\Delta T}{2} < T < T_{pc}+\frac{\Delta T}{2} \\ 1 & if T > T_{pc}+\frac{\Delta T}{2} \end{cases}$	**(A19)**
$C_p = \frac{1}{\rho}\left(\theta_l\rho_l C_{p,l}+\theta_s\rho_s C_{p,s}\right)+C_L(T)$	**(A20)**
$\rho = (\theta_l\rho_l+\theta_s\rho_s)$ and $\theta_l+\theta_s = 1$	**(A21)**
$\int_{T_{pc}-\frac{\Delta T}{2}}^{T_{pc}+\frac{\Delta T}{2}} C_L(T)dT = L\int_{T_{pc}-\frac{\Delta T}{2}}^{T_{pc}+\frac{\Delta T}{2}}\frac{d\alpha_m}{dT}dT$	**(A22)**
$\alpha_m = \frac{1}{2}\frac{\theta_s\rho_s-\theta_l\rho_l}{\rho}$	**(A23)**
$\rho_l = \rho_s = \rho$	**(A24)**

References

1. Saini, D.K., Jha, P.K.: Fabrication of aluminum metal matrix composite through continuous casting route: a review and future directions. J. Manuf. Process. **96**, 138–160 (2023). https://doi.org/10.1016/j.jmapro.2023.04.041
2. Dhindaw, B., Singh, S., Mandal, A., Pandey, A.: Modelling and experimental characterization of processing parameters in vertical twin roll casting of aluminium alloy A356. Arch. Foundry Eng. **20**(4), 121–132 (2020). https://doi.org/10.24425/afe.2020.133358
3. Kumar, R., Jha, P.K.: Numerical simulation for EMS induced solidification and inclusion behavior in bloom caster CC mold with bifurcated SEN. J. Manuf. Process. **81**, 396–405 (2022). https://doi.org/10.1016/j.jmapro.2022.06.061
4. Kano, T., Harada, H., Nishida, S., Watari, H.: Production of copper alloy sheet by twin-roll casting. In: 8th Pacific Rim International Congress on Advanced Materials and Processing 2013, PRICM 8, vol. 3, pp. 2587–2591 (2013). https://doi.org/10.1002/9781118792148.ch320
5. Saini, D., Jha, P.K.: Numerical modelling and experimental study of solidification behaviour of Al-Mg2Si composite sheet fabricated using continuous casting route. J. Heat Mass Transf. (2023). https://doi.org/10.1115/1.4062758
6. Patil, Y.G., Shukla, A.K.: Numerical simulation of the effect of process parameters on cooling rate and secondary dendrite arm spacing in high-speed twin roll strip casting of al-15 wt % Cu alloy. J. Heat Transf. **141**(10) (2019). https://doi.org/10.1115/1.4044108
7. Barenji, R.V.: Casting fluidity, viscosity, microstructure and tensile properties of aluminum matrix composites with different Mg 2 Si contents. https://doi.org/10.1007/s12598-017-0923-8

Influence of Sodium on the Microstructure of Laser Rapid Manufactured NiCrSiBC Hardface Alloy Coatings

Akash Singh[1(✉)], T. N. Prasanthi[1], R. Punniyamoorthy[2], S. Ravishankar[2], C. Sudha[1], C. P. Paul[3], V. Srihari[4], and S. Chandramouli[2]

[1] Metallurgy & Materials Group, Indira Gandhi Centre for Atomic Research, Kalpakkam 603102, India
akash@igcar.gov.in
[2] Reactor Design and Technology Group, Indira Gandhi Centre for Atomic Research, Kalpakkam 603102, India
[3] Laser Material Processing Division, Raja Ramanna Centre for Advanced Technology, Indore 452013, India
[4] High Pressure and Synchrotron Physics Division, Bhabha Atomic Research Centre, Mumbai 400085, India

Abstract. To prevent high temperature wear and galling, core structural steels in fast breeder reactors are hardfaced with Ni-based alloys. To overcome cracking and dilution related issues, the hardfaced alloy with a nominal composition of Ni-13.25Cr-4Fe-4Si-2.3B-0.5C (wt. %) was deposited on 304L SS build plate by laser rapid manufacturing (LRM) process. As-deposited specimen had the following precipitates with unique morphology and microchemistry: CrB, Cr_2B, Cr_5B_3, Cr_7C_3 and $M_{23}C_6$, which are responsible for its excellent wear resistance. Microstructure of as-deposited specimen was taken as the reference to infer changes in the microstructure due to sodium exposure. Sodium testing was done under static conditions with argon cover gas at temperature of 823 K for 500, 1000, 2000 and 4000 h. Phase identification in sodium exposed sample was done using XRD and microstructure, morphology and microchemistry of the precipitate phases were analyzed using electron microprobe. Samples in as-deposited, thermally aged and sodium exposed conditions were compared to delineate the influence of sodium from thermal ageing.

Keywords: Hardface alloy coatings · Additive manufacturing · Sodium exposure · Microstructure

1 Introduction

To prevent high temperature wear and galling structural steels are generally hardfaced with Ni, Co and Fe-based alloys. Among them, Ni-based alloys are preferred for nuclear applications. The improved wear behaviour of the hardfacing alloy is due to solid solution

S. Patra et al. (Eds.): METCENT 2023, *Proceedings of the International Conference on Metallurgical Engineering and Centenary Celebration*, pp. 199–207, 2024.
https://doi.org/10.1007/978-981-99-6863-3_20

strengthening of alloying elements and presence of high volume fraction of borides and carbides. Hardness, wear resistance and effective thickness of the hardface deposit are influenced by the extent of dilution of the deposit by the substrate material [1]. Hardfacing of Ni-based alloys is done by welding methods like Plasma Transferred Arc (PTA), Gas Tungsten Arc (GTA), and Shielded Metal Arc (SMA) and laser processes. The work reported by Das et al., [1] on the fabrication of wear resistant Colmonoy 6 (Ni-13.56Cr-4.75Fe-4.25Si-2.5B-0.6C; wt.%) bushes through GTA welding shows about 2.5 mm thick dilution layer with low hardness of 350 VHN, when compared to that of undiluted deposit, which is about 600 VHN. However dilution can be further controlled by the optimization of welding process and processing conditions. NiCrBSi coating achieved through laser cladding had higher hardness and improved wear resistance due to uniform distribution of precipitates [2]. Laser cladding has further advantages such as minimum dilution, low distortion of the work piece and good metallurgical bond between the deposit and the substrate material. Also, the rapid solidification after cladding leads to more refined microstructures with improved mechanical and corrosion properties [3]. High temperature (~623 K) oxidative wear behavior under high contact stress makes NiCrFeBSiC as a candidate hard facing material for nuclear power plant applications. In general, the NiCrFeBSiC hardface coatings consists of (i) γ-Ni solid solution (ii) Cr rich CrB, Cr_5B_3 and Cr_7C_3 type precipitates and (iii) Ni-B-Si binary and ternary eutectics containing Ni_3B, Ni_2B and Ni_3Si [4, 5]. However, the morphology, volume fraction, sequence of formation of these phases differs with the method of deposition and its related parameters. The influence of aging at 923 K on the microstructure, wear and corrosion of a nickel based Colmonoy 5 hardfacing alloy has been investigated by Keshvan et. al and they reported coarsening of γ-nickel dendrites, $Cr_{23}C_6$ precipitates, and the transition of Ni_3Si precipitates from the near spherical to aligned cuboidal morphology [8].

In this study, in order to reduce dilution, Ni-13.25Cr-4Fe-4Si-2.3B-0.5C (in wt.%) hardfacing alloy was deposited on 304L SS build plate through laser rapid manufacturing, a 3D printing method for applications related to liquid sodium cooled fast breeder reactors [6, 7]. To check for sodium compatibility and effect of thermal aging, sodium testing under static conditions with argon cover gas and thermal aging were carried out at temperature of 823 K for up to 4000 h. The aim of the present study was to understand the microstructural and microchemical modification introduced, if any, on the surface of the hardfacing alloy deposit due to long term sodium exposure and thermally aging.

2 Experimental

Laser rapid manufacturing technique was used to deposit Ni-13.25Cr-4Fe-4Si-2.3B-0.5C; wt.%) alloy on 304L SS build plate. Thickness of the deposit was 6–8 mm. For static sodium exposure NiCrFeSiBC/SS 304 specimens with dimensions of 15 mm × 15 mm × 10 mm were put in a SS wire mesh separately and tied with a stainless steel hanging rod which is welded with the top flange of test pot (Fig. 1(a)). The pot is provided with surface heater, thermocouple and insulation. The height of the pot is 1000 mm and it is filled with pure sodium up to a height of 500 mm with a cover gas. The specimens were kept inside the test pot and pot temperature was kept at 393K. Pressure hold test of the test pot was carried out for 24 h and then the temperature was gradually increased up to

823 K. The specimens were taken out after 500 h, 1000 h, 2000 h and 4000 h respectively. Figure 1(a) shows the experimental set up where samples after static sodium experiments can be seen. Each sample was cleaned using a cotton cloth. Figure 1(b) shows a cleaned sample after sodium exposure. For long term thermal aging experiments samples were kept in a box type furnace having a temperature accuracy of ± 2 °C and aged up to 4000 h. Samples were taken out after 500 h, 1000 h, 2000 h and 4000 h respectively and air cooled. Visual examination of the samples after Na exposure and long term aging experiments showed that there is no damage to the sample surface. To study the effect of sodium, specimens were extracted from the cross section which were ground and polished as per standard metallographic procedures. To reveal the microstructure the specimen was etched with Murakami reagent (10 g of $K_3Fe(CN_6)$ and KOH each in 100 ml distilled H_2O). X-ray diffraction (XRD) patterns were recorded in a XRG-3000 diffractometer (INEL, France) fitted with a curved position sensitive detector at a glancing angle of incidence (ω) of 6° using Co $K\alpha_1$ ($\lambda \sim 1.7889$ Å) radiation. Data was recorded in the 2θ range of 10 to 100° with a step size of 0.023°. Back Scattered Electron (BSE) imaging and microchemical analysis was carried out using SX-FIVE EPMA operating at an accelerating voltage and current of 20 kV and 20 nA respectively. For diffracting $K\alpha$ X-rays the crystals used were LiF for Fe, Ni and Cr, TAP for Si and PC2 for B and C. For the set acquisition parameters the X-ray generation volume will be ~ 2–3 μm^3. For accurate chemical composition analysis ZAF correction procedure was adopted [9].

3   Results and Discussion

Figure 2 shows a comparison of the XRD patterns obtained from as-deposited, thermally aged and sodium exposed specimens. XRD analysis of aged sample showed the presence of CrB, Cr_2B, Cr_2B_3, Cr_5B_3, Cr_7C_3 and $Cr_{23}C_6$ phases along with Ni matrix in the deposit. Whereas, strong peak corresponding to Cr_5B_3, Cr_7C_3, $Cr_{23}C_6$ phases were observed in the as deposited and sodium exposed samples. The intensity ratio of Ni in as deposited and aged sample was also found to be lower compare to sodium exposed sample. Annealed texture could be a reason for the change in the intensity after 4000 h of aging and sodium exposure. The presence of Ni_3B phase was noticed at 2θ values of 34.9, 43 and 67.2° respectively for aged sample only.

3.1   Effect of Thermal Aging on Microstructure

Figure 3(a) and (b) shows the BSE image of the microstructure of as deposited and aged NiCrFeBSiC deposit respectively. Different phases having different morphologies were observed in the deposit and it is labeled as 'A', 'B', 'C' and 'D'. Based on the morphology and contrast difference, these labeled phases were designated as follows and marked in Figure: Dark chunky – A; Grey coloured flower - B; Grey colored butterfly irregular shaped – C; Dark fine Irregular shaped– D. The size and shape of the phases and precipitates depends on the processing parameters of LRM such as deposition rates, cooling rates, temperature and the aspects of phase formations from liquid/solid [10]. However, the high amount of Cr, Fe, B, and C of Deloro 50® deposit during solidification preferably gives rise to boride and carbide precipitation, which are distinguished

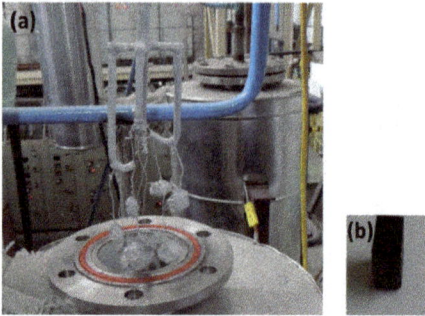

Fig. 1. (a) Experimental set up; (b) Sample after sodium exposure

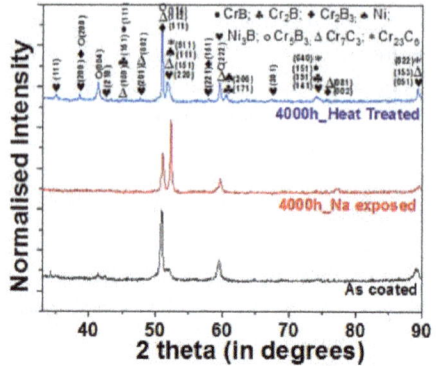

Fig. 2. XRD patterns of as deposited, 4000 h Na exposed and thermally aged samples

by dark and bright contrast in BSE imaging. Based on the C, B and Cr content and the temperatures of the phase reactions involved during LRM process, the first precipitates should be chromium carbides which formed with nearly globular/bulky type of morphology. In laser cladding, in general orthorhombic M_7C_3 type of carbides exhibits dendritic/irregular shaped morphology as observed in the present study. At increasing Cr/C ratios it is reported that the formation of fcc structured $M_{23}C_6$ is easier than M_7C_3. Although, it is not so clear on the shape possessed by $M_{23}C_6$, the distinct flower type morphology of this phase can be attributed to the $M_{23}C_6$ due to the orientation relationship with the γ-Ni matrix. BSE image and X-ray elemental mapping was also taken from the 4000 h thermally aged NiCrFeBSiC deposit and it is shown in Fig. 4(a) and Fig. 4(b–g) respectively. From the elemental maps of Ni, Cr, Fe, Si, B, and C it is evident that A and D phases are rich in boron while B and C phases (marked in Fig. 3(b)) are rich in carbon. However Cr enrichment indicated that all precipitates (A, B, C and D) are either Cr rich borides or carbides. In addition to this it was also observed that the matrix phase is predominately rich in Fe and Ni. Preferential Si enrichment was also noticed in the matrix phase. Table 1 shows composition of CrB, $Cr_{23}C_6$, Cr_7C_3 and Cr_2B phases present in the deposit after long term aging of 4000 h. Similar phases have been earlier reported in as deposited and aged sample [5, 8].

3.2 Effect of Long Term Static Sodium Exposure on Microstructure

Figure 5(a) and (b) shows the BSE image of 4000 h Na exposed NiCrFeBSiC deposit near the surface and far from surface region respectively. Figure 5(b) shows the presence of carbide and borides phases in the deposit with unique morphologies. In addition to this, surface modified layer of about 94 μm was also observed near the surface after long term sodium exposure (Fig. 5(a)). This Surface modified layer shows higher volume fraction of M_7C_3 type precipitate. It was also noticed that the modified layer is not uniform and ranging from 23 to 94 μm.

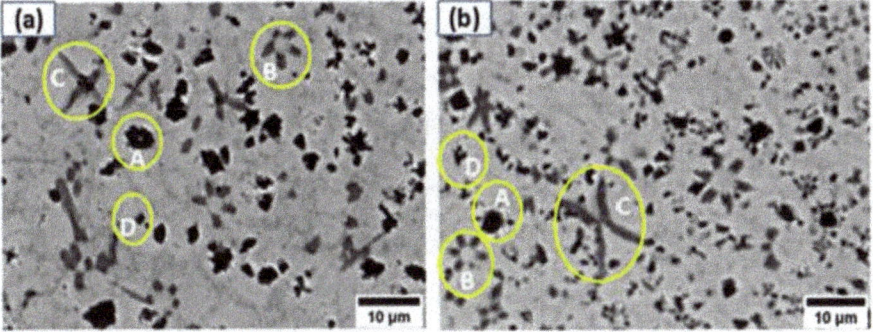

Fig. 3. BSE image of (a) as deposited and (b) 4000 h aged NiCrFeBSiC deposit

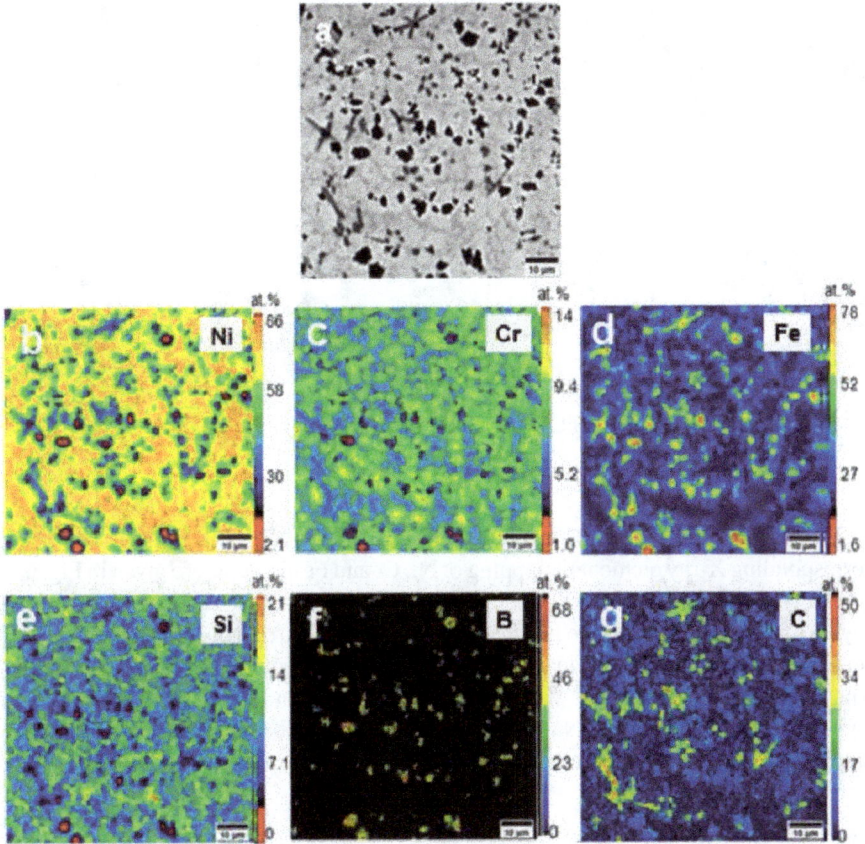

Fig. 4. (a) BSE image obtained from aged NiCrFeBSiC deposit, corresponding X-ray elemental maps for (b) Ni, (c) Cr, (d) Fe, (e) Si, (f) B and (g) C

Table 1. EDS analysis of phases present in 4000 h aged NiCrFeBSiC deposit (at.%)

Labeled phase	at.%						Assigned Phase
	B	C	Fe	Cr	Ni	Si	
A	25.3	4.2	2.7	55.4	11.7	0.7	CrB
B	3.2	20.5	11.05	53.52	10.42	1.45	$M_{23}C_6$
C	2.8	15.73	8.16	59.27	15.55	0.41	M_7C_3
D	7.4	4.2	2.4	4.7	75.7	5.5	Cr_2B and Matrix phase

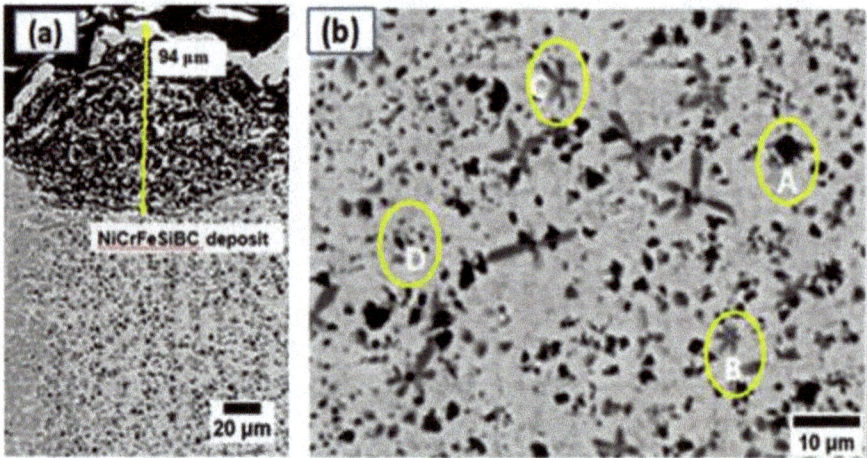

Fig. 5. BSE image of NiCrFeBSiC deposit after the 4000 h static Na exposure (a) near surface; (b) far from surface

Figure 6(a) shows the BSE image of sodium exposed NiCrFeBSiC deposit surface. Corresponding X-ray elemental mapping of Ni, Cr and Fe elements are given in Fig. 6(b), (c) and (d) respectively. The X-ray mapping of Ni shows the leaching of Ni into Na (Fig. 6(b)). To confirm this line profile analysis was also carried out and shown in Fig. 7. This clearly indicates the Ni leaching near the surface region after long term sodium exposure. Table 2 shows the at. % of CrB, $Cr_{23}C_6$, Cr_7C_3 and Cr_2B phases present in the deposit after 4000 h of static Na exposure. Based on the observations listed in Tables 1 and 2, it is found that the volume fraction of the fine precipitates was lower in sodium exposed samples in comparison with the aged samples.

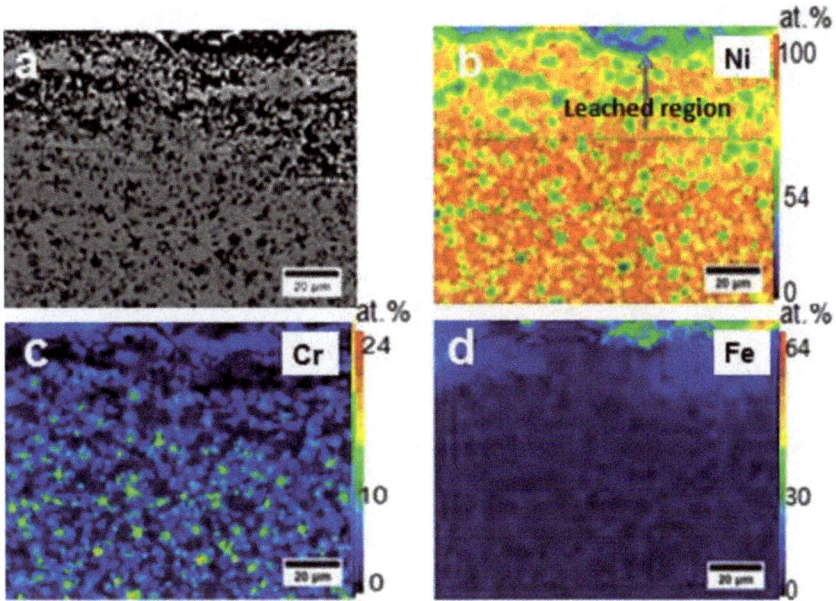

Fig. 6. (a) BSE image obtained from Na exposed NiCrFeBSiC deposit, corresponding X-ray elemental maps for (b) Ni, (c) Cr and (d) Fe

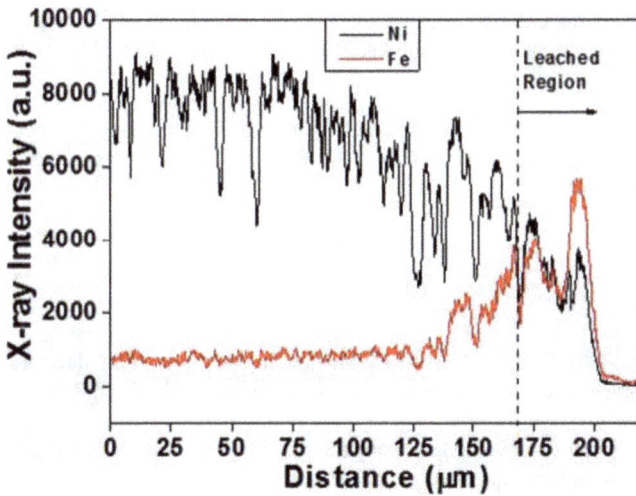

Fig. 7. Line profile analysis across the interface region

Table 2. EDS analysis of phases present in 4000 h Na exposed NiCrFeBSiC deposit (at.%)

Labeled phase	atm.%						Assigned Phase
	B	C	Fe	Cr	Ni	Si	
A	24	3.8	3.0	56	12.4	0.8	CrB
B	4	21	13.1	52	8.9	1	$M_{23}C_6$
C	3.2	15.8	10.2	54.6	15.8	0.4	M_7C_3
D	5.1	2.4	3.0	3.7	84.6	1.2	Cr_2B and Matrix phase

4 Conclusion

Hardfaced alloy with a nominal composition of Ni-13.25Cr-4Fe-4Si-2.3B-0.5C (wt. %) was deposited on 304L SS build plate by laser rapid manufacturing. NiCrFeBSiC deposit samples were subjected to long term aging and static sodium exposure at 823 K and up to 4000 h.

- Long term aging experiment of 4000 h showed the presence of Ni_3B phase along with chromium borides and carbides. Change in the intensity ratio was also noticed in comparison with as deposited samples, which could be due to annealing texture during long term aging.
- In addition to borides and carbides precipitates 4000 h sodium exposed samples showed the presence of non uniform modified layer close to surface which was limited to about 94 μm. Line profile analysis of modified layer shows Ni leaching in Na.

Acknowledgement. The authors would like to express their sincere thanks to Director IGCAR and Director, Metallurgy and Materials Group, IGCAR for their encouragement and support during the course of this work.

References

1. Das, C.R., Albert, S.K., Bhaduri, A.K., Kempulraj, G.: Technology development for hard-facing of large components using Ni base hardfacing alloys. J. Mater. Process. Technol. **141**, 60–66 (2003)
2. Gómez-del Río, T., Garrido, M.A., Fernández, J.E., Cadenzas, M., Rodríguez, J.: Influence of the deposition techniques on the mechanical properties and microstructure of NiCrBSi coatings. J. Mater. Process. Technol. **204**, 304–312 (2008)
3. Ming, Q., Lim, L.C., Chen, Z.D.: Microstructures of laser-clad nickel-based hardfacing alloys. Surf. Coat. Technol. **106**, 182–192 (1998)
4. Hemmati, I., Ocelík, V., De Hosson, J.: Effects of the alloy composition on phase constitution and properties of laser deposited Ni-Cr-B-Si coatings. Phys. Procedia **41**, 302–311 (2013)
5. Sudha, C., Shankar, P., Subba Rao, R.V., Thirumurugesan, R., Vijayalakshmi, M., Raj, B.: Microchemical and microstructural studies in a PTA weld overlay of Ni–Cr–Si–B alloy on AISI 304L stainless steel. Surf. Coat. Technol. **202**, 2103–2112 (2008)

6. Bhaduri, A.K., Indira, R., Albert, S.K., Rao, B.P.S., Jain, S.C., Asokkumar, S.: Selection of hardfacing material for components of the Indian Prototype Fast Breeder Reactor. J. Nucl. Mater. **334**, 109–114 (2004)
7. Zhang, H., Shi, Y., Kutsuna, M., Xu, G.J.: Laser cladding of colmonoy 6 powder on AISI316L austenitic stainless steel. Nucl. Eng. Des. **240**, 2691–2696 (2010)
8. Kesavan, D., Kamaraj, M.: Influence of aging treatment on microstructure, wear and corrosion behavior of a nickel base hardfaced coating. Wear **272**, 7–17 (2011)
9. Scott, V.D., Love, G., Reed, S.J.B.: Quantitative Electron-probe Microanalysis. Ellis Horwood, Chichester (1995)
10. Li, Q., Zhang, D., Lei, T., Chen, C., Chen, W.: Comparison of laser clad and furnace melted Ni based alloy microstructures. Surf. Coat. Technol. **137**, 122–135 (2001)

Study of Geometry Modulated Magnetoelectric Composite Structure

S. Sai Harsha[1], P. Kondaiah[1(✉)], and K. Deepak[2]

[1] Department of Mechanical Engineering, Mahindra University, Hyderabad 500043, Telangana, India
{harsha21pmee005,Kondaiah.p}@mahindrauniversity.edu.in
[2] Department of Metallurgical Engineering, Indian Institute of Technology (BHU), Varanasi 221005, India
Deepak.met@iitbhu.ac.in

Abstract. The promising potential of Magneto-Electric (ME) composites in sensing and energy harvesting applications has sparked significant interest in both experimental and theoretical characterization because of its multiferroic nature. This work examines the position of the piezoelectric layer in an unsymmetrical tri-layer laminate ME composite. A coupling of three-dimensional structure of ME composite is modelled and investigated using finite element method. The proposed model incorporates the nonlinearity of magnetoelectric coupling and magnetostrictive material. The study investigates the characterisation of non-uniform magnetization distribution, strain, and magnetoelectric coefficient of the ME composite material. Additionally, a parametric study also performed with the aim of high ME coefficient by optimizing the length of the piezoelectric layer with respect to applied DC bias field. The results shows that the ME coefficient is increased when the piezoelectric layer is positioned in the middle of the tri-layer composite structure, rather than the spanning its entire length, leading to a noticeable reduction in material usage.

Keywords: Magneto-Electric composite · Unsymmetric structure · Piezoelectric · Magnetostrictive · Finite Element Method

1 Introduction

Multiferroic materials or multi-functional materials are ever increasing attention over the last few years due to combining two or more ferroic orders in one system and the number of potential multifunctional applications. Magnetoelectric materials are a class of multiferroic materials that usually have two or more ferroic orders, including ferromagnetic, ferroelectric, and ferro-elastic in the same phase and exhibits linear and nonlinear coupling. The magnetoelectric coupling is a coupling of piezoelectric material and magnetostrictive materials, this coupling enables the combination of mechanical, electric, and magnetic phases in the host composite which is not possible with piezoelectric and magnetostrictive material individual phases. In direct magnetoelectric coupling, change the

S. Patra et al. (Eds.): METCENT 2023, *Proceedings of the International Conference on Metallurgical Engineering and Centenary Celebration*, pp. 208–223, 2024.
https://doi.org/10.1007/978-981-99-6863-3_21

polarization in piezoelectric phase by applying magnetic field in magnetostrictive phase [1, 2]. Conversely, changing of magnetization in magnetostrictive phase by applying electric field in piezoelectric phase this phenomenon is called as converse magnetoelectric effect [3]. In order to combine these piezoelectric and magnetostrictive phases there are different connectivities for ME composite i.e., particulate (0–3) [4–6], cylindrical matrix type (1–3) [7], laminated or layered type (2–2) [2, 8], and fibre type (1–1) [9]. The various configurations available for composites each possess their own set of advantages, disadvantages, and applications. However, when considering the magneto-electric (ME) coefficient, it is evident that layered composites exhibit a significant advantage over other configurations. This can be attributed to their excellent mechanical coupling and efficient stress transfer mechanisms [26], less current leakages [cylindrical layer] resulting in a remarkably high ME coefficient. By using these magnetoelectric ferroic orders, the ME devices are used in antennas [10–12], current sensors [13–15], field sensors [16, 17], energy harvesting [18–21], gyrators [22], resonators [23], and memory devices [24, 25] applications.

In the instant of direct magnetoelectric effect (DME) and converse magnetoelectric effect (CME) response in any configuration is characterized by the magnetoelectric (ME) coefficient (α_{ME}), which defines the capability of output voltage with respect to the applied magnetic field externally [5]. Up to now, there are several investigations are proposed to improve the ME coefficient such as varying the materials [8, 27], changing the size of the geometry [28, 31] connection methods [29], boundary conditions [30] etc. Upon careful examination and comparison of these aspects, it becomes apparent that the structural variation plays a pivotal role in enhancing the magneto-electric (ME) coefficient of composites. The ME coefficient can be significantly increased by strategically modifying the structure of the composite material.

Dong et al., 2005 introduced a quasi-ring type magnetoelectric laminated composite which can work in C-C mode and obtained a high ME coefficient $5.5V/cm.Oe$. Then, Leung et al., 2010 proposed and developed a ring shaped laminated magnetoelectric current sensor for vortex magnetic field detection. In this, piezoelectric ring is sandwiched between two specially designed magnetostrictive rings and the theoretical results shows that ME voltage coefficient is 32.16 mV/Oe at non resonant frequency and 201.2 mV/Oe at 67 kHz. Song et al., 2016 developed a hallow multi-layered ME composite in order to decrease the weight and cost effective such as P-T-P, T-P, T-P-T-P structures. Results shown that P-T-P structure deliver the high ME coefficient $4.8V/cm.Oe$ at resonant frequency $39.1kHz$ compared to other structures because this structure has optimal volume fraction and effective working surface area (A_{eff}) is large.

Zhang et al., 2020 developed self-biased magnetoelectric composite with the help of high magnetic conductivity material using finite element method. The self-biased ME coefficient is observed $9mV/Oe$ accounting for 47% of the maximum ME coefficient. Similarly, Song et al., 2022 developed and modelled a cylindrical layered structures as P/T and T/P ME composite. The results indicated that P/T structure has high ME coefficient and less optimum bias field compared to T/P structure. Ren et al., 2022 numerically developed magnetoelectric composite for energy harvesting application for cantilever structure with full length beam of magnetostrictive layer and half-length is the piezoelectric layer and in the comparison of Galfenol and Ni materials, Galfenol composite has delivered the high output in both ac and dc magnetic fields. Magnetoelectric composite carries high output density at the resonant frequency, in this scenario, Gao *et al.,* proposed numerically and experimentally, an advanced hybrid energy harvester which has the capability of capturing both energies in terms of mechanical vibrations and magnetic fields. This magnetoelectric composite has three piezoelectric plates on the surface of magnetostrictive cantilever beam. Wen *et al.,* 2017 developed numerical approach for a 3D finite element model of magnetoelectric tri-layer laminate structure for dynamic response, and observed that shear stress and transverse normal stress is large when the resonance mode occurred. Sudersan *et al.,* 2018 developed a finite element formulation for the analysis of unsymmetric magnetoelectric laminated structures by incorporating of nonlinear constitutive relations for static and resonating conditions. The results demonstrated the importance of considering the coupling stiffness and simultaneous axial and bending deformations in the structure when subjected to an applied magnetic field.

In retrospect, the variation of structural geometry in multi-layered ME composite is playing a crucial role to increase the ME coefficient. Motivated by this observation, this paper will focus on modifying the geometry for ME composite by changing the length of the piezoelectric layer. This approach involves taking into account the multi-field coupling characteristics of the magneto-electric (ME) composite and integrate the nonlinear behaviour of the constituent materials such as ferromagnetic and piezoelectric materials to establish a comprehensive theoretical model. To solve these theoretical model equations, finite element method based COMSOL is used because, the Finite Element Method (FEM) offers adaptability to meet specific accuracy requirements, thereby reducing the need for physical prototypes. It provides highly accurate solutions with a high degree of precision.

2   Working Principle and Modelling

2.1   Working Principle

The principle for layered ME composite is strain mediated mechanism illustrates the in Fig. 1. The ME composite component in this study consists of three layers: two magnetostrictive layers polarized magnetically along the length direction, and a piezoelectric layer polarized electrically along the thickness direction. When a magnetic field is applied to the ME component, the molecular dipoles and magnetic field boundaries in the magnetostrictive layers align with the field, resulting in magnetostrictive strain. This strain is then transmitted to the piezoelectric layer through the coupling interface, generating electrical charges through the piezoelectric effect. In this way, magnetic field energy is converted to electrical energy.

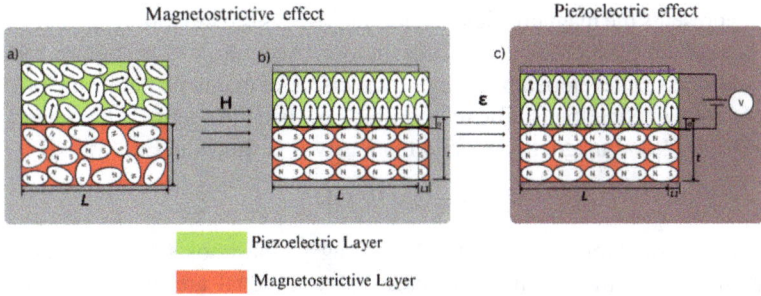

Fig. 1. Strain Mediated Mechanism in layered Magnetoelectric composite. (a) Random arrangements of domains before applying magnetic field, (b) turn over the domains and strain developing in magnetostrictive layer under the external magnetic field (H), (c) output voltage (V) is developed in piezoelectric layer under the magnetostrictive strain (ε)

2.2 Modelling

The present work investigates an unsymmetrical tri-layer magnetoelectric structure consisting of Terfenol-D and PZT-5A. It is assumed that the piezoelectric and magnetostrictive layers are ideally bonded together. The dynamic equilibrium equation of motion for the magnetoelectric layered cantilever beam structure is shown in Eq. (1) [27]

$$^i\rho \frac{\partial^{2i}u}{\partial t^2} - \nabla^i\sigma + F_v = 0, \ (i = m, p) \tag{1}$$

where the superscript i represents the material variation like $i = m$ for magnetostrictive material and $i = p$ for piezoelectric material. $^i\rho$ represents the density of the material, ^iu is the mechanical displacement vector $(^iu = [\ ^iu\ ^iv\ ^iw]^T)$ $^i\sigma$ is the stress tensor, F_v is the external force applied. In this study the laminate ME structure is affected by magnetic field without any external force i.e., $F_v = 0$.

By applying external magnetic field in laminate ME composite, the magnetostrictive material shows the nonlinear response of magnetization and magnetostrictive strain. The nonlinear magnetization of magnetostrictive material is expressed in Eq. (2) [27, 31, 37]

$$M = M_s L(|H_{eff}|) \frac{H_{eff}}{|H_{eff}|} \tag{2}$$

Here M_s is the saturation magnetization with units of $\frac{A}{m}$, $L(|H_{eff}|)$ represents the argument of Langevin function L is the norm of H_{eff}. The Langevin function is expressed as χ in Eq. (3) [27, 31]

$$L = \coth\left(\frac{3\chi_0|H_{eff}|}{M_s}\right) - \frac{M_s}{3\chi_0|H_{eff}|} \tag{3}$$

Here χ_0 is the initial magnetic susceptibility which is the degree of magnetization of the material in response to the externally applied magnetic field and it is dimensionless. And H_{eff} is the effective magnetic field in the material expressed in Eq. (4) [27]

$$H_{eff} = H + \frac{3\lambda_s}{\mu_0 M_s^2}\sigma_{dev}M \tag{4}$$

where H represents the magnetic field intensity ($\frac{A}{m}$), λ_s is the saturation magnetostriction (ppm). σ_{dev} is the elastic deviatoric stress tensor. According to the villari effect, the mechanical stress is developed by applying the effective magnetic field which indicated the second term in equation (). The relationship between the magnetic field intensity H, magnetization M and magnetic induction intensity B is shown in Eq. (5) [37]

$$B = \mu_0(H + M) \tag{5}$$

Here μ_0 is the vacuum permeability. According to the hook's law, the relationship between stress and strain in the magnetostrictive material are mentioned in Eq. (6) [31]

$$\sigma^m = C_E^m[\varepsilon_{el} - \varepsilon_{me}(M)] \tag{6}$$

where C_E^m is the stiffness matrix, in the present model assuming that the magnetostrictive material is isotropic, the stiffness matrix can be represented by the parameters young's modulus E and poisson's ratio ϑ. The strain divided into the elastic strain tensor ε_{el}, and magnetostrictive strain tensor means the strain caused by magnetostrictive effect and it is the function of magnetization M which can be expressed as [27, 31, 37]

$$\varepsilon_{me} = \frac{3}{2} \frac{\lambda_s}{M_s^2} dev(M \otimes M), \tag{7}$$

In the above equation, $dev(M \otimes M)$ is the deviatoric tensor of $(M \otimes M)$, in which the tensor product of two vectors is defined as [31]

$$(M \otimes M)_{ij} = M_i M_j \tag{8}$$

In the laminate magnetoelectric composite, the magnetostrictive strain is developed because of external magnetic field. This strain will transfer to the piezoelectric material which is limited to the linear elastic range. The constitutive relations for the piezoelectric material in the form of strain-charge mode are mentioned in Eqs. (9) and (10) [1]

$$\varepsilon^p = S^p \sigma^p + d^T E \tag{9}$$

$$D = d\sigma^p + k^p E \tag{10}$$

where ε^p, σ^p are the stress and strain of the piezoelectric material, S^p, d and k^p represents the stiffness matrix, piezoelectric coefficient matrix and dielectric constant matrix for piezoelectric material, E and D are the electric field and electric displacement respectively. In the same way, as the material deformation remains within the linear elastic range, it is possible to express the relationship between displacement and strain of the magneto-electric (ME) component as [31]

$$\varepsilon = \frac{1}{2}\left[\left(\nabla^i u\right)^T + \nabla^i u\right] \qquad (i = m, p) \tag{11}$$

In addition, according to the gauss law, the governing equations for electric displacement D and electric field E are shown as [31]

$$\nabla.D = \rho_v \tag{12}$$

$$E = -\nabla V \tag{13}$$

where ρ_v is the special charge density, V is the electric potential and ∇ is the gradient operator.

Under the external magnetic field, the ME structure is unconstrained in mechanical and electrical fields. So, the mechanical boundary condition is that one end is fixed for only two magnetostrictive layers and other end is free, the electric boundary conditions for piezoelectric material is open circuit boundary condition that means the upper surface of piezoelectric layer is connected to charge and lower surface is grounded.

$$u^p|_{x=0} = 0 \tag{14}$$

$$\sigma^i|_{x=l} = 0 \qquad (i = p, m) \tag{15}$$

$$D_z = 0 \tag{16}$$

Then the magnetoelectric coupling coefficient is calculated by using the relation as follows [27, 30]

$$\alpha_{ME} = \frac{\partial V}{\partial H} \tag{17}$$

In the form of electric field, the ME coefficient is defined as [30]

$$\alpha_{ME} = \frac{E_{avg}}{H_0} = \frac{V}{H_0 \times t_p} \tag{18}$$

Here E_{avg} is the normalized electric field ($\frac{V}{m}$), H_0 indicates the amplitude of the magnetic field, and t_p is the thickness of piezoelectric layer thickness. The procedure for solving the ME coefficient by implementing three different physics interfaces using above equations is shown in the Fig. 2.

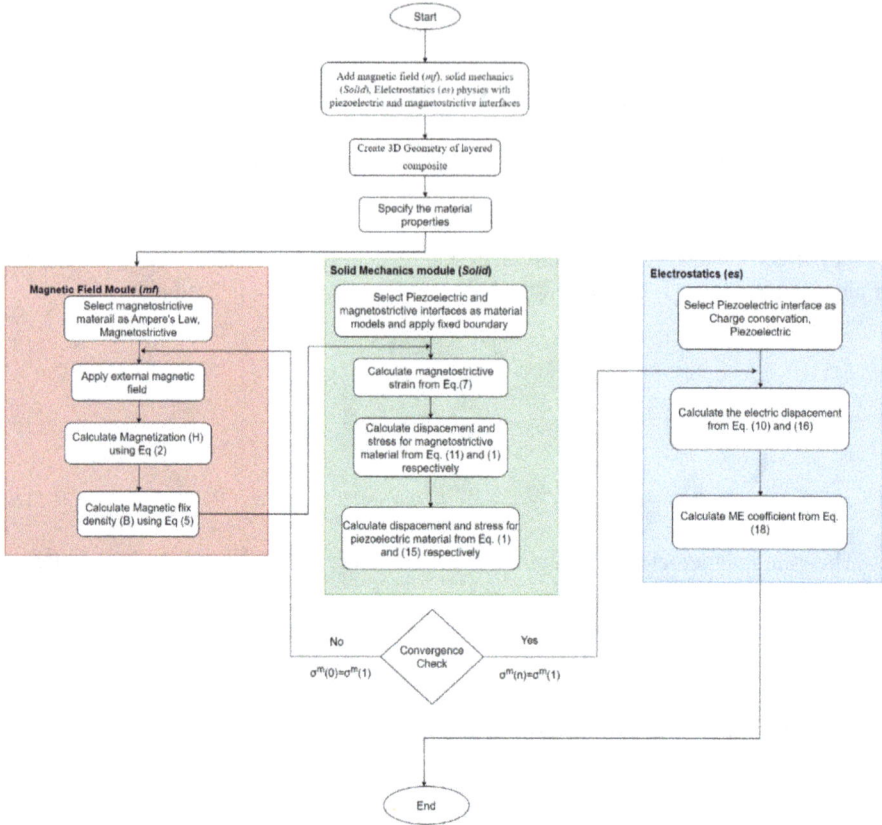

Fig. 2. Procedure for implementing the different physics interfaces in COMSOL Multiphysics

3 Finite Element Simulation Procedure

In the present study, the ME properties of unsymmetrical magnetoelectric composite structure is calculated using finite element software COMSOL Multiphysics v6.0 the tri-layer ME composite has two different thickness of magnetostrictive layers and sandwiched piezoelectric layer by varying its length from the midpoint as shown in the Fig. 3. And this structure is unsymmetric about x-axis. The materials of the ME composite are Terfenol-D as magnetostrictive layer and PZT-5A as the piezoelectric layer. The material properties and the dimensions of these two materials are shown in Table 1. The ME structure is surrounded by air domain to simulate the structure in real working conditions and facilitate the application of external magnetic field. The results are accurate as big as the air domain but at the same time the computation time also will increase, the spherical air domain is used in the present study with radius of 40mm as shown in the Fig. 4.

To simulate the magnetoelectric coupling in COMSOL Multiphysics, magnetic field module (mf), solid mechanics (solid), and electrostatic (es) module. The procedure for calculating the ME composite is shown in the Fig. 3. The implementation of these modules in COMSOL as following;

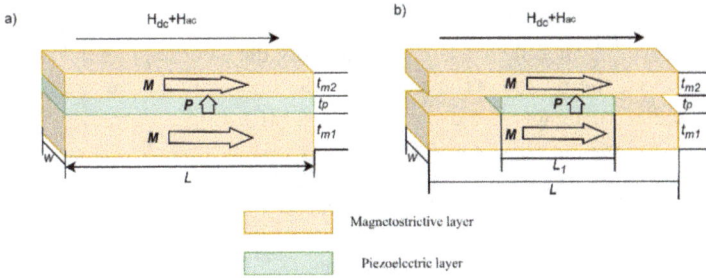

Fig. 3. Illustration of unsymmetric tri-layer ME composite, (a) ME structure with full length piezoelectric layer, (b) ME structure with varying of piezoelectric layer

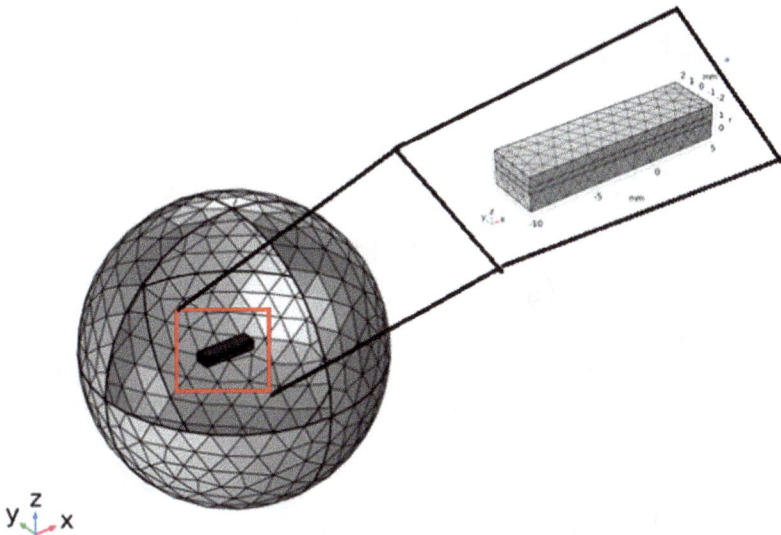

Fig. 4. Simulation model blocks in COMSOL and zoomed view of ME tri-layer

Magnetic field *(mf)* module:

In this module, a magnetic field is applied parallel to the length of the surface in the air domain. To confine the magnetic induction (B) to the in-plane directions, the *Magnetic Insulation* boundary condition is applied to all external boundaries. The magnetostrictive material is chosen for its non-linear magnetization and is utilized in the Ampere's law equation for the magnetostrictive node.

Solid mechanics *(solid)* module:

This module is used for modelling the magnetostrictive strain and elastic behaviour for magnetoelectric composite. Two types of material models are added that is piezo-electric material and magnetostrictive material. For magnetostrive material, the isotropic nonlinear magnetoelastic properties are chosen. Similarly, for piezoelectric material, the strain-charge form is selected. And the fixed constraint boundary condition is applied for magnetostrictive material for one end fixed, and other nodes are free.

Electrostatics *(es)* module:

This electrostatics module is used for measuring the electric potential or normalised electric field from the piezoelectric material. The open circuit boundary condition is used for piezoelectric material as described in Eq. (16).

Table 1. Material parameters and properties for magnetostrictive [35, 38] and piezoelectric [37]

Parameters	Terfenol-D	PZT-5A
Length of the layer *(L)*	15 mm	15 mm (initially)
Width *(w)*	4 mm	4 mm
Thickness (t_p, t_m)	$t_{m1} = 1.25$ mm $t_{m2} = 0.75$ mm	0.5 mm
Young's modulus *(E)*	110 *GPa*	A1
Poisson's ratio (ϑ)	0.3	–
Density (ρ)	9230 Kg/m^3	7500 Kg/m^3
Electrical conductivity (σ)	1 S/m	1 S/m
Relative permittivity (ε_r)	1	A3
Saturation magnetization (M_s)	0.821 T	–
Initial magnetic susceptibility (χ_0)	27	–
Saturation magnetostriction (λ_s)	718.2 ppm	–
Piezoelectric coefficient (d_{ij})	–	A2

4 Results and Discussion

In order to create and model the correct simulation results, calculated the ME coefficient for layered ME composite by applying bias magnetic field at the resonant frequency of the structure at *162* Hz is replicated and compared with the literature [27] as shown in the Fig. 5. The results seems that the validation results are agree well with the Ren *et al.* (2022) work. Based on the current verification, the mechanical, magnetic and electric field distributions of layered ME composite is analysed in the present study as precise and effectively.

In the present study, a three-dimensional unsymmetrical tri-layer magnetoelectric composite structure is considered and the geometrical parameters and material parameters are described in Table 1. The magnetostrictive parameters taken from [35, 38] and for piezoelectric material properties are considered from COMSOL inbuilt properties. In the tri-layer composite, all dimensions are fixed throughout the entire simulation procedure and the length of the piezoelectric layer is initially fixed as 15 mm and then varied accordingly. Finer Tetrahedral mesh is employed for ME composite and normal mesh is applied for the air domain because to reduce the computation time and the major

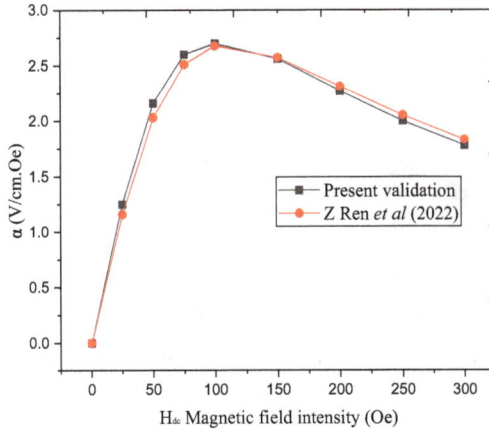

Fig. 5. Validation of ME coefficient simulation results

results concentrated on the laminated structure. The meshed model of the ME structure is shown in the Fig. 2 which has the domain elements 16,034 included.

In order to solve ME composite while applying the DC bias field with small AC signal of magnetic fields, the study step called small-signal (harmonic) analysis is considered which has the combination of stationary study with varying dc magnetic field and frequency domain perturbation by varying the frequency. In solver settings, segregated solver with Anderson acceleration is used and for frequency domain perturbation fully coupled approach is used. Based on these study solver settings, the convergence solutions are obtained for varying dc bias magnetic field under small AC signal field are determined.

4.1 Distribution of Magnetization of Tri-Layer ME Composite Under Varying Magnetic Field

The static analysis is performed to understand the distribution of magnetization in unsymmetrical laminate composite by varying the H_{dc} magnetic field at small signal of 1 kHz. The external magnetic field is starting from 0 Oe and gradually increased up to it reaches its saturation level. The magnetization first increased linearly with respect to the applied DC magnetic field and then it gets saturated at 8000 Oe. Figure 6(a) shows the distribution of magnetization by applying dc magnetic field. The presence of boundary effects causes the component to exhibit higher magnetization in the middle while experiencing lower magnetization as it approaches both ends. Furthermore, as the magnetic field increases, the degree of magnetization within the Terfenol-D becomes more uniform. Figure 6(b) represents the distribution of magnetic flux density with respect to the length at two different thickness that is 0.625 mm and 2.125 mm respectively. It is observed that the magnetic flux density is equally distributed and high magnetic field intensity at middle of the structure.

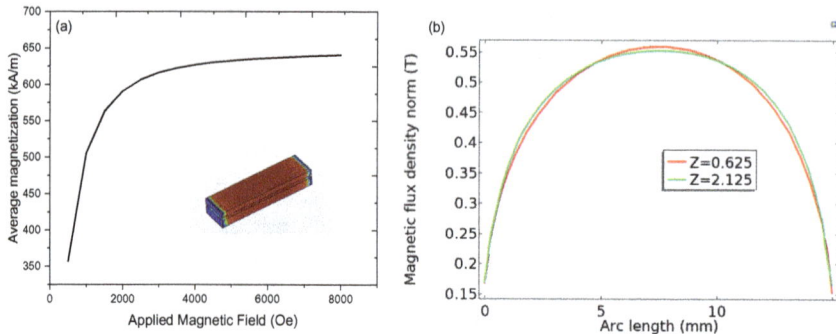

Fig. 6. (a) M-H curve of the Terfenol-D component, (b) distribution of magnetic field density in two magnetostrictive layers

From Eq. (8), it is observed that within a specific range, higher magnetization results in greater magnetostrictive strain. Figure 7(a) illustrates the variations in magnetostrictive strain in X direction for different feature points on the Terfonol-D component as the H_{dc} magnetic field increases. Notably, within this range, the equivalent strain of each point steadily increases with the growing H_{dc} magnetic field until it eventually reaches saturation.

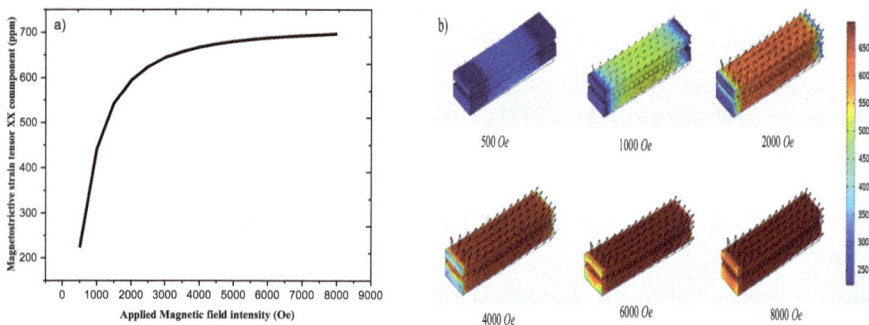

Fig. 7. (a) Magnetostrictive strain under different magnetic fields (b) Contour of magnetostrictive strain developed at different magnetic fields

4.2 Variation of Magnetoelectric Coefficient

In this section, the variation of magnetoelectric coupling coefficient by varying the length of the piezoelectric layer for varying of dc bias magnetic field is discussed. Figure 7 shows the variation of ME coefficient for tri-layer structure and the initial piezoelectric layer length is 15 mm, the maximum ME coefficient is obtained is $2.99 \, V/cm.Oe$ at $700 Oe$ at 1 kHz. The ME coefficient is linearly increased by increasing the magnetic field till it reaches the optimum magnetic field $700 Oe$ and then it decreased and reached the saturation state (Fig. 8).

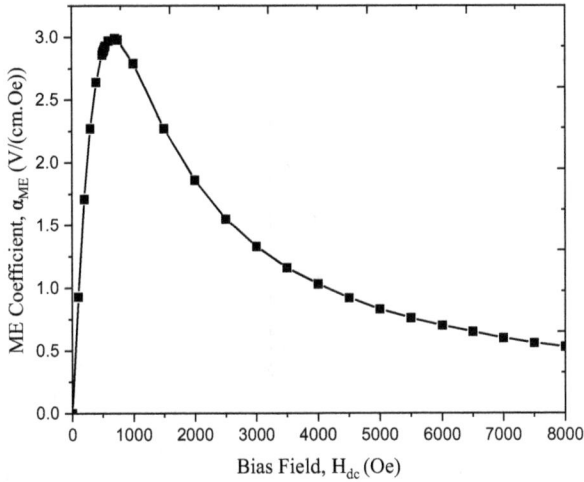

Fig. 8. Magnetoelectric coefficient for variation of H_{dc}

In addition, the piezoelectric layer length is also affecting the ME coefficient. The piezoelectric layer length is gradually decreased from 15 mm to 5 mm length. The ME coefficient is increasing till it reaches the optimum magnetic field while decreasing the length of the piezoelectric layer up to the length of 9 mm, because as discussed in Sect. 4.1, the magnetization and magnetostrictive strains are developing at the middle of the magnetostrictive layer. So, the rate of strain transfer is high at middle compared to both ends. This strain will distribute the entire piezoelectric layer and generates the output voltage. Figure 9 shows the variation of ME coefficient from the length 14 mm to 9 mm and the highest ME coefficient is obtained for 9 mm length is $3.24V/cm.Oe$ at $700\ Oe$. After 9 mm length, the ME coefficient is decreasing while decreasing of length of piezoelectric layer because, the output voltage is depending on the material quantity. Figure 10 shows the variation of ME coefficient from 8 mm to 5 mm in the decreasing manner and the Coefficient for 5 mm length is $2.98V/cm.Oe$ which is approximately equal to 15 mm length of the piezoelectric layer. The variation of ME coefficient with respect to the length of the piezoelectric layer is shown in the Fig. 11.

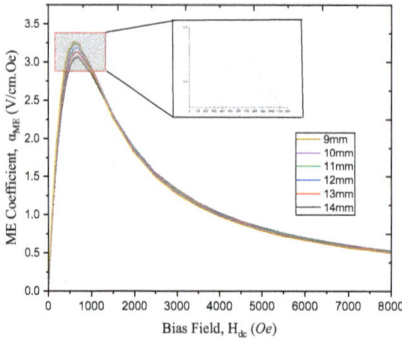

Fig. 9. Magnetoelectric coefficient for piezoelectric layer length 14 to 9 mm

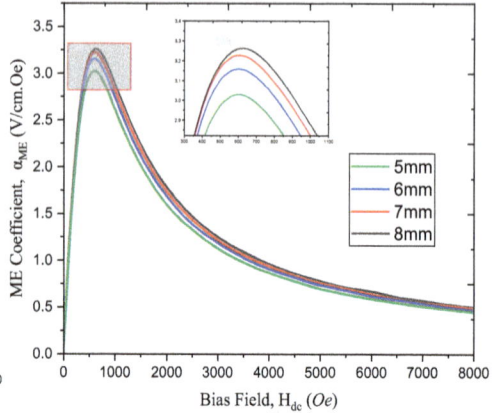

Fig. 10. Magnetoelectric coefficient for piezoelectric layer length 8 to 5 mm

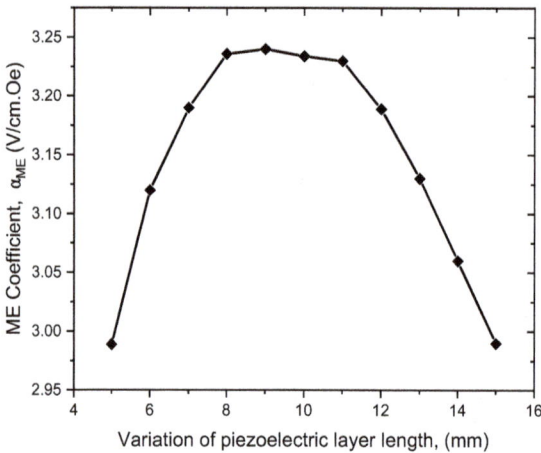

Fig. 11. Variation of ME coefficient with respect to the length of piezoelectric layer

5 Conclusions

In the present study, the unsymmetric magnetoelectric tri-layer composite with Terfenol-D and PZT-5A is built and investigated using finite element software COMSOL Multiphysics. The non uniform characteristics of magnetostrictive material and magnetoelectric coupling are explored under the external dc bias magnetic field. The magnetization, magnetic field density and the magnetostrictive strains are developed in the middle of the structure compared to the both ends. Consequently, the piezoelectric layer length is symmetrically varied from the centre and observed the magnetoelectric coefficient. The predicted Magnetoelectric coefficient is $2.99 V/cm.Oe$ at $700 Oe$ at 15 mm length, and it increased to $3.24 V/cm.Oe$ for 9 mm length. And then it decreased again up to

$2.98\ V/cm.Oe$ which is approximately equals to the 15 mm length ME coefficient. This indicates that instead of placing the full length of piezoelectric layer in multilayer, it would be a beneficial that placing at the centre in layered ME composite this enroute to light weight and cost-effective with high performance in the applications like energy harvesting and magnetic field sensors etc.

Acknowledgement. The Authors acknowledges the COMSOL Multiphysics software facility accessed through I-STEM (Indian Science Technology and Engineering facilities Map) portal, supported by Office of Principal Scientific Adviser to the Govt. of India.

Appendix

The strain-charge form of piezoelectric material parameters is defined as;

Compliance matrix, pC_E

$$
= \begin{bmatrix}
1.64e-11 & -5.74e-12 & -7.22e-12 & 0 & 0 & 0 \\
-5.74e-12 & 1.64e-11 & -7.22e-12 & 0 & 0 & 0 \\
-7.22e-12 & -7.22e-12 & 1.88e-11 & 0 & 0 & 0 \\
0 & 0 & 0 & 4.75e-11 & 0 & 0 \\
0 & 0 & 0 & 0 & 4.75e-11 & 0 \\
0 & 0 & 0 & 0 & 0 & 4.43e-11
\end{bmatrix} 1/Pa \quad (A1)
$$

Piezoelectric coupling matrix, d

$$
= \begin{bmatrix}
0 & 0 & 0 & 0 & 5.84e-10 & 0 \\
0 & 0 & 0 & 5.84e-10 & 0 & 0 \\
-1.71e-10 & -1.71e-10 & 3.74e-10 & 0 & 0 & 0
\end{bmatrix} C/N \quad (A2)
$$

$$
\text{Relative permittivity, } p\varepsilon_r = \begin{bmatrix}
1730 & 0 & 0 \\
0 & 1730 & 0 \\
0 & 0 & 1700
\end{bmatrix} \quad (A3)
$$

References

1. Sunar, M., Al-Garni, Z., Ali, M.H., Kahraman, R.: Finite element modeling of thermopiezo-magnetic smart structures. AIAA J. **40**(9), 1846–1851 (2002)
2. Chu, Z., PourhosseiniAsl, M., Dong, S.: Review of multi-layered magnetoelectric composite materials and devices applications. J. Phys. D Appl. Phys. **51**(24), 243001 (2018). https://doi.org/10.1088/1361-6463/aac29b
3. Dinesh Kumar, S., Gupta, S., Swain, A.B., Subramanian, V., Padmanabhan, M.K., Mahajan, R.L.: Large converse magnetoelectric effect in Sm doped Pb(Mg1/3Nb2/3)-PbTiO3 and NiFe2O4 laminate composite. J Alloys Compounds **858**, 157684 (2021). https://doi.org/10.1016/j.jallcom.2020.157684
4. Bok, I., Haber, I., Qu, X., et al.: In silico assessment of electrophysiological neuronal recordings mediated by magnetoelectric nanoparticles. Sci. Rep. **12**, 8386 (2022). https://doi.org/10.1038/s41598-022-12303-4

5. Fiocchi, S., et al.: Modeling of core-shell magneto-electric nanoparticles for biomedical applications: Effect of composition, dimension, and magnetic field features on magnetoelectric response. PLoS ONE **17**(9), e0274676 (2022). https://doi.org/10.1371/journal.pone.0274676

6. Kaviraj, P., Pramanik, R., Arockiarajan, A.: Influence of individual phases and temperature on properties of CoFe2O4-BaTiO3 magnetoelectric core-shell nanocomposites. Ceram. Int. **45**(9), 12344–12352 (2019). https://doi.org/10.1016/j.ceramint.2019.03.153

7. Jian, G., et al.: Orientation dependence of magnetoelectric coefficient in 1–3–type BaTiO3/CoFe2O4. J. Magn. Magn. Mater. **449**, 263–270 (2018). https://doi.org/10.1016/j.jmmm.2017.10.046

8. Xavier, D., Dinesh Kumar, S., Subramanian, V.: Significant magnetoelectric coupling in P(VDF-TrFE)/48%NiFe bilayer laminate composite for energy harvesting applications. J. Phys. D: Appl. Phys. **55**(30), 305502 (2022). https://doi.org/10.1088/1361-6463/ac6b64

9. Chu, Z., et al.: Enhanced resonance magnetoelectric coupling in (1–1) connectivity composites. Adv. Mater. **29**(19), 1606022 (2017). https://doi.org/10.1002/adma.201606022

10. Chen, H., Liang, X., Sun, N., Sun, N.-X., Lin, H., Gao, Y.: An ultra-compact me antenna design for implantable wireless communication. In: 2020 IEEE International Symposium on Antennas and Propagation and North American Radio Science Meeting, pp. 655–656. IEEE Montreal (2020). https://doi.org/10.1109/IEEECONF35879.2020.9329973

11. Das, D., et al.: A radio frequency magnetoelectric antenna prototyping platform for neural activity monitoring devices with sensing and energy harvesting capabilities. Electronics **9**(12), 2123 (2020). https://doi.org/10.3390/electronics9122123

12. Nan, T., et al.: Acoustically actuated ultra-compact NEMS magnetoelectric antennas. Nat. Commun. **8**(1), 296 (2017). https://doi.org/10.1038/s41467-017-00343-8

13. Leontiev, V., Sokolov, O., Tatarenko, A., Petrov, R., Bichurin, M.: Magnetoelectric current sensor based on MEMS technology. In: 2019 PhotonIcs & Electromagnetics Research Symposium - Spring (PIERS-Spring), pp. 2587–2592. IEEE, Rome (2019). https://doi.org/10.1109/PIERS-Spring46901.2019.9017218

14. Petrov, R.V., Yegerev, N.V., Semenov, G.A., Petrov, V.M., Bichurin, M.I., Aleksic, S.: Current sensor based on magnetoelectric effect. In: 2014 18th International Symposium on Electrical Apparatus and Technologies (SIELA), pp. 1–4. IEEE, Bourgas (2014). https://doi.org/10.1109/SIELA.2014.6871882

15. Bichurin, M., Petrov, R., Leontiev, V., Semenov, G., Sokolov, O.: Magnetoelectric current sensors. Sensors **17**(6), 1271 (2017). https://doi.org/10.3390/s17061271

16. Duc, N.H., Tu, B.D., Ngoc, N.T., Lap, V.D., Giang, D.T.H.: Metglas/PZT-Magnetoelectric 2-D Geomagnetic device for computing precise angular position. IEEE Trans. Magn. **49**(8), 4839–4842 (2013). https://doi.org/10.1109/TMAG.2013.2241446

17. PourhosseiniAsl, M.J., Chu, Z., Gao, X., Dong, S.: A hexagonal-framed magnetoelectric composite for magnetic vector measurement. Appl. Phys. Lett. **113**(9), 092902 (2018). https://doi.org/10.1063/1.5022094

18. Zhu, Y., Zu, J.W.: A Magnetoelectric generator for energy harvesting from the vibration of magnetic levitation. IEEE Trans. Magn. **48**(11), 3344–3347 (2012). https://doi.org/10.1109/TMAG.2012.2199289

19. Dong, S., Zhai, J., Li, J.F., Viehland, D., Priya, S.: Multimodal system for harvesting magnetic and mechanical energy. Appl. Phys. Lett. **93**(10), 103511 (2008). https://doi.org/10.1063/1.2982099

20. Gao, J., Hasanyan, D., Shen, Y., Wang, Y., Li, J., Viehland, D.: Giant resonant magnetoelectric effect in bi-layered Metglas/Pb(Zr, Ti)O 3 composites. J. Appl. Phys. **112**(10), 104101 (2012). https://doi.org/10.1063/1.4765724

21. Narita, F., Fox, M.: A review on piezoelectric, magnetostrictive, and magnetoelectric materials and device technologies for energy harvesting applications. Adv. Eng. Mater. **20**(5), 1700743 (2018). https://doi.org/10.1002/adem.201700743

22. Leung, C.M., et al.: A dual-output magnetoelectric gyrator. J. Phys. D Appl. Phys. **52**(6), 065003 (2019). https://doi.org/10.1088/1361-6463/aaf18c
23. Hayes, P., et al.: Converse magnetoelectric composite resonator for sensing small magnetic fields. Sci. Rep. **9**(1), 16355 (2019). https://doi.org/10.1038/s41598-019-52657-w
24. Li, X., Lee, A., Razavi, S.A., Wu, H., Wang, K.L.: Voltage-controlled magnetoelectric memory and logic devices. MRS Bull. **43**(12), 970–977 (2018). https://doi.org/10.1557/mrs.2018.298
25. Hu, J.-M., Li, Z., Chen, L.-Q., Nan, C.-W.: High-density magnetoresistive random access memory operating at ultralow voltage at room temperature. Nat. Commun. **2**(1), 553 (2011). https://doi.org/10.1038/ncomms1564
26. Viehland, D., Wuttig, M., McCord, J., et al.: Magnetoelectric magnetic field sensors. MRS Bull. **43**, 834–840 (2018). https://doi.org/10.1557/mrs.2018.261
27. Ren, Z., Tang, L., Zhao, J., Zhang, S., Liu, C., Zhao, H.: Comparative study of energy harvesting performance of magnetoelectric composite-based piezoelectric beams subject to varying magnetic field. Smart Mater. Struct. **31**(10), 105001 (2022). https://doi.org/10.1088/1361-665X/ac798c
28. Wen, J., Zhang, J., Gao, Y.: A coupling finite element model for analysis the nonlinear dynamic magnetoelectric response of tri-layer laminate composites. Compos. Struct. **166**, 163–176 (2017). https://doi.org/10.1016/j.compstruct.2017.01.056
29. Zhai, J., Xing, Z., Dong, S., Li, J., Viehland, D.: Magnetoelectric laminate composites: an overview. J. Am. Ceram. Soc. **91**(2), 351–358 (2008). https://doi.org/10.1111/j.1551-2916.2008.02259.x
30. Sudersan, S., Maniprakash, S., Arockiarajan, A.: Nonlinear magnetoelectric effect in unsymmetric laminated composites. Smart Mater. Struct. **27**(12), 125005 (2018). https://doi.org/10.1088/1361-665X/aae858
31. Song, R., Zhang, J., Weng, G.J.: Three-dimensional finite element simulation of magnetic-mechanical-electrical coupling in layered cylindrical multiferroic structures. Mech. Mater. **175**, 104476 (2022). https://doi.org/10.1016/j.mechmat.2022.104476
32. Song, Y., Pan, D., Xu, L., Liu, B., Volinsky, A.A., Zhang, S.: Enhanced magnetoelectric efficiency of the Tb1−xDyxFe2−y/Pb(Zr, Ti)O3 cylinder multi-electrode composites. Mater. Des. **90**, 753–756 (2016). https://doi.org/10.1016/j.matdes.2015.11.031
33. Dong, S., et al.: Circumferential-mode, quasi-ring-type, magnetoelectric laminate composite—a highly sensitive electric current and/or vortex magnetic field sensor. Appl. Phys. Lett. **86**(18), 182506 (2005). https://doi.org/10.1063/1.1923184
34. Leung, C.M., Or, S.W., Zhang, S., Ho, S.L.: Ring-type electric current sensor based on ring-shaped magnetoelectric laminate of epoxy-bonded Tb0.3Dy0.7Fe1.92 short-fiber/NdFeB magnet magnetostrictive composite and Pb(Zr, Ti)O3 piezoelectric ceramic. J. Appl. Phys. **107**(9), 09D918 (2010). https://doi.org/10.1063/1.3360349
35. Zhang, J., Du, H., Xia, X., Fang, C., Weng, G.J.: Theoretical study on self-biased magnetoelectric effect of layered magnetoelectric composites. Mech. Mater. **151**, 103609 (2020). https://doi.org/10.1016/j.mechmat.2020.103609
36. Elakkiya, V.S., Sudersan, S., Arockiarajan, A.: Stress-dependent nonlinear magnetoelectric effect in press-fit composites: a numerical and experimental study. Eur. J. Mech. A. Solids **93**, 104536 (2022). https://doi.org/10.1016/j.euromechsol.2022.104536
37. COMSOL multiphysics ® v.6.0. COMSOL multiphysics reference manual. https://doc.comsol.com/6.0/docserver/#!/com.comsol.help.comsol/helpdesk/helpdesk.html
38. Sudersan, S., Arockiarajan, A.: Thermal and prestress effects on nonlinear magnetoelectric effect in unsymmetric composites. Compos. Struct. **223**, 110924 (2019). https://doi.org/10.1016/j.compstruct.2019.110924

Development and Characterization
of Advanced Steels

Development of Third Generation Advanced High Strength Steel

J. N. Mohapatra$^{(\boxtimes)}$ and Satish Kumar Dabbiru

JSW Steel Ltd., Toranagallu, Bellary 583275, Karnataka, India
jitendra.mohapatra@jsw.in

Abstract. Present study intends to develop third generation Advanced High Strength Steel (AHSS) from a first generation lean alloyed TRIP assisted steel with Nb and Ti micro alloying by innovative heat treatments such as Transformation Induced Plasticity (TRIP) aided Bainitic Ferrite (TBF), Dual Stabilization (DS), Harden and Temper (H&T), Quench and Partitioning (Q&P), Austenite Reverted Transformation (ART). All the steels showed tensile toughness greater than 20 GPa. %, higher than 25 GPa% for TBF, DS, Q&P after inter critical (IC) - austenitization, whereas for ART it is found to be $\geq$ 30 GPa. %. A versatile range of properties was achieved in the single steel by changing the heat treatment conditions to vary the microstructure such as ferrite, bainite, martensite and retained austenite. The retained austenite with the TRIP effect further enhanced the strength-elongation product to third generation AHSS regime. Due to the higher martensite content in the Q&T and Q&P after austenitization above Ac_3 temperature, the steel showed very high level of strength (1300–1450 MPa) with relatively lower elongations (14.5–17%). In the TBF, DS and Q&P after inter critical annealing give ultra high strength (994–1067 MPa) with very good ductility (25–29%) due to the presence of bainite and retained austenite. The ART showed relatively lower strength (750–767 MPa) with excellent elongation (39.5–47%) due to the presence of more ferrites compared to the other heat treatment cycles. The third generation AHSS was developed by varying heat treatment cycles from a first generation TRIP assisted AHSS.

Keywords: TRIP Assisted Steel · Innovative Heat Treatment · Microstructure · Mechanical Properties · Third Generation AHSS

1 Introduction

Third generation advanced high strength steel (AHSS) development emerges in the automotive industry for the achievement of high strength and elongations compared to the first generation AHSS and lower cost and better weldability compared to the second generation AHSS [1–3]. The steels are growing demand for their excellent combination of strength and elongation to give tensile toughness in the range of 20GPa. % to 40GPa. %. Such properties helps in the weight reduction of the automotive components to improve fuel efficiency, reduce CO_2 emission with high crash resistance properties to

S. Patra et al. (Eds.): METCENT 2023, *Proceedings of the International Conference on Metallurgical Engineering and Centenary Celebration*, pp. 227–236, 2024.
https://doi.org/10.1007/978-981-99-6863-3_22

give safety to the passengers. Steel makers produce third generation AHSS primarily by the use of lean chemistry with the start of art annealing practices for the cost saving and achieving the desired microstructure responsible for the beneficial properties [4]. The microstructure of third generation AHSS constitutes retained austenite in a bainite or martensite matrix and potentially some amount of ferrite and/or precipitates, all in specific proportions and distributions, to develop these enhanced properties [5]. Although lot of studies can be found on the development of third generation AHSS with different chemical compositions and processing routes, a few study may be found on the use of a single steel for the utilization of various processing conditions for the improvement of mechanical properties.

2 Experimental

A hot rolled TRIP assisted steel containing (wt. %) 0.17C-1.35Si-1.72Mn-0.0.2Nb-0.04Ti of 3mm thickness manufactured at JSW Steel Ltd. Vijayanagar Work was used in the present study. The A_1 (731 °C), A_3 (875 °C), B_s (557 °C) and M_s(417 °C) temperatures were obtained through thermocalc (2020b) software and Cambridge University software MUCG83 respectively [6]. The steel was subjected to various heat treatment cycles such as TRIP aided Bainitic Ferrite (TBF), Dual Stabilization (DS), Harden and Temper (H&T), Quench and Partitioning (Q&P), Austenite Reverted Transformation (ART) as shown in Fig. 1. The H&T and Q&P were conducted both after austenitization above A_3 and in the inter-critical (IC) region followed by further treatments whereas for TBF it is only austenitized above A_3 and for DS only austenitized in the IC zone. The ART cycle follows full austenitization followed by bainitic holding and water quenched followed by holding at two different inter critical temperatures. The details of the processes conditions is summarized in Table 1.

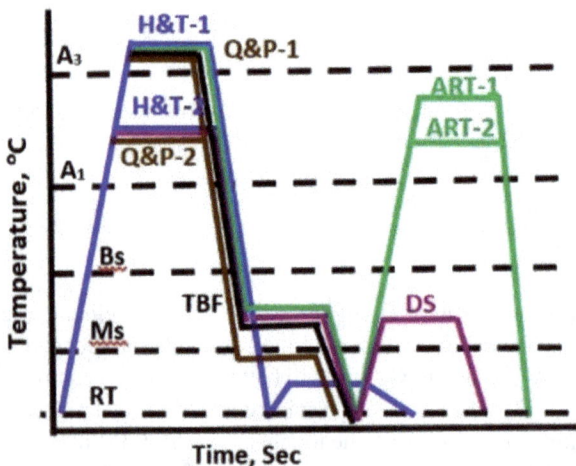

Fig. 1. Heat Treatment Cycle (TBF, DS, H&T, Q&P, ART).

The heat treatments were conducted in laboratory muffle furnace with salt bath arrangements. After heat treatment the samples were cut to ASTM E8M standard sub size specimens for tensile testing on a 250 kN Zwick/Roell make tensile testing machine at a standard strain rate of 0.008/s. Samples were also cut for the microstructure analysis using Olympus make optical and Hitachi make Scanning Electron Microscopy (SEM). Panlalytic make XRD was used to evaluate the retained austenite content in the steel. Thermocalc software was used to estimate the phase fraction at the intercritical temperatures.

Table 1. Heat treatment condition.

Heat Treatment	Aust^z. Temp., $^\circ$C	Aust^z.Time, Min	Aust^P./Tempering/Partitioning Temp., $^\circ$C	Aust^P./Tempering/Partitioning Time, Min	Re-Aust^z. Temp. $^\circ$C	Re-Aust^z. Time, Min	Cooling
TBF	930	20	425	5	–	–	WQ
DS	810	5	450WQ	0.5	450	0.5	WQ
H&T-1	920	20WQ	150	15	–	–	AC
H&T-2	810	20WQ	150	15	–	–	AC
Q&P-1	900	5	387	1	–	–	WQ
Q&P-2	830	5	362	5	–	–	WQ
ART-1	900	2	450WQ	5	750	5	WQ
ART-2	900	2	450WQ	5	780	5	WQ

3 Results and Discussion

3.1 Microstructure and Phases

The optical microstructure and SEM micrograph of the steel in TBF heat treatment cycle with the demarcation of phases is shown in Fig. 2. The microstructure of the TBF steel constitutes bainitic matrix with the martensite and retained austenite (RA) phase as reported earlier [7]. In the microstructure film type RA can be seen and it was estimated to be around 3% by XRD. The martensite/austenite (M/A) helps in suppressing a fatigue crack initiation and propagation [8]. Due to the bainitic ferrite matrix with the TRIP effect from the transformation of RA to martensite provides the excellent combination of strength-elongation product in the TBF steel.

DS heat treatment is made to increase the RA content in the steel. The RA on TRIP effect gives excellent combination of strength and ductility. The optical microstructure and SEM micrograph of the dual stabilized steel is shown in Fig. 3. It was observed that inter critical austenitization resulted in ferrite content of 47% in the steel for the adequate ductility. The RA content in the DS steel was estimated to be 7% by XRD. The retained austenite content can be further improved up to 30% in steels by increasing the C & Mn content to achieve tensile strength of 1650 MPa with 20% elongation reported earlier [9].

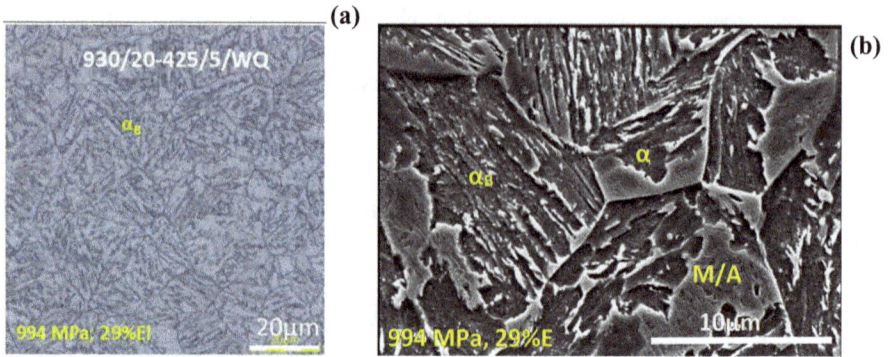

Fig. 2. (a) Optical microstructure and (b) SEM micrograph of the steel with TBF heat treatment cycle.

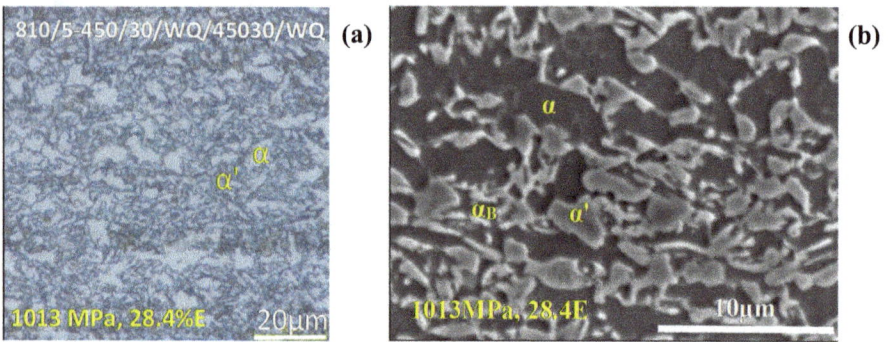

Fig. 3. (a) Optical microstructure and (b) SEM micrograph of the steel with DS heat treatment cycle.

The optical and SEM micrograph of the harden and temper steel after austenitization above A_3 (a & c) and inter critical austenitization (b & d) is shown in Fig. 4. Lathe type martensite is found in the full austenitization condition whereas intercritical ferrite with colony of martensites are seen in the inter critical austenitization condition. The martensites after tempering improved ductility in the steel to give third generation AHSS properties. Very high level of strengths were found in H&T condition as reported earlier [10, 11].

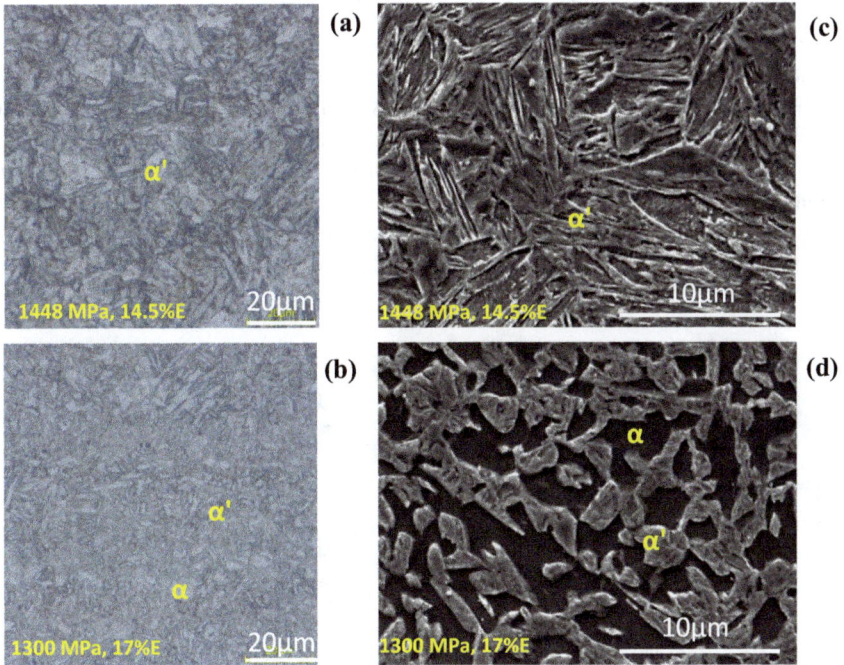

Fig. 4. Microstructure of the steel with H&T heat treatment cycle **(a)** after full austenitization and **(b)** intercritical austenitization, **(c)** & **(d)** are their corresponding SEM micrographs.

The microstructure of the Q&P steel after austenitization above A_3 temperature and with intercritical austenitization is shown in Fig. 5. Tempered martenstitic microstructure with retained austenite is seen in the fully austentitized Q&P steel whereas bulk type martensite with inter critical ferrite (21%) and retained austenite is found in the inter critical austenitization condition. Approximately 12% retained austenite is estimated in both the steels by XRD. Higher strength with lower ductility is expected in the fully austenitized steel followed by Q&P due to presence of higher amount of martensite. Whereas, relatively lower strength and higher ductility is expected in the inter critical austenitized steel followed by Q&P due to presence of lower martensite content and presence of inter critical ferrite [12, 13].

Austenite reverted transformation annealing is mostly conducted in medium manganese steel for the enrichment and stabilization of retained austenite [14, 15]. In the present study, the concept has been used for the utilization for a lean alloyed TRIP steel. The microstructure of the austenite reverted transformation (ART) at higher inter-critical and lower intercritical annealed steel is shown in Fig. 6. The microstructure of the steel shows ferrite, bainite with retained austenite in the steel. The retained austenite was estimated through XRD and found to be quite lower in the range of 2–4%. Hence it is mostly of ferrite and bainite in the steel resulting very good ductility with reasonably good strength to achieve third generation AHSS properties.

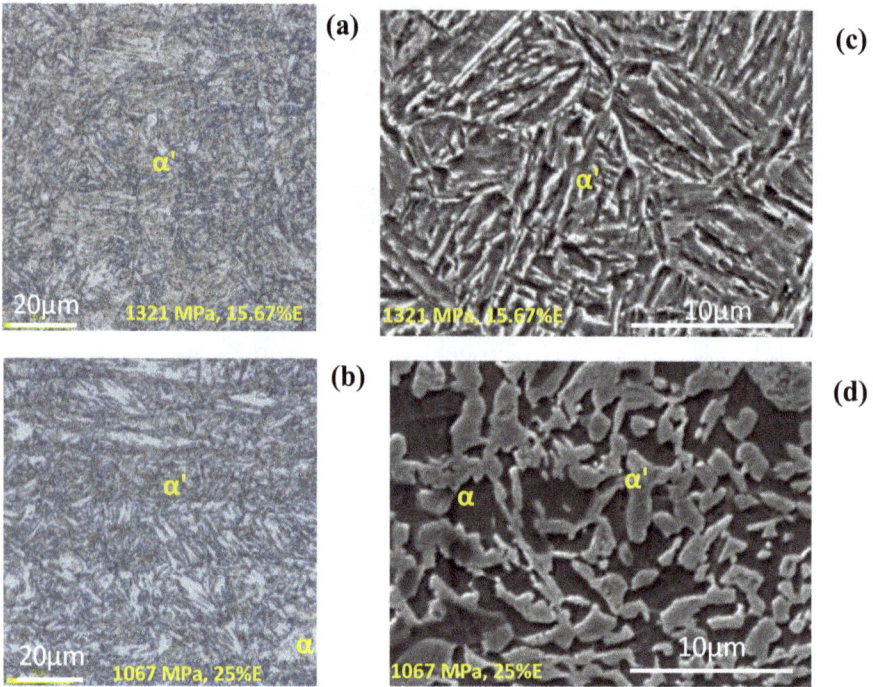

Fig. 5. (a) Optical microstructure of the steel with Q&P heat treatment cycle after full austenitization and (b) intercritical austenitization, (c) & (d) are their corresponding SEM micrographs.

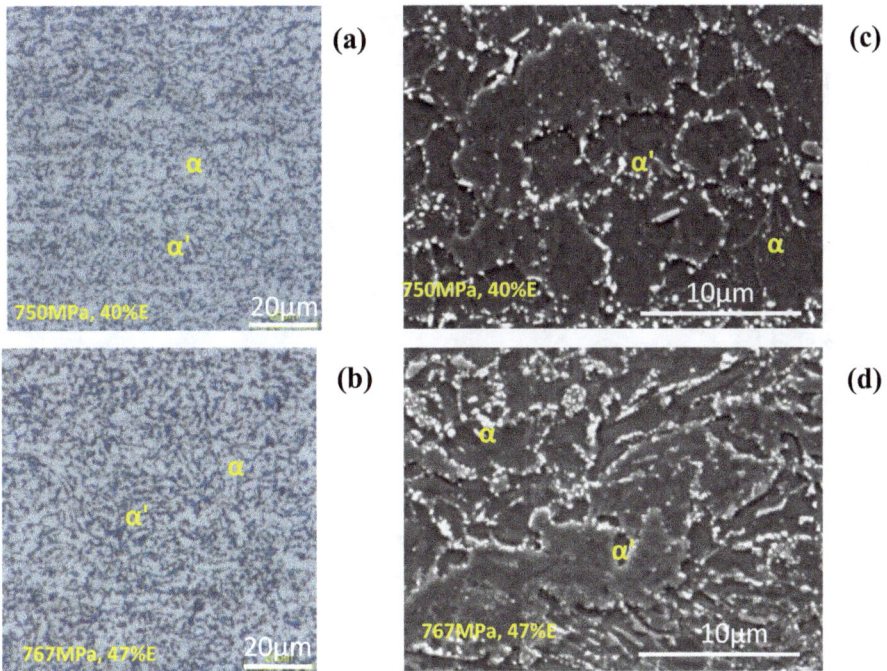

Fig. 6. Optical microstructure of the steel with ART heat treatment cycle after full austenitization followed by higher intercritical annealing temperature (**a**) and lower inter critical annealing temperature (**b**), (**c**) & (**d**) are their corresponding SEM micrographs.

3.2 Mechanical Properties

The stress strain diagram of the steel at various heat treatment cycles is shown in Fig. 7. The heat treatment condition and mechanical properties of the steel is summarized in Table 2. The ART steels are having lower strength with high elongations due to the presence of higher amount of ferrites bainite microstructure. H&T and Q&P-1 steels are having higher strength with lower elongations due to the presence of higher quantity of martensite. The TBF, DS and Q&P-2 are having excellent combinations of strength and elongations due to the presence of complex microstructure in addition to the TRIP effect for the transformation of RA to martensite during deformation.

The plot of ultimate tensile strength with total elongation with the classification of different generation of AHSS is shown in Fig. 8. It can be seen that all the studied heat treated steels are falling in the third generation AHSS regime. Hence the steels can be used in the automotive crash resistance applications where third generation AHSS properties are desired.

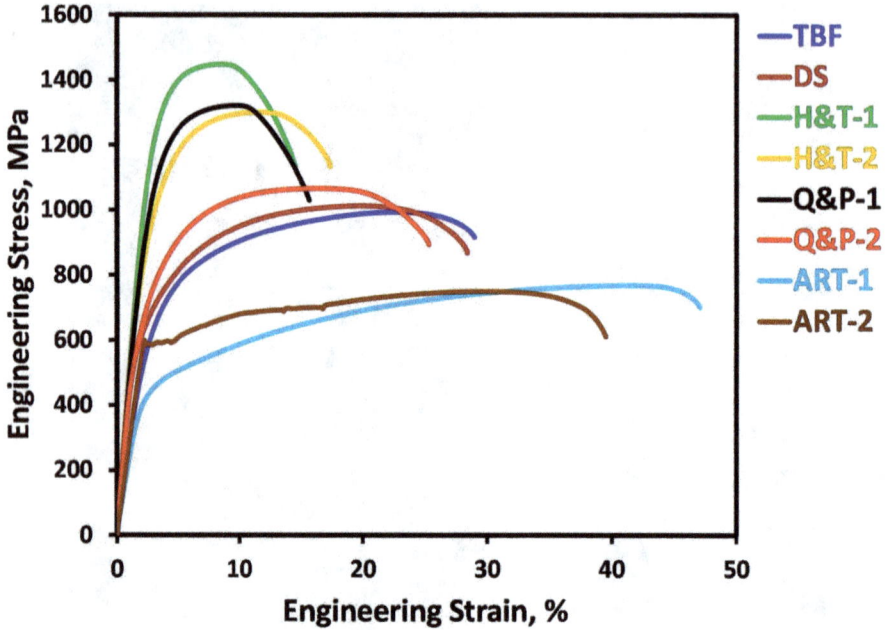

Fig. 7. Engineering stress-strain diagram of the steel at various heat treatment cycles.

Table 2. Heat treatment condition and mechanical properties.

	Heat Treatment Condition	YS, MPa	UTS, MPa	TE, %	YR	UTS*TE, GPa.%
TBF	Austenitized Above A_3	447	994	28.97	0.45	28.79
DS	IC-Annealed	479	1013	28.40	0.47	28.78
H&T-1	Austenitized Above A_3	870	1448	14.50	0.60	21.08
H&T-2	IC-Annealed	595	1300	16.96	0.46	22.05
Q&P-1	Austenitized above A_3	661	1321	15.67	0.50	20.70
Q&P-2	IC-Annealed	463	1067	25.28	0.43	26.97
ART-1	Austenitized above A_3 with higher IC partitioning temperature	511	767	47.05	0.66	36.08
ART-2	Austenitized above A_3 with lower IC partitioning temperature	578	750	39.50	0.77	29.62

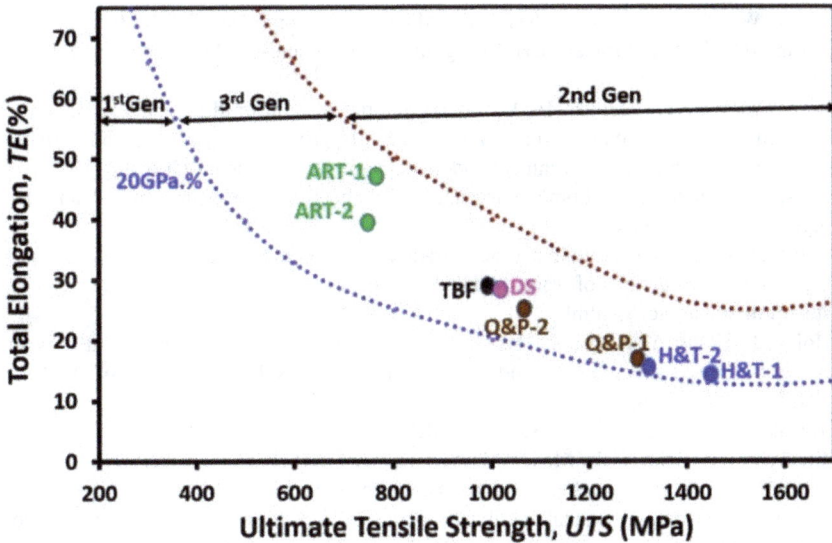

Fig. 8. Mechanical properties of the steel studied falling in the banana curve.

4 Conclusions

A versatile range of mechanical properties fulfilling the requirements of third generation AHSS were developed with the various processing routes such as Trip aided Bainitic Ferrite (TBF), Dual Stabilization (DS), Harden and Temper (H&T), Quench and partitioning (Q&P) and Austenite reverted Transformation (ART) cycles.

For all the cycles the ultimate tensile strength (UTS) was found to be greater than 1000 MPa indicating ultra high strength steels except ART cycle where the UTS is close to 750 MPa with exceptional high elongation ($\geq$40%).

Highest Strength (UTS) with lower elongation was found for the harden and temper condition. Elongation close to 28% were observed for the TBF, DS and Q&P Steel with inter critical austenitization.

For all the steels the tensile toughness was found to be greater than 20 GPa. %, higher than 25 GPa% for TBF, DS, Q&P after IC- austenitization, whereas for ART it is found to be $\geq$ 30 GPa. %.

It is inferred that by changing the heat treatment cycle in innovative way microstructure of a TRIP assisted steel can be modified to achieve a versatile range of mechanical properties in the third generation AHSS regime for automotive applications.

References

1. Noder, J., Gutierrez, J.E., Zhumagulov, A., Dykeman, J., Ezzat, H., Butcher, C.: A comparative evaluation of third-generation advanced high-strength steels for automotive forming and crash applications. Materials **14**(4970), 1–37 (2021)

2. Bleck, W., Bruhl, F., Ma, Y., Sasse, C.: Materials and processes for the third-generation advanced high-strength steels. BHM Berg- und Huttenmannische Monatshefte **164**, 466–474 (2019)
3. Bo, L., Yuehong, D., Xiaoyan, H., Yue, W.: Development of the third generation advanced high strength steel for automobile. In: Advances in Engineering Research, 163, 7th International Conference on Energy, Environment and Sustainable Development (ICEESD 2018)
4. https://www.thefabricator.com/thefabricator/article/bending/third-generation-advanced-high-strength-steel-emerges
5. https://ahssinsights.org/metallurgy/steel-grades/3rd-generation-steels/
6. Cambridge University software MUCG83. https://www.phasetrans.msm.cam.ac.uk/map/steel/programs/mucg83.html
7. Hojoa, T., Ukaib, Y., Akiyamaa, E.: Effects of hydrogen on tensile properties at slow strain rate of ultra high-strength TRIP-aided bainitic ferrite steels. Procedia Eng. **207**, 1868–1873 (2017)
8. Yoshikawa, N., Kobayashi, J., Sugimoto, K.I.: Notch-fatigue properties of advanced TRIP-aided bainitic ferrite steels. Metall. Mater. Trans. A **43**, 4129–4136 (2012)
9. Qu, H., Michal, G.M., Heuer, A.H.: Third generation 0.3 C-4.0 Mn advanced high strength steels through a dual stabilization heat treatment: austenite stabilization through paraequilibrium carbon partitioning. Metall. Mater. Trans. A **45**(6), 2741–2749 (2014)
10. Sahoo, G., Kumar, P., Nageswaran, K, Singh, K.K., Kumar, V., Dhua, S.K.: High-Si-content quenched and tempered steel for high-strength application. J. Mater. Eng. Perform. **30**(12), 9290–301 (2021)
11. Horn, R.M., Ritchie, R.O.: Mechanisms of tempered martensite embrittlement in low alloy steels. Metall. Trans. A **9**(8), 1039–1053 (1978)
12. Wang, L., Speer, J.G.: Metallogr. Microstruct. Anal. **2**, 268 (2013). https://doi.org/10.1007/s13632-013-0082-8
13. Fekhreddine, M.R., Baibhaw, K., Tisza, M.: Comparison study between TBF and Q&P steels in sheet metal forming: an overview. Int. Res. J. Eng. Technol. (IRJET) **8**(3), 3131–3134 (2021)
14. Xu, Z., Shen, X., Allam, T., Song, W., Bleck, W.: Austenite transformation and deformation behavior of a cold-rolled medium-Mn steel under different annealing temperatures. Mater. Sci. Eng. A **829**(1), 142115 (2022)
15. Li, Z., et al.: Effect of austenite reverted transformation-annealing heat treatment on microstructure and mechanical properties of Fe–0.14C–5Mn–1Al–Ce steel. Proc. Inst. Mech. Eng. Part E J. Process Mech. Eng. **235**(6) (2021). https://doi.org/10.1177/0954408921103 36225

Cracking Problem During Room Temperature Cold Rolling of Three High Al Low C Ferritic Low Density-Steels

Vinit Kumar Singh[1]([envelope]) and Amrita Kundu[2]

[1] Department of Metallurgical and Materials Engineering, IIT Kharagpur, Kharagpur 721302, West Bengal, India
`vksmme@iitkgp.ac.in`

[2] Department of Metallurgical and Material Engineering, Jadavpur University, Kolkata 700032, West Bengal, India

Abstract. High Al low C ferritic low-density steels exhibit a remarkable reduction in density, around 8 to 15%, compared to conventional steels. However, in high Al, Fe-Al-Mn-C system, during room temperature cold rolling process, a challenging issue arises in the form of center and edge cracking, which negatively impacts the ductility. This study is aimed to investigate the effects of cold, warm, and hot rolling of three high Al low-carbon low-density steels with 6.8 wt.%, 9.1 wt.%, and 9.7 wt.% Al. The focus is on the resulting microstructural features, including the presence of brittle Fe_3Al intermetallic and κ-carbide within the ferrite matrix. During cold rolling, centre and edge cracks were observed in the 9.1 wt.% and 9.7 wt.% Al steels due to brittle Fe_3Al intermetallic and κ-carbide, higher solid solution hardening and hardness. In contrast, the 6.8 wt.% Al steel does not exhibit any cracks during cold rolling. Therefore, to prevent cracking problems in high Al steels, warm rolling was carried out at intermediate rolling temperatures range of 200 °C and 500 °C.

Keywords: Cold/warm rolling · Center and edge cracking · Ferritic low-density steels

1 Introduction

The low-density steels have more than 6 wt. % Al gained significant attention as potential replacements for conventional steels like HSLA, DP, TWIP and TRIP steels due to their environmental benefits, particularly in reducing greenhouse gas emissions through weight reduction, especially in automotive applications [1]. For instance, every 1 wt.% increase in aluminium (Al) leads to 1.3% reduction in density [1, 2]. The development of low-density steels with a ferritic microstructure began in the late 1990s, pioneered by Baligidad and colleagues [3]. Recently, low-carbon ferritic low-density steels with Al content ranging from 6 to 10 wt.% have emerged as promising materials, offering an excellent combination of strength and ductility [1–6]. These steels possess good

S. Patra et al. (Eds.): METCENT 2023, *Proceedings of the International Conference on Metallurgical Engineering and Centenary Celebration*, pp. 237–247, 2024.
https://doi.org/10.1007/978-981-99-6863-3_23

weldability due to their low carbon (C) content, eliminating the need for additional heat treatments. However, the addition of Al beyond 7 wt.% adversely affects the formability of the steels [7]. It has been observed that the formation of secondary phases such as Fe_3Al intermetallic or κ-carbide ($(Fe,Mn)_3AlC$ perovskite structure) can induce cracking during cold rolling in high-Al (> 7 wt.%) ferritic low-density steels [8]. The size, fraction and distribution of Fe_3Al intermetallic/κ-carbides activated the voids and cracks during cold rolling, harmful for ductility. Therefore, to prevent cracking in high-Al ferritic low-density steels, it is necessary to heat to intermediate temperatures and then rolled, which may vary based on alloying elements, and morphology, fraction of secondary phases [9]. Additionally, warm rolling of high-Al steels can enable the production of crack-free sheets of low-density steels, thereby improving their mechanical properties. In this study, three high Al low C ferritic low-density steels were examined. Cold, warm, and hot rolling techniques were applied to these steels to validate the strategies of cracking followed by evaluating their microstructural characteristics, effects of Al, hardness and tensile properties.

2 Experimental Details

The chemical compositions of three steels (S1, S2, and S3) are presented in Table 1. The densities of S1, S2, and S3 were measured using Archimedes principle. The processing routes for all steels are illustrated in a schematic diagram shown in Fig. 1, explained elsewhere [4].

Initially, hot-rolled (HR) plates of 3 mm thickness were denoted as HRS1, HRS2, and HRS3, respectively. These plates underwent a 60% cold rolling (CR) process at room temperature (RT) in eight passes, with a 7.5% thickness reduction in each pass, resulting sample name as CRS1, CRS2, and CRS3. However, cracking issues were encountered in CRS2 and CRS3 samples. To address this, HRS2 and HRS3 samples were subjected to homogenization at 300 °C for 0.5 h and 500 °C for 1 h in muffle furnace. Subsequently, warm rolling (WR) was performed in eight pass to achieve a final thickness of 1.2 mm, name as WRS2 and WRS3. The finished rolling temperatures (FRT) was measured using a laser pyrometer and found to be 250 °C ± 4 °C for WRS2 and 450 °C ± 7 °C for WRS3, then WR plates were air-cooled to room temperature. Thermodynamic calculations of binary phase diagram using the TCBIN1.1 database was conducted to predict the expected equilibrium phases present at room temperature. To identify the phase present in initial HRS1, HRS2 and HRS3 samples, X-Ray diffraction technique was performed in D8 Advance, Bruker X-Ray diffractometer with Co-K$_\alpha$ source at scan speed of 0.5°/min, scan step of 0.02° with operating current (40 mA) and voltage (40 kV). Metallographic polishing of all steels sample was carried out in the rolling direction (RD) and normal direction (ND), following standard polishing techniques until achieving a 0.05-micron colloidal silica finish. The optical microstructures were examined using a Leica DM 2500M optical microscope after etching with 5% nital for 10–15 s. Scanning electron microscopy (SEM) was performed using a Zeiss Merlin FEG-SEM model. The macro Vickers hardness of the hot/warm/cold rolled samples were measured using a LECO LV700 instrument with a diamond pyramid indenter at a load of 10 kg and a dwell time of 15 s. Tensile samples were machined in the RD according to the ASTM

E8M standard and tested at a strain rate of 1×10^{-3} s^{-1} at room temperature using a servo-electric universal testing machine with a capacity of 100 kN (INSTRON 8862 model).

Table 1. Chemical composition of steels.

composition (wt.%)	C	Al	Mn	Ti	P	S	Fe
S1	0.0034	6.8	...	0.101	<0.01	<0.01	balance
S2	0.0400	9.7	...	0.090	<0.01	<0.01	balance
S3	0.1340	9.1	5.3	...	<0.01	<0.01	balance

3 Results and Discussion

3.1 Density Measurement

The main objective in developing low-density steels for automotive applications is to achieve a reduction in density lower than pure Fe. This is often accomplished by introducing alloying elements with lower densities than iron (Fe, 7.87 g/cm^3), such as aluminium (Al, 2.7 g/cm^3), silicon (Si, 2.3 g/cm^3), and manganese (Mn, 7.21 g/cm^3), into Fe-C system [1, 2]. These elements not only alter the lattice parameter of the steels but also contribute to density reduction due to their lower atomic masses [1]. Among these elements, Al is particularly effective in reducing the density of low-density steels. The densities of S1, S2, and S3 steels were measured using the Archimedes principle and are depicted in Fig. 2. The measured densities for S1, S2, and S3 steels are approximately 7.14, 6.81, and 6.71 g/cm^3, respectively. Notably, S3 steel exhibits the highest density reduction, with a maximum reduction of 14% compared to pure iron, followed by S2 and S1 steels.

3.2 Phase Diagram and XRD

Figure 3a illustrate the thermodynamic calculation of the binary Fe-Al phase diagram. The composition ranges of Al in the S1, S2, and S3 steels confirm the presence of body-cantered cubic (BCC, ferrite(α)) phase at room temperature. Figure 3b represents the XRD plot of HR samples, the peaks corresponding to ferrite(α) were observed in HRS1, HRS2, and HRS3. The inset in Fig. 3b reveals the peak corresponding to the κ-carbide in the HRS3 sample.

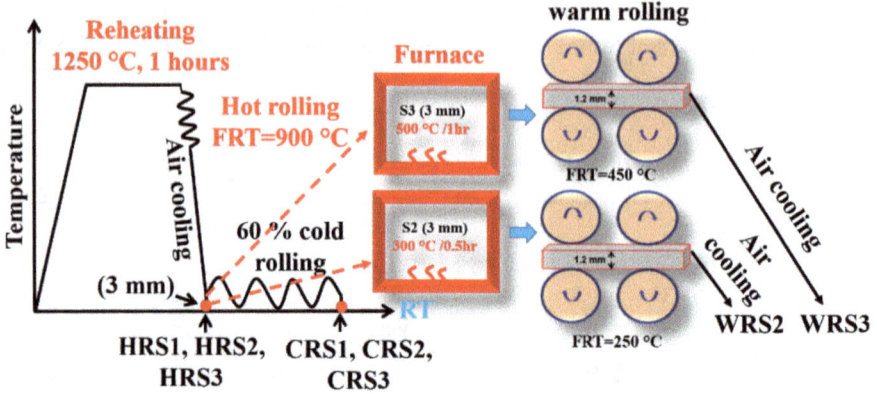

Fig. 1. Schematic diagram of hot/cold/warm rolling of S1, S2 and S3 steels. (RT: room temperature)

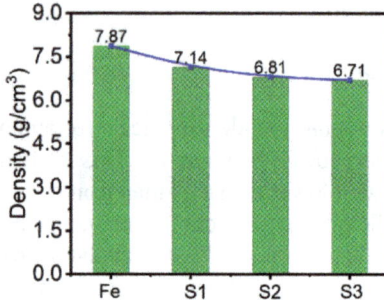

Fig. 2. Measured density of S1, S2 and S3 steels.

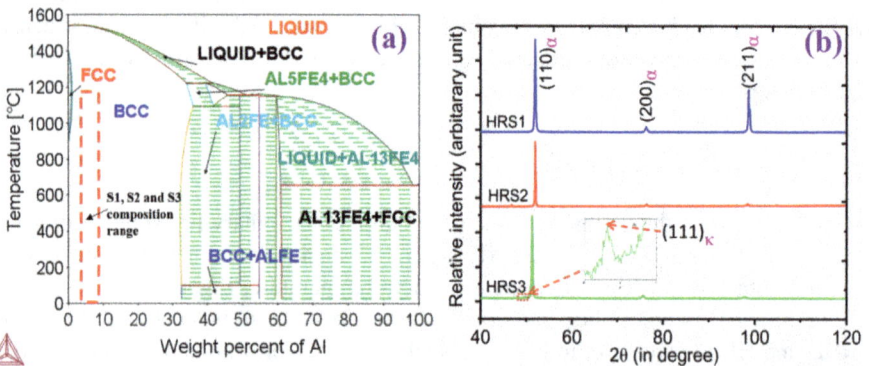

Fig. 3. (a) Thermodynamic calculation of binary phase diagram of Fe-Al system, (b) XRD profile of HRS1, HRS2 and HRS3 samples.

3.3   Microstructural Features

The optical micrographs of the hot-rolled (HR), cold-rolled (CR), and warm-rolled (WR) samples of S1, S2, and S3 steels are shown in Fig. 4. Figure 4a–c display the optical micrographs of the hot-rolled samples (HRS1, HRS2, and HRS3) of S1, S2, and S3 steels, respectively. Following hot rolling and subsequent air cooling to room temperature, the microstructure consists of ferrite in HRS1 (Fig. 4a) and ferrite and Fe_3Al intermetallic in HRS2 (Fig. 4b) samples. However, in the case of HRS3 (Fig. 4c) sample, ferrite islands containing lamellar κ-carbide are observed. The ferrite grains in HRS1 and HRS2 are equiaxed, with mean diameter of 477 μm and 406 μm, respectively. The presence of Fe_3Al intermetallic in HRS2 was explained in previous study [4]. However, in the HRS3 alloy, point energy dispersive spectroscopy (EDS) analysis (Fig. 5) shows the κ-carbide due to the presence of Al, Mn, and higher C content. The estimated area fraction of Fe_3Al intermetallic (HRS2) and κ-carbide given in Table 2 are 2.38 and 12.5%. Observations of the κ-carbide on a larger scale (shown in SEM image Fig. 5) indicate their alignment parallel to the RD. This morphology is partly attributed to micro-segregations of Mn/Al [4]. Upon cold rolling (Fig. 4d–f), the ferrite grains in CRS1 and CRS2 samples become flattened and elongated in the RD, while in CRS3 sample (Fig. 4f), the κ-carbide align along the RD. Pramanik and Suwas [10] observed the presence of lamellar bands in the rolling directions of the Fe-Al system during cold rolling, due to segregation of Al, C and anisotropy tensile properties of rolled sheets. In current study, we have examined the ferrite matrix of CRS1 and CRS2 samples and found lamellar bands aligned with the rolling direction. However, in the CRS3 sample, κ-carbide bands were observed in RD. In the warm-rolled samples (WRS2 and WRS3), the ferrite phase with lamellar bands were observed, while κ-carbide still appear in the WRS3 sample. Warm rolling of the HRS1 sample was not performed in this work, as due to successful cold rolling was achieved.

3.4   Hardness

The macro-Vickers hardness (HV_{10}) values, measured across the entire cross-sectional surface (RD-ND), are presented in Fig. 6. It can be observed that the hardness increases with Al concentration, as well as the presence of Mn in the S3 steel (specifically in the HRS3 sample), which promotes the formation 12.5% area fraction of κ-carbide. Consequently, the HRS3 ($265HV_{10}$) exhibits higher hardness compared to the HRS1 ($167HV_{10}$) and HRS2 ($216HV_{10}$). The hardness further increases with cold rolling due to the accumulation of strain in the ferrite matrix. The maximum hardness is reported in the CRS3 ($437HV_{10}$), primarily due to the strong influence of high Al content and the presence of higher amount of κ-carbide. Subsequently, the hardness value experiences a significant decrease in the WRS2 ($308HV_{10}$), followed by a slight increase in the WRS3 ($366HV_{10}$) than CRS3.

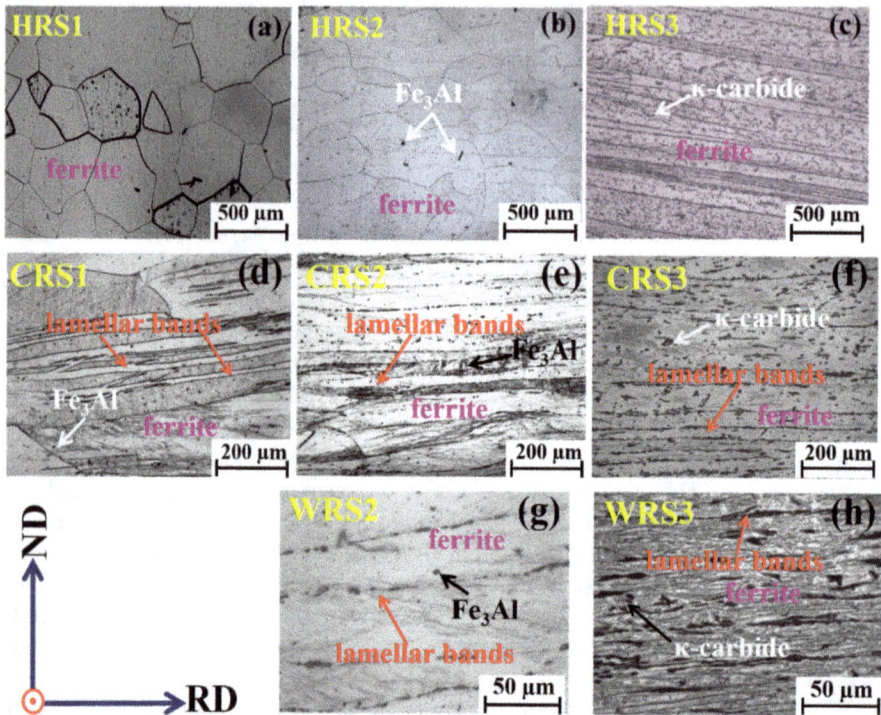

Fig. 4. Optical microstructures of (a) HRS1, (b) HRS2, (c) HRS3, (d) CRS1, (e) CRS2, (f) CRS3, (g) WRS2 and (h) WRS3 samples.

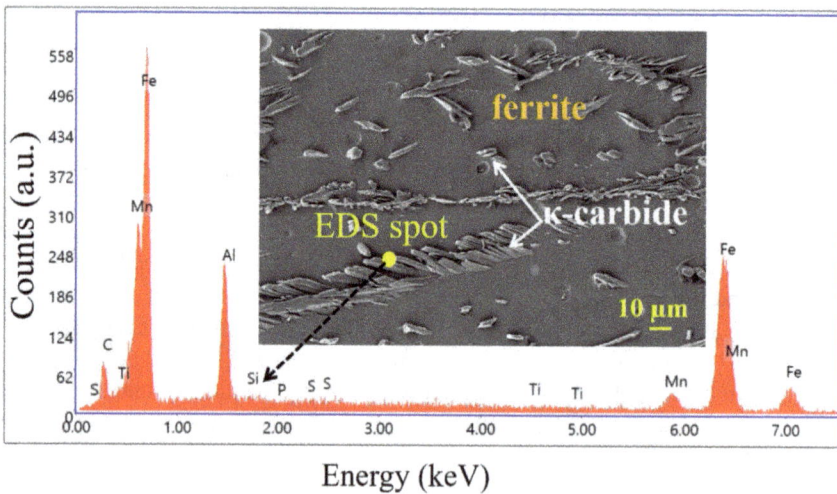

Fig. 5. The EDS profile of κ-carbides of HRS3 sample.

Table 2. Area fraction (%) of ferrite, Fe₃Al, and κ-carbide of HRS1, HRS2 and HRS3 samples.

	area fraction (%)		
	ferrite	Fe$_3$Al	κ-carbide
HRS1	100	...	...
HRS2	97.59(Bal.)	2.38 ± 0.10	...
HRS3	87.50(Bal.)	...	12.50 ± 3.00

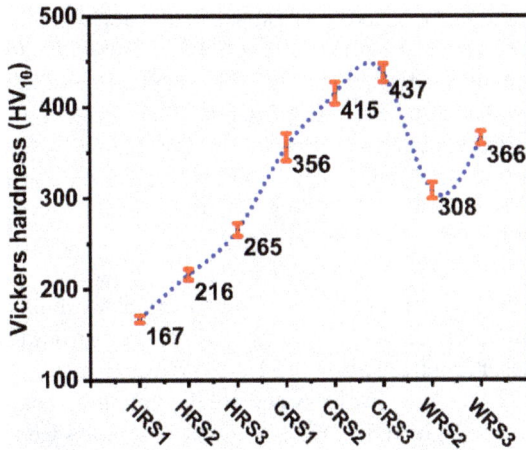

Fig. 6. Vickers hardness of hot/cold/warm rolled sample of S1, S2, and S3 steels.

3.5 Solid Solution Hardening Effect Due to Al

The Eq. (1) [11] describe the solid solution hardening effect resulting from the presence of Al in the S1, S2, and S3 steels:

$$\Delta\sigma_{ss} = M_{SS}C_{Al} \tag{1}$$

Here, $\Delta\sigma_{ss}$ (in MPa) represents the contribution of solid solution hardening, M_{SS} (in MPa/at.%) is the solid solution hardening coefficient (which combines the effects of both modulus misfit and size misfit and is equal to 9 MPa/at.%), and C_{Al} (in at.%) denotes the concentration of Al solute atoms. After converting the weight percentage (wt.%) of Al into atomic percentage (at. %) for the S1, S2, and S3 steels, the values are determined as follows: 13.12 at.% for S1, 18.18 at.% for S2, and 17.99 at.% for S3. The calculated values of $\Delta\sigma_{ss}$ are presented in Table 3, illustrating the solid solution effect of Al in S1, S2, and S3 steels are 118.17 MPa, 163.62 MPa, and 161.91 MPa, respectively. These values explain the higher solid solution hardening effects in S2 followed by S3 then S1 steels.

Table 3. Solid solution hardening contribution due to Al in S1, S2, and S3.

	S1	S2	S3
$\Delta\sigma_{ss}$ (MPa)	118.17	163.62	161.91

3.6 Effects of Al on Cracking Behaviour During Cold Rolling

D. Raabe [12] reported in the Fe-Al-Mn-C system, the addition of Al greater than 7 wt. % poses a challenge to alloy design and thus application in the automotive industry. Figure 4 shows the microstructures of three high Al low-density steels in their hot-rolled (HR), cold-rolled (CR), and warm-rolled (WR) states. Table 2 provides the area fraction of ferrite, Fe_3Al intermetallic and κ-carbide and Table 3 explain the solid solution hardening effect resulting from Al addition, while Fig. 6 displays the corresponding hardness values of specimens. These data explain the appropriate rolling strategy for S1, S2, and S3 steels. Figure 7 shows the rolled steel plate images after cold rolling and warm rolling. Steel S1, which underwent 60% cold rolling at RT (Fig. 7a), did not exhibit any cracking. However, S2 (Fig. 7b) and S3 (Fig. 7c) steels are higher Al contents of approximately 9.7 wt.% and 9.1 wt.%, showed cracking at both the center and edges during cold rolling. This can be attributed due to the higher area fraction of secondary phase, bulk hardness and solid solution hardening in S2 and S3 steels. Cold cracking can be also due to frictions between roller and steels plates. In contrast, S1, with 6.8 wt. % Al, demonstrated relatively absence of secondary phase, lower bulk hardness and solid solution hardening (118.17 MPa). The cracking problem in high Al, i.e., HRS2 and HRS3 samples were avoided by warm rolling at temperatures ranging from 200 °C to 500 °C, as depicted in schematic diagram shown in Fig. 1. By performing rolling at 200 °C to 500 °C temperature, the hardening effects due to Al were minimized, and the elongation and thickening of the secondary phase were reduced. This enabled simultaneous deformation of both ferrite and the hard secondary phase. These measures proved necessary to mitigate the detrimental effects of cracking on the materials properties effectively [13]. The impact of coarse secondary phases on cold rolling in S2, and S3 steels were interrelated to the weakest link theory model [14]. The cracking probability of the hard secondary phase (P_{cr}) is described by the Eq. (2) [14]:

$$P_{cr} = 1 - exp\left\{-\left(\frac{d}{d_n}\right)^3\left(\frac{\sigma - \sigma_{min}}{\sigma_0 - \sigma_{min}}\right)^{m\prime}\right\} \quad (2)$$

In Eq. (2), d, d_n, σ, σ_0, σ_{min}, and $m\prime$ are the precipitates size, normalized precipitate size, tensile stress, normalise stress, minimum fracture stress and Weibull factor. Equation (2) indicates that the cracking probability of the secondary phase is directly proportional to the precipitate size, d. Therefore, microscopically, CRS2 and CRS3 samples exhibit higher cracking probabilities (P_{cr}) due to their larger d values of secondary phases compared to CRS1. This observation aligns with the findings of Lee et al. [15] who investigated cracking during rolling in ferrite-pearlitic steels. Under applied stress, cracks propagate and fracture along the grain boundaries due to higher stress accumulation. However, in pearlite colonies, crack initiation and propagation are avoided due to the higher density of fine lamellar carbides. Thus, during cold rolling, cracks can be mitigated by refining the size of precipitates and adjusting their distribution within the ferrite matrix.

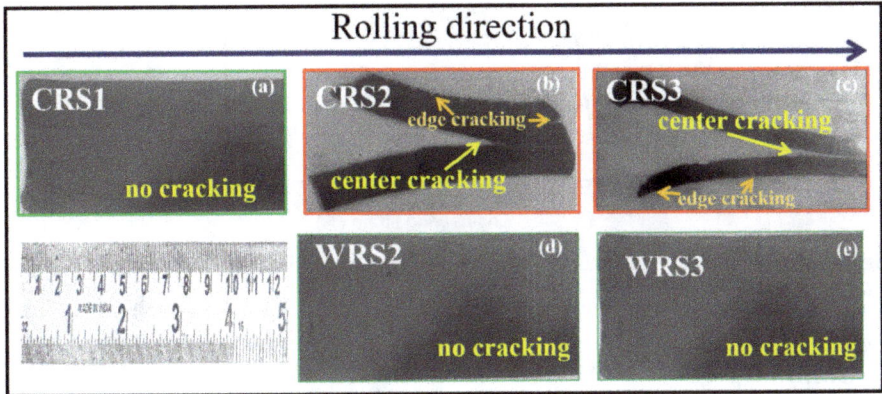

Fig. 7. The image of cold rolled steel plates of (a) CRS1, (b) CRS2, (c) CRS3 and warm rolled plates of (d) WRS2 and (e) WRS3 samples in rolling direction.

4 Tensile Curve

In the present study, it was observed that the HRS1, HRS2, and HRS3 samples experienced a decrease in tensile elongation, which can be attributed to the high Al content of 6.8 wt.%, 9.7 wt.%, and 9.1 wt.%, respectively. The presence of high Al content led to the segregation of Al, creating vulnerable sites that were prone to cracking during deformation. To address this reduction in ductility, an alternative approach could be to decrease the Al content while maintaining the density by incorporating other lightweight elements instead of Al. However, it was found that even a 5 wt.% Al addition could be sufficient to mitigate the loss of tensile elongation. Figure 8 illustrates the tensile properties of the HR and WR samples, wherein less than 2% elongation was observed due to the high Al contents. The strength values ranged from 300 MPa to 935 MPa. Notably, the tensile strength of the WRS2 and WRS3 samples were higher than that of the HRS1, HRS2, and HRS3 samples (Fig. 8).

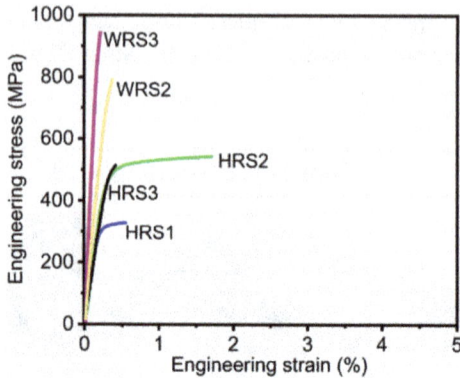

Fig. 8. Room temperature engineering stress-strain curves of HRS1, HRS2, HRS3, WRS2 and WRS3 samples

5 Summary

Based on the findings of the current investigation, the measured density of the S1, S2, and S3 steels are 8 to 15% lower than that of pure Fe. The microstructures of the HRS1 and HRS2 alloys primarily consist of ferrite, while HRS2 contains an additional 2.38% area fraction of Fe_3Al intermetallic. In HRS3, a brittle κ-carbide phase is present in the ferrite matrix, constituting 12.5% of the area fraction. After a 60% cold rolling of the HRS2 and HRS3 samples, cracks appeared at the center and edges due to the higher Al content, solid solution hardening and microhardness of HRS2 and HRS3 are 163.62 MPa, 161.91 MPa, and $216HV_{10}$ and $265HV_{10}$. To prevent cracking issues and minimize the effects of hardness and solid solution hardening, warm rolling of the HRS2 and HRS3 samples was carried out at rolling temperatures of 250 °C and 450 °C, respectively. The tensile results of the HR, and WR samples of the alloys show elongation values less than 2%.

Acknowledgement. Amrita Kundu acknowledges the financial support for this work from the Science and Engineering Research Board, Department of Science and Technology, Government of India under core research grant (File No.: CRG/2020/001511). The authors are grateful to Prof. P. C. Chakraborti from Jadavpur University for providing the experimental facilities and useful discussion. Special thanks to Prof. Shiv Brat Singh from IIT Kharagpur and Dr. Radhakanta Rana from Tata Steel for their valuable contributions in conceptualization, formal analysis, review, and discussions.

References

1. Rana, R., Liu, C., Ray, R.K.: Low-density low-carbon Fe-Al ferritic steels. Scr. Mater. **68**(5), 354–359 (2013)
2. Chen, S., Rana, R., Haldar, A., Ray, R.K.: Current state of Fe-Mn-Al-C low density steels. Prog. Mater. Sci. **89**, 354–359 (2013)

3. Baligidad, R.G.: Effect of niobium on microstructure and mechanical properties of high carbon Fe – 10.5 wt .% Al alloys. Mater. Sci. Eng. A **368** 131–138 (2004)
4. Singh, V.K., Rana, R., Singh, S.B., Kundu, A.: Effects of temperatures of rolling and annealing on microstructures and tensile properties of low carbon ferritic low density steels. Isij Int. **63**(5), 930–940 (2023)
5. Mishra, B.: Microstructure and deformation behaviour of austenitic low-density steels: the defining role of B2 intermetallic phase. Materialia **20**, 101198 (2021)
6. Morris, D.G., Munoz-Morris, M.G.: Work hardening in Fe–Al alloy. Mater. Sci. Eng. A **460**, 163–173 (2007)
7. Zuazo, I.: Complex metallurgy for automotive applications. Jom **66**(9), 1747–1758 (2014)
8. Sohn, S.S., Lee, B.J., Lee, S., Kwak, J.H.: Effects of aluminum content on cracking phenomenon occurring during cold rolling of three ferrite-based lightweight steel. Acta Mater. **61**(15), 5626–5635 (2013)
9. Xu, J., Wang, Z., Yan, Y., Li, J., Wu, M.: Effect of hot/warm rolling on the microstructures and mechanical properties of medium-Mn steels. Mater. Charact. **170**, 110682 (2020)
10. Pramanik, S., Suwas, S.: Low-density steels: the effect of Al addition on microstructure and properties. Jom **66**(9), 1868–1876 (2014)
11. Lu, Q., Xu, W., Van Der Zwaag, S.: Designing new corrosion resistant ferritic heat resistant steel based on optimal solid solution strengthening and minimisation of undesirable microstructural components. Comput. Mater. Sci. **84**, 198–205 (2014)
12. Raabe, D., Gutierrez-Urrutia, I.: Influence of Al content and precipitation state on the mechanical behaviour of austenitic high-Mn low-density steels. Scr. Mater. **68**(6), 343–347 (2013)
13. Han, S.Y., Shin, S.Y., Lee, S., Kim, N.J., Kwak, J.H., Chin, K.G.: Effect of carbon content on cracking phenomenon occurring during cold rolling of three light-weight steel plates. Metall. Mater. Trans. A **42**(1), 138–146 (2011)
14. Lanzillotto, C.A.N., Pickering, F.B.: Structure-property relationships in dual-phase steels. Met. Sci. **16**(8), 371–382 (1982)
15. Lee, S., Lee, C.G., Kwon, D., Park, S.H., Kim, N.J.: Analysis and prevention of vertical cracking phenomena during deep drawing of hot-rolled SG295 steel strips. Metall. Mater. Trans. A **27**(5), 1241–1250 (1996)

Effect of Cu on the Microstructure and Properties of Hot Rolled Low Carbon Steels

Kapil Dev Sharma[1], Arnab Sarkar[2], Sudipta Patra[3], and Anish Karmakar[1 (✉)]

[1] Department of Metallurgical and Materials Engineering, Indian Institute of Technology,
Roorkee 247667, India
`anish.karmakar@mt.iitr.ac.in`
[2] Department of Metallurgical and Materials Engineering, National Institute of Technology,
Rourkela 769008, India
[3] Department of Metallurgical Engineering, Indian Institute of Technology, BHU,
Varanasi 221005, Uttar Pradesh, India

Abstract. This study explores the impact of Cu addition on the mechanical properties of hot rolled low carbon steel. LCu and HCu steels with 0.6wt% Cu and 1.1 wt% Cu, respectively, are subjected to specific heat treatments that involves soaking the as-cast steels at 1250 °C for 2 h, followed by hot rolling at 1050 °C with 60% reduction in thickness, and thereafter air cooled to room temperature. A novel combination of ferrite and pearlite volume %, resulting from a carefully designed processing schedule, leads to a remarkable enhancement in the strength of HCu steel. Interestingly, HCu steel excels in hardness and yield strength, while LCu steel outperforms in ductility and strain hardening. Overall, this study firmly establishes the relationship between microstructural characteristics and mechanical properties of the investigated steels.

Keywords: Cu added steel · Precipitate · Tensile Property

1 Introduction

Low-carbon hot rolled steels are widely used in various industries such as automotive, construction machineries, and structural applications due to their favourable combination of strength, toughness, formability, and cost-effectiveness [1]. The reduced carbon content in this type of steel results in improved weldability and machinability [2]. The demand for high strength steels with superior toughness and excellent weldability ignited the extensive research on continuously cooled microstructures of low-carbon steels [3].

In the same aspect, Cu modified steels are extensively used due to its important role in manoeuvring solid solution and precipitation strengthening [4, 5]. Cu is highly soluble in austenite while being almost insoluble in ferrite. Additionally, Cu being an austenite stabilizer, enhances hardenability and reduces the austenite-ferrite transformation temperature, resulting in a finer ferrite microstructure. Apart from Cu, elements like Ti produces stable carbides and nitrides at high temperature, refining the austenite grain

S. Patra et al. (Eds.): METCENT 2023, *Proceedings of the International Conference on Metallurgical Engineering and Centenary Celebration*, pp. 248–254, 2024.
https://doi.org/10.1007/978-981-99-6863-3_24

size and resulting in increment in overall strength [6]. However, achieving the desired microstructural properties can be facilitated by optimizing the composition, thermomechanical processing and controlled cooling. Therefore, the present study aims to sheds some light on the role of Cu addition on the microstructure and mechanical properties of low carbon hot rolled steels.

2 Experimental Procedure

In this study, we investigated the hot rolled (HR) structures of two steel compositions with varying Cu content, denoted as LCu (0.6 wt.% Cu) and HCu (1.1 wt.% Cu) in Table 1. The Cu added low carbon steel were prepared through melting and casting route in a vacuum melting furnace. The resultant cast ingots underwent homogenization at 1250 °C for 2 h in a programmable muffle furnace. Subsequently, the thickness was reduced till 60% via hot rolling at 1050 °C, and thereafter air cooled to room temperature. The schematics of the thermomechanical processing are in Fig. 1(a).

Table1. Composition of both the LCu and HCu steels being investigated in this study.

Element	C	Si	Mn	Cr	Ni	Cu	Ti	Fe
LCu	0.2	0.06	0.6	0.5	0.7	0.6	0.14	Bal
HCu	0.2	0.07	0.6	0.5	0.7	1.1	0.14	Bal

The microstructural observation of both the hot rolled steels were examined using an optical microscope (Leica 2500DM), scanning electron microscope (SEM, Zeiss EVO18), and high-resolution transmission electron microscope (HR-TEM, JEM 3200FS). We utilized a Vickers hardness tester (FIE-VM50PC 07/2004–681) to quantify both microhardness and macrohardness, with a load of 50 gf and 10 Kgf respectively, and a dwell time of 10 s. Subsequently, uniaxial tensile tests was performed using an INSTRON-5982 universal testing apparatus equipped with 100 KN load cell along with INSTRON 2630–100 series extensometer. Specimen for tensile test were prepared in conformity with the ASTM E8 standards meticulously having gauge length, gauge width, thickness, overall length, fillet radius, and grip length of 50 mm, 12.5 mm, 5 mm, 200 mm, 12.5 mm, and 50 mm respectively. To ensure data consistency, three tests for each specimen were carried out with a strain rate of 0.001 s^{-1}. Ultimately, to evaluate the fracture behaviour and establish its correlation with microstructures, fractographs of tensile specimens were recorded using SEM.

3 Result

3.1 Microstructural Characterization

The optical and SEM micrographs in Fig. 1(b–e) depicts the complete dominance of pearlitic microstructure, alongside the ferrite phase in HR samples. The corresponding volume fraction of each phase are estimated from EBSD or SEM analysis and shown in Table 2.

Table 2. Volume % of different phases in hot rolled structures.

Phase	LCu	HCu
Ferrite	$(47 \pm 2)\%$	$(13 \pm 1)\%$
Pearlite	$(51 \pm 3)\%$	$(16 \pm 1)\%$
Degenerate Pearlite	$(2 \pm 0.5)\%$	$(71 \pm 3)\%$

It is observed that the ferrite content is significantly higher in LCu steel (~47 ± 2%) than HCu steel (~13 ± 1%). The elongated and continuous nature along the interface of the adjacent pearlitic boundaries indicates the allotriomorphic morphology of ferrite (Fig. 1(b–c)). Allotriomorphic ferrite is formed at the vicinity of pearlites which is a high energy region. The growth of this BCC phase has one-dimensional thickening normal to the boundary planes resulting in its rapid formation along the grain boundaries [7, 8]. Along with allotriomorphic ferrite, minimal amount of idiomorphic ferrite is also present. From Fig. 1(b–c), the refinement of pearlite colonies and adjacent ferrite grains is more visible in HCu steels (grain diameter ≈21 ± 2 μm) compared to the LCu one (grain diameter ≈ 30 ± 3 μm). Further, the high magnification SEM micrographs in Fig. 1(d) also exhibited a greater proportion of lamellar pearlite (~ 51 ± 3 vol.%, Table 2) in LCu steel. Conversely, the HCu steel predominantly showcases a degenerated pearlitic structure (Fig. 1(e)), constituting approximately (~ 71 ± 3 vol.%, Table 2).

An array of earlier literatures reported that the strength-ductility correlation is primarily influenced by the number density, morphology, and size of the precipitates [9, 10]. In the current study, we found TiC and Cu-rich precipitates of varying sizes, plays a vital role in controlling the strength-ductility combination. The bright field TEM image of HCu samples in Fig. 2(a) shows of the existence of TiC precipitate embedded between the adjacent ferrite grains, as marked with a yellow circle. The corresponding fast Fourier transform (FFT) pattern in Fig. 2(b) confirms the probable planes and directions of TiC precipitates lying along the [100] zone axis of ferrite.

In Fig. 2(c-f), through rigorous EDS mapping, we observed various Cu rich precipitates scattered in the grain boundaries, and grain interiors. The size of Cu precipitates in LCu specimen is around ≈ 80 ± 20 nm, while in the case of the HR HCu specimen, the Cu precipitate size is ≈ 53 ± 15 nm. Predominantly, the coarser precipitates are found at high-energy regions, i.e., grain boundaries or interphases in both the steels [11].

3.2 Evaluation of Mechanical Properties

To assess the initial mechanical response, macro hardness measurements were obtained from both hot rolled steels. Figure 3(a) shows that HCu steel exhibits a higher macro-hardness (~ 424 ± 18 VHN) than LCu steel (~ 260 ± 10 VHN). Furthermore, the engineering stress-strain curve of HR steels as the function of Cu concentration is presented in Fig. 3(b). From the stress-strain behavior, HCu specimens in the hot rolled condition show superior yield and tensile strength while the LCu specimen features better ductility. Stress-strain curves of both the steels exhibit

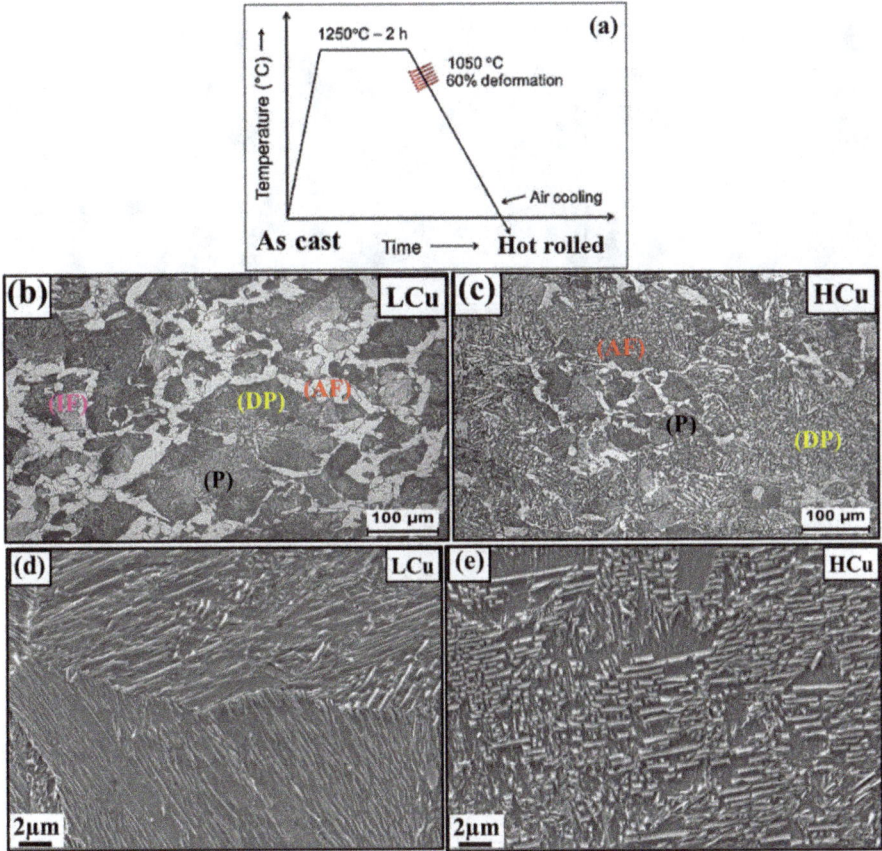

Fig. 1. (a) Processing schedule followed for both the investigated steels, (b, c) Optical microstructure of LCu and HCu respectively, representing allotriomorphic ferrite (AF) (red), idiomorphic ferrite (IF) (Magenta), pearlite (P) (blue) and degenerate pearlite (DP) (yellow), (d, e) SEM micrograph from LCu and HCu respectively, representing lamellar pearlite and degenerate pearlite respectively.

a smooth elasto-plastic transition. Figure 3(c) illustrates the variation of the strain hardening rate $\left(\frac{d\sigma}{d\varepsilon}\right)$ with true strain (ε) for both steels. Notably, it becomes evident that HCu steel possesses a greater strain hardening capability compared to LCu steel. The observed strain hardening during tensile deformation is a result of the deformation of various microstructural constituents.

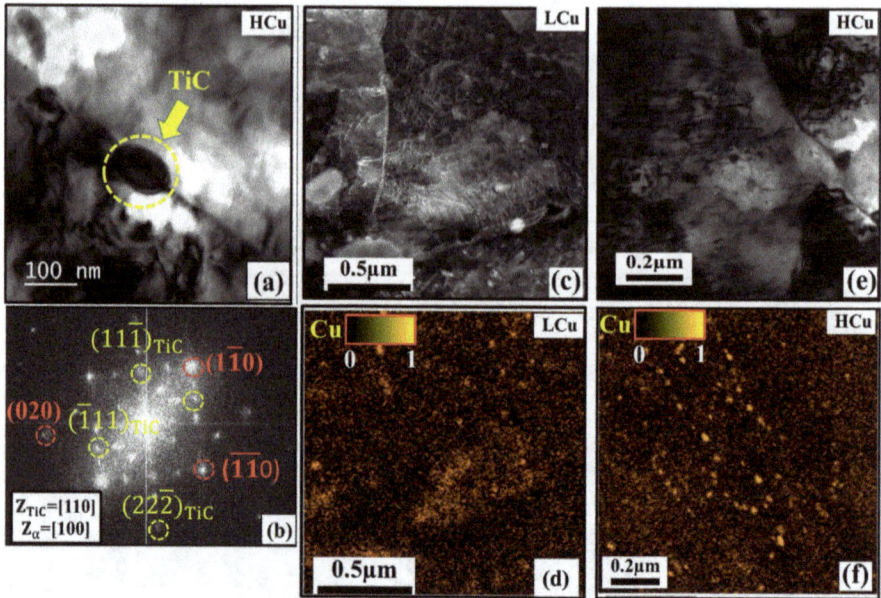

Fig. 2. (a) Bright field (BF) TEM micrograph showing TiC precipitate embedded between the two ferrite grains in HCu specimen, (b) FFT of the precipitate along with the matrix shown in Fig. (a), (c, d) Bright field TEM image of ferritic matrix and corresponding edx mapping of HR LCu specimen respectively, (e, f) Bright field TEM image of ferritic matrix and corresponding edx mapping of HR HCu specimen respectively.

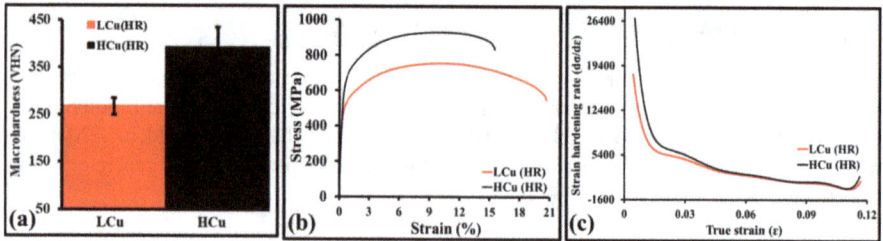

Fig. 3. (a) Macrohardness of hot rolled (HR) structure of LCu and HCu, (b) Engineering stress-strain curves for the hot rolled structure of both LCu and HCu steels, (c) Variation of strain hardening rate as a function of the true strain for the hot rolled structure of both LCu and HCu steels.

4   Discussion

In HCu steel, finer Cu precipitates have the tendency to pin down the grains resulting in the refinement of grain sizes. Therefore, the refined microstructure of HCu having average grain diameter $\approx 21 \pm 2\,\mu$m along with a substantial volume fraction of degenerate pearlite, leads to enhanced hardness and superior yield and tensile strength in the above mentioned steels. Fractographic analysis of the hot rolled LCu and HCu samples in Fig. 4(a, b) reveals the presence of void coalescence, river-like patterns, cleavage faces, and pullouts kind of morphology on fractured surfaces.

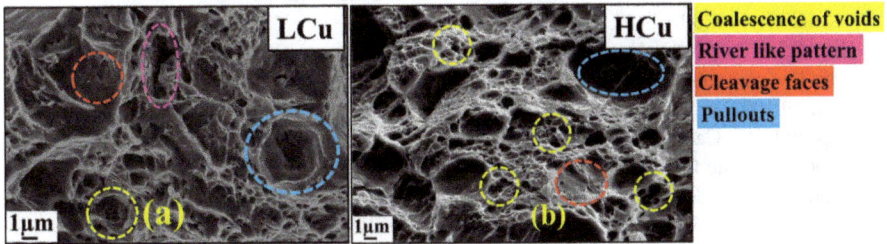

Fig. 4. (a, b) Fractographs of the fractured tensile specimens captured using SEM for both LCu and HCu specimens respectively, in hot rolled condition.

The river-like patterns observed are the consequence of delamination of pearlite colonies. High number density of fine Cu precipitates and smaller grain size in HCu specimen reduces ductility by accelerating crack growth. Conventionally, refined microstructure as well as smaller second phase particles causes the dislocations pile up during plastic deformation leading to the initiation of microcrack and its associated brittle behavior [12]. Even the better strain hardening behavior is the consequence of enhanced dislocation density occurs due to finer precipitates and smaller grain size in HCu specimen. Additionally, larger precipitates in LCu steels ($\sim 80 \pm 20$ nm) contribute to the formation of deeper pullouts by dislodging at the precipitate-matrix interface.

5   Conclusions

Both investigated steels show primarily pearlitic structures with allotriomorphic ferrite and a few traces of idiomorphic ferrite in the HR condition. In terms of mechanical performance, the HCu steel stands out with a dominant degenerate-type pearlite morphology in comparison to LCu specimen. Cu precipitates in HCu specimen are finer than their LCu counterpart. Therefore, large fraction of degenerate pearlite along with dominance of finer Cu precipitates and refined grain size are responsible for enhancing the strength, hardness and strain hardening behaviour in HCu specimen with reasonable amount of ductility.

References

1. Makhatha, M.E.: Effect of titanium addition on sub-structural characteristics of low carbon copper bearing steel in hot rolling. AIMS Mater. Sci. **9**, 604–616 (2022). https://doi.org/10.3934/MATERSCI.2022036
2. Kong, H., Liu, C.: A review on Nano-scale precipitation in steels. Technologies **6**, 36 (2018). https://doi.org/10.3390/technologies6010036
3. Reed-Hill,R.E.: Physical Metallurgy Principles, n.d. https://books.google.co.in/books/about/Physical_metallurgy_principles.html?id=_plTAAAAMAAJ

4. Shahriari, B., Vafaei, R., Sharifi, E.M., Farmanesh, K.: Aging behavior of a copper-bearing high-strength low-carbon steel. Int. J. Miner. Metall. Mater. **25**, 429–438 (2018)

5. Zhao, N., et al.: Strengthening-toughening mechanism of cost-saving marine steel plate with 1000 MPa yield strength. Mater. Sci. Eng. A **831**, 142280 (2022). https://doi.org/10.1016/j. msea.2021.142280

6. Kim, S.I., Chin, K.G., Yoon, J.B.: Precipitation hardening with copper sulfide in Cu bearing extra low carbon steel sheet. ISIJ Int. **49**, 109–114 (2009). https://doi.org/10.2355/isijinternat ional.49.109

7. Capdevila, C., Caballero, F.G., Garc, C., Andre, D.: Austenite grain size effects on isothermal Allotriomorphic ferrite formation. Mater. Trans. **44**, 1085–1095 (2003)

8. Babu, S.S., Bhadeshia, H.K.D.H.: A direct study of grain boundary allotriomorphic ferrite crystallography. Mater. Sci. Eng. A **142**, 209–219 (1991). https://doi.org/10.1016/0921-509 3(91)90660-F

9. Sarkar, A., Modak, P., Mandal, A., Chakrabarti, D., Karmakar, A.: Correlation between microstructure and tensile properties of low-carbon steel processed via different thermo-mechanical routes. J. Mater. Eng. Perform. 30–33 (2023). https://doi.org/10.1007/s11665-023-08492-2

10. Kisku, N., Sarkar, A., Ray, K.K., Mandal, S.: Development and characterization of a novel Ti-modified high-Si medium-Mn steel possessing ultra-high strength and reasonable ductility after hot rolling. J. Mater. Eng. Perform. **27**, 4077–4089 (2018). https://doi.org/10.1007/s11 665-018-3480-x

11. Stechauner, G., Kozeschnik, E.: Thermo-kinetic modeling of Cu precipitation in α-Fe. Acta Mater. **100**, 135–146 (2015). https://doi.org/10.1016/j.actamat.2015.08.042

12. Dieter, G.E., Bacon, D., Copley, S.M., Wert, C.A., Wilkes, G.L.: Adapted by (1988)

Role of Cerium on High-Temperature Oxidation Behaviour of Low-Carbon Steel

Chetan Kadgaye[1], Rachit Trivedi[1], Sudipta Patra[2], and Anish Karmakar[1(✉)]

[1] Department of Metallurgical and Materials Engineering, Indian Institute of Technology, Roorkee 247667, India
anish.karmakar@mt.iitr.ac.in

[2] Department of Metallurgical Engineering, Indian Institute of Technology, BHU, Varanasi 221005, Uttar Pradesh, India

Abstract. This study investigates the effect of Cerium (Ce) on the oxidation behaviour of low-carbon steels. Different compositions of the Ce-containing steels have been microstructurally examined through scanning electron microscopy (SEM) and energy dispersive spectroscopy (EDS) methods after isothermal holding at 700 °C for 120 h, simulating the oxidation phenomenon during various metallurgical processes. Characterization of the oxidized surface regarding their peculiar morphology and appearance has also been carried out. The inhibitory role of Ce on the kinetics of high-temperature corrosion and oxidation has been enumerated in this study. The insightful effect of the segregated and dissolved Ce, along with the different Ce-precipitates towards the oxidation behaviour, has eventually justified the resistance to high-temperature corrosion.

Keywords: Ce-modified steel · High-temperature corrosion · Oxidising mechanisms · Oxide scales · Electron microscopy

1 Introduction

Around the globe, ongoing efforts are being made to create engineering alloys, especially steels, that are affordable and effective for a particular application [1–5]. The low-carbon steels are frequently employed in various high-temperature metallurgical applications due to their reduced cost and consistent property [2–4, 6]. Strong efforts are being made to improve the mechanical properties of these materials by changing the alloy chemistry and microstructure, ranging from the development of basic carbon-manganese (C-Mn) to advanced high-strength steels (AHSS) [2, 3, 6]. These results in the development of rare-earth metal (REM) added steels, which have the potential for better high-temperature properties and oxidation resistance [4, 6].

Equipment failure results in direct or indirect economic losses and disruptions to human productivity and daily living [3, 6]. One of the leading causes of material failure is high-temperature oxidation, a highly intricate process [6, 7]. According to several research, REM substantially impacts steel's ability to oxidize at high temperatures [6].

S. Patra et al. (Eds.): METCENT 2023, *Proceedings of the International Conference on Metallurgical Engineering and Centenary Celebration*, pp. 255–260, 2024.
https://doi.org/10.1007/978-981-99-6863-3_25

In general, research on rare earth steel's ability to resist oxidation at high temperatures is examined principally from the perspectives of oxide shape, oxide growth rate, oxide phase composition, and oxide film structure [3, 7]. Additionally, the presence of rare earth elements might result in the selective oxidation of some oxides [1, 4]. Few researchers investigated the oxidation resistance of rare earth steel under high-temperature steam conditions, primarily utilized in power plant applications [7, 8]. Certain reports addressed the effect of REM on the cyclic oxidation behavior of steels at 1100 °C [4]. These ideas primarily claimed that REM were abundant near grain boundaries, which inhibited the diffusion of anions and cations [6, 9]. The oxidation rate may be slowed down while the adhesion of the REM oxides could be boosted during the high-temperature process [1, 3, 7].

In this study, the impact of Ce addition on high temperature (700°C) oxidation resistance properties of low carbon steels has been summarized. The reported experimental effort seeks to recommend an ideal REM addition for this steel to achieve a good trade-off between the quantity of large inclusions, precipitates, and their superior characteristics.

2   Experimental Methods

The chemical composition of the three hot-rolled low-carbon steels for the given investigation is shown in Table 1. The oxidized specimens with 10 mm × 10 mm × 5 mm dimensions were grounded, polished, and dried.

Table 1. Chemical Composition of the Three Investigated Steels in Weight Percent

Sample Name	C	Si	Mn	O	Cr	Ce	P	S	N	Ni	Cu	Mo
X65	0.05	0.23	1.57	–	–	–	0.015	0.003	0.0065	–	–	–
LCe	0.05	0.08	1.21	0.0088	0.03	0.03	0.017	0.007	0.0076	0.03	0.01	0.075
HCe	0.05	0.14	1.36	0.029	0.05	0.6	0.02	0.004	0.0185	0.04	0.01	0.08

The oxidation experiments were carried out in a resistance box furnace. The samples were exposed at 700 °C for 120 h, followed by air cooling to room temperature. The weight measurements were taken using an analytical scale with a 0.1 mg precision before and after the oxidation durations. After 120 h, the measurement time was completed. The study was conducted using scanning electron microscopy (SEM) with energy dispersive spectroscopy (EDS) to describe the microstructure of the specimens' oxidized surfaces.

3   Results and Discussion

The initial microstructures of the hot rolled blocks are shown in Fig. 1, together with the quantification of ferrite grain sizes and second phases.

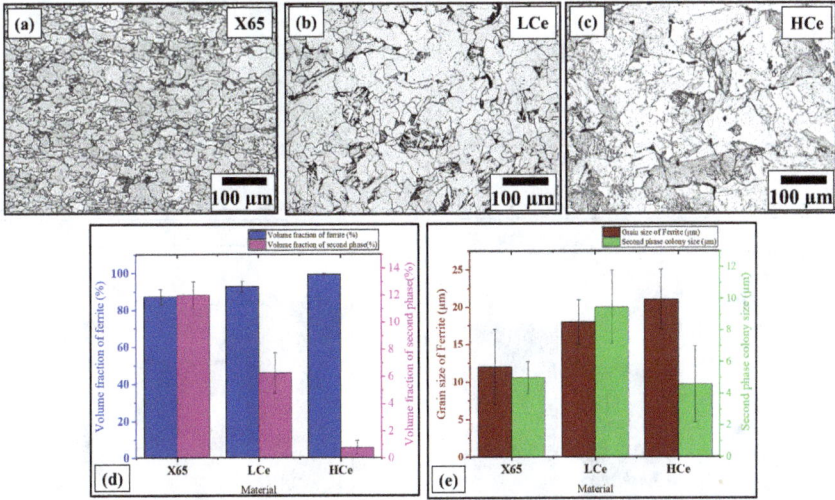

Fig. 1. (a–c) Optical micrographs of the various steel samples under investigation as shown in the images and (d & e) plot quantifying several microstructural characteristics in these steels.

The structures of all three materials are primarily ferritic. However, only the X65 and LCe steels exhibit pearlitic features. The HCe, on the other hand, emphasizes the presence of the Ce-rich intermetallics [6]. The X65 steel has the smallest grain size, followed by the LCe steel. According to Fig. 1(a, b, & d), the second phase's volume fraction was more significant for the X65 steel and smaller for the HCe steel [2, 6]. However, it was discovered that the X65 and HCe steel had comparable average second-phase colony sizes. In contrast, the LCe steel had the highest average size for the second phase colony, as shown in Fig. 1e.

The processing plan for the 120 h studied isothermal oxidation test at 700 °C is shown in Fig. 2a. The chart in Fig. 2(b–c) lists the mass ruptured, mass gain, and the oxidation rate, respectively, after the investigated test schedule. Figure 2 dictates that oxidation happened least for the LCe steel and highest by X65 steel. The oxidized surfaces of the experimental materials are shown in Fig. 3(a–c). The surface morphology of the oxidized specimens differs considerably for various steel compositions, Fig. 3(a–c). The polishing traces have not been seen, and the surfaces of the experimental steels are mainly covered in oxides with diverse morphologies. This demonstrates that the oxide films are sufficiently thick to cover exposed surfaces with continuous growth.

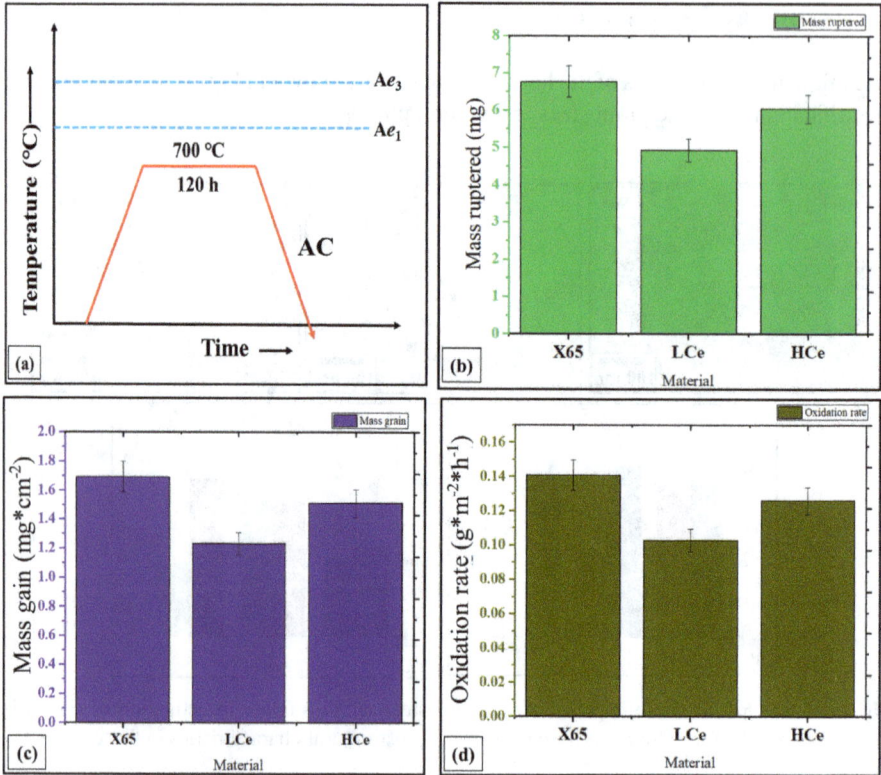

Fig. 2. (a) Processing schedule undertaken for isothermal oxidation test and associated (b) mass ruptured, (c) mass gain, and (d) the oxidation rate for the investigated steels.

As seen in Fig. 3a, the surface of the X65 steel has a nodular-like oxide layer and is inhomogeneous in size. Additionally, it is seen that there are numerous signs of shedding area on the surface of the X65 sample, Fig. 3a. However, the equal distribution of acicular-like oxide is more common in Ce-modified steels, Fig. 3(b, c). These acicular oxides fill the spaces between the spinel oxide and the denser surface oxides. The dense oxide coating can increase the antioxidant capacity of materials and reduce the inward diffusion of oxygen ions [3, 9].

An in-depth investigation of the surface oxides of the X65 steel revealed the presence of some minute oxidation cracks and fissures, making them loose and porous, Fig. 3a. Additionally, the oxides are very badly separated, and prismatic oxides may be seen in the shaded areas, Fig. 3a. Moreover, due to the higher fraction of the grain boundary areas, X65 steel is more prone to oxidation [2]. On the other hand, the acicular oxides present on the Ce-modified steel's surface impart flaky-shaped morphology. The morphologies of the oxides vary greatly with varying Ce content. Flaky-like formations are more common in LCe steel contrasted with HCe one, Fig. 3(b, c). Furthermore, the surfaces of Ce-modified steel contain numerous fine granular oxides scattered around the spinel oxide. These tiny granular oxides fill the spaces left by the spinel oxides, making a

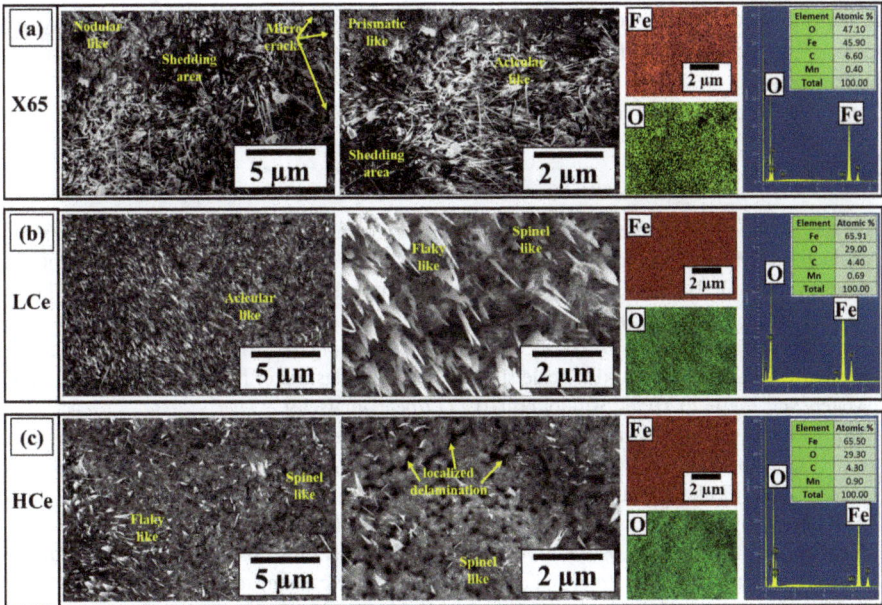

Fig. 3. Illustration of the various morphologies of the surface layers of specimens of (a) X65, (b) LCe, and (c) HCe steel that were oxidised at 700 °C for 120 h.

more compact oxidized surface. This is due to the fact that at high temperatures, the rare earth element can encourage the production of spinel oxides in steel surface, which are proven to be more effective in stopping future oxidation of the steel's interior [1, 3, 5]. However, the segregation of Ce at different interfaces and grain boundaries along with Ce-rich precipitates influenced the quest for better oxidation performance by inhibiting the diffusion of anions and cations [6].

The oxidation resistance also depends on the relationship between the thermal expansion coefficient of the various oxides produced on the steel's surface [4]. The distinct thermal expansion coefficient of oxides of X65 steel further promotes scale fragmentation with the development of microcracks and shaded areas. Additionally, similar thermal expansion coefficients of various oxides and matrix enhance the adhesion of the scale with the matrix [3, 9]. In the present case, compared to others, the LCe steel exhibits good oxidation resistance from the perspective of the oxidation rate and oxide layer properties. However, the insufficient Ce in LCe composition prevents it from acceptable oxidative characteristics related to the scale behaviour. Nonetheless, a value between 0.03 and 0.6 wt. % will probably be the critical Ce concentration at which the best resistant oxidative characteristics may be attained.

4 Conclusions

After continuous oxidation at 700 °C for 120 h., the Ce-added steels have shown less mass gain with a slower oxidation rate compared to X65 steel. Ce addition encourages the production of surface spinel oxide and increases its compactness, which may encourage the formation of interior oxides at the oxide/matrix interface. These interior oxides increase the surface area of the oxide film, improve the film's adhesion, and prevent the oxide shedding.

References

1. Wang, S., et al.: High temperature oxidation behavior of heat resistant steel with rare earth element Ce. Mater. Res. Express. **7**, 016571 (2020). https://doi.org/10.1088/2053-1591/ab692d
2. Kadgaye, C., Godase, S., Karani, A., Barat, K., Chakrabarti, D., Karmakar, A.: Correlating in-plane strength anisotropy with its microstructural counterpart for a hot rolled line pipe steel. Int. J. Press. Vessel. Pip. **199**, 104753 (2022). https://doi.org/10.1016/j.ijpvp.2022.104753
3. Wang, X., et al.: Effect of oxide scale structure on shot-blasting of hot-rolled strip steel. PeerJ Mater. Sci. **2**, e9 (2020). https://doi.org/10.7717/peerj-matsci.9
4. Yan, J., et al.: Effect of yttrium on the cyclic oxidation behaviour of HP40 heat-resistant steel at 1373K. Corros. Sci. **53**, 329–337 (2011). https://doi.org/10.1016/j.corsci.2010.09.039
5. Huntz, A.M., et al.: Oxidation of AISI 304 and AISI 439 stainless steels. Mater. Sci. Eng. A **447**, 266–276 (2007). https://doi.org/10.1016/j.msea.2006.10.022
6. Kadgaye, C., Hasan, S.M., Patra, S., Ghosh, M., Nath, S.K., Karmakar, A.: Role of cerium on transformation kinetics and mechanical properties of low carbon steels. Metall. Mater. Trans. A **52**, 3978–3995 (2021). https://doi.org/10.1007/s11661-021-06358-7
7. Suárez, L., Rodríguez-Calvillo, P., Houbaert, Y., Colás, R.: Oxidation of ultra low carbon and silicon bearing steels. Corros. Sci. **52**, 2044–2049 (2010). https://doi.org/10.1016/j.corsci.2010.02.001
8. Liang, Z., Zhao, Q., Singh, P.M., Wang, Y., Li, Y., Wang, Y.: Field studies of steam oxidation behavior of austenitic heat-resistant steel 10Cr18Ni9Cu3NbN. Eng. Fail. Anal. **53**, 132–137 (2015). https://doi.org/10.1016/j.engfailanal.2015.02.019
9. Suarez, L., Coto, R., Vanden Eynde, X., Lamberigts, M., Houbaert, Y.: High temperature oxidation of ultra-low-carbon steel. In: Defect Diffusion Forum, vol. 258–260, pp. 158–163 (2006). https://doi.org/10.4028/www.scientific.net/DDF.258-260.158

High Temperature Oxidation of T91 Alloy

Ashish Jain[✉], S. Maharajan, P. Logaraj, and A. Senthamil Selvam

Materials Chemistry and Metal Fuel Cycle Group, Indira Gandhi Centre for Atomic Research, Kalpakkam 603102, Tamilnadu, India
ashish@igcar.gov.in

Abstract. Ferritic-martensitic steels are extensively used in thermal and nuclear power plants due to their higher thermal conductivity and lesser thermal expansion coefficient. T91 alloy (Ferritic-martensitic steel) is proposed as material of clad for metal fuel fast breeder reactors. The choice is based on characteristics viz. Low swelling (BCC structure) and minimum DBTT (Ductile Brittle Transition Temperature) shift. The possibility of oxidation of this alloy under the influence of high temperature and exposure time cannot be ruled out. The studies on high temperature oxidation of T91 will be useful in understanding the underlying mechanism of oxidation and for further improvement in the oxidation resistance of T91. In the present work the high temperature oxidation of T91 alloy was studied using thermogravimetry technique. Both isothermal and non-isothermal methods were used. A multistep oxidation reaction was observed. Change in slope of the curve in temperature interval (1073–1673 K) showed the simultaneous oxidation of constituent elements of alloy T91 which is due to difference in the tendency of constituent elements to form respective oxides and is governed by the standard Gibbs energy of formation of respective oxide and kinetics of the reaction.

Keywords: T91 alloy · Kinetics · Oxidation · Isoconversional method

1 Introduction

T91 alloy, a Ferritic / martensitic (F/M) steel is considered as preferred structural material for advanced energy generating systems viz. Gen IV nuclear power plants and ultra-supercritical coal-fired power-generating units [4]. It has good creep rupture strength, steam oxidation resistance, modest corrosion resistance, good thermal conductivity, low coefficient of thermal expansion, low swelling and activation under irradiation, low susceptibility to inter granular stress corrosion cracking (IGSCC) [1, 8].

T91 steel is having excellent physical and mechanical properties, the long term exposure of T91 steel at high temperature may lead to oxidation. The susceptibility towards oxidation and high temperature may result in the formation of thick oxide scale. Scales formed inside the tube is the major cause of their failure. Heat transfer rate across the tube also decreases due to the accumulated scales inside the tube. It will increase the metal temperature which result in corrosion and creep degradation. High temperature

S. Patra et al. (Eds.): METCENT 2023, *Proceedings of the International Conference on Metallurgical Engineering and Centenary Celebration*, pp. 261–268, 2024.
https://doi.org/10.1007/978-981-99-6863-3_26

oxidation and analysis of products of oxidation have been studied by several authors. A brief description of the work reported so far is given in next few paragraphs.

Pantip et al. [1] studied the oxidation behavior of (F–M) alloys in supercritical water (SCW) to evaluate their suitability for Gen IV supercritical water reactor (SCWR) concept. The oxidation was studied in the temperature range 673–873 K and oxygen concentration was controlled to 10 - 300 ppb. Authors observed the dependence of reaction rate on temperature and it followed exponential rate law. The oxide formed on the alloy surface consisted of two layers. An outer layer of porous magnetite (Fe_3O_4) and an inner layer of iron chromium oxide, $(Fe-Cr)_3O_4$ was observed along with a transition region beneath the inner oxide. An oxidation mechanism based on the short circuit oxygen diffusion to the oxide–metal interface was inferred by the authors.

Dinesh Gound et al. [2] studied the oxidation behavior of T91 and T-22 steel in air under isothermal conditions at a temperature of 900 °C in a cyclic manner using thermogravimetry technique. Authors reported that the oxidation of T91 and T-22 followed somewhat linear rate and parabolic rate of oxidation respectively. Products of oxidation were analyzed by XRD, SEM and EDAX analysis.

Das et al. [3] reported the oxidation of T91 alloy in dry air at 873 K and above for 1000 h. It was observed that oxidation rate was influenced by temperature and exposure time and follows a parabolic rate law. Authors also observed formation of duel oxide layer as reported by Pantip et al. Inner layer was primarily rich in Fe and Cr and the outer layer only in Fe. The developed kinetic equation could be used to predict the thickness of oxide layer for a given temperature and exposure time.

Zhongfei et al. [4] reported the characterization of oxidation product formed on T91 ferritic/martensitic steel in oxygen saturated liquid lead-bismuth eutectic (LBE) at 823K at the nanoscale using focused-ion beam and transmission electron microscope. Authors reported that during the oxidation process, the diffusion of Cr occurred at micro and nano-scale. The micron-scale diffusion of Cr ensured the continuous advancement of IOZ (internal oxidation zone) and inner oxide layer, and nano-scale diffusion of Cr gave rise to the typical appearance of the IOZ.

ZHANG et al. [5] studied the oxidation of 9Cr-1Mo steel in (Ar + 10% H_2O) atmosphere at 600 °C, 650 °C and 700 °C. During oxidation for 10h the oxidation kinetics followed parabolic law at 600 °C, but two–stage parabolic law at 650 °C and 700 °C, in which the rate constant at the initial stage was higher than that at the second stage. The oxide scale with three layers (top to inner layer) was composed of Fe_2O_3, Fe_3O_4 and $(Fe,Cr)_3O_4$. Meanwhile the internal oxidation was also observed, the corresponding internal oxides were Cr_2O_3 and FeO.

Xiangyu et al. [6] studied the characteristic of oxide scales on T91 superheater tube after long-term service in an ultra-supercritical coal power plant by using x-ray diffraction and microscopy (electron and optical). Authors reported that the scale formed had multilayer structure. The outer layer consisted of Fe_3O_4 and Fe_2O_3 with some pores and tended to exfoliate. The inner layer consisted of Cr-rich spinel. Mechanism of oxidation and exfoliation was also described in the work.

Seifallah Fetni et al. [7] studied the corrosion of T91 tube used in subcritical conditions in an oil power plant. Authors reported the nature and morphology of the oxide scale multilayer structure (hematite, magnetite and spinel) which was depended on the environment and exposure time.

Most of the studies reported in the literature were carried out at a constant temperature. The information about the onset temperature of the oxidation reaction, total mass increase during the oxidation reaction, quantitative treatment of the data to obtain kinetic parameters (rate constant and activation energy) is scarce. In view of the above, it was decided to carry out studies on the oxidation of T91 alloy. This study explores isothermal and non-isothermal kinetics of oxidation of T91 alloy using isoconversional method.

2 Experimental

2.1 Materials

T91 alloy tubes were received from Nuclear Fuel Complex, Hyderabad. High purity oxygen and argon gas (purity > 99.99 wt. %) were procured from M/s. Inox Air Products Ltd, Chennai.

2.2 Experimental Procedure

Characterization

The T91 alloy was characterized for its chemical composition, impurity content and phase composition. The carbon content was determined by oxidizing the sample in a stream of oxygen in an induction furnace (T > 2500 K) and measuring the carbon dioxide evolved by an infrared detector [8]. The oxygen and nitrogen content in T91 alloy was determined by inert gas fusion technique [9].

The x-ray powder diffraction pattern was obtained by using an x-ray diffractometer (Explorer System procured from M/S. G N R s r l, Italy), employing filtered Cu Kα radiation in $\theta - \theta$ geometry, with a 2θ-scan step of 0.05°. National Institute of Standards and Technology (NIST) certified standard silicon powder was used for θ – calibration.

The thermogravimetry and differential thermal analyses was carried out by using a TGA / DSC 3 + combined differential thermal analysis and thermogravimetric analyzer supplied by M/S. Mettler Toledo, Switzerland. The temperature calibration of thermogravimetric analyzer was carried out by the method of fixed melting points employing International Confederation for Thermal Analysis and Calorimetry (ICTAC) recommended standards such as indium, aluminum and gold. This equipment has a mass sensitivity of 0.1 μg and a temperature sensitivity of 0.01 K. An alumina crucible of 70 μl capacity was used as the sample container. For each experiment 5 mg of the T91 alloy was used as sample. The temperature was measured using a Pt- Pt-10% Rh thermocouple which was in firm contact with the sample holder on which the alumina sample container was placed. The inner muffle of the furnace was constantly purged with oxygen, the flow of which was controlled by a mass flow controller. A constant gas flow of 20 sccm and a pre-determined rate of heating was used to carry out dynamic experiments.

A program with single segment was used for this study. Samples were heated from 1073 K to 1673 K at various heating rates viz. 1, 3, 5, 7 and 10 K min^{-1}. A fresh sample was used for each experiment. Reproducibility of the results obtained from different measurements was verified by carrying out measurements in duplicate or triplicate. The results were found to be in agreement with each other. Isothermal experiments were carried out by heating the sample at pre-determined temperatures viz. 1173, 1223, 1273 and 1323 for 360 min or till measurable change in mass was observed.

3   Results and Discussion

3.1   Characterization of T91 Alloy

The concentration of residual non-metallic impurities viz. Oxygen, nitrogen, carbon and sulphur is given in Table 1. The chemical composition of the T91 alloy under study is within the chemical composition of ASME standard T91 alloy [10].

Table 1. Elemental (Non metallic) composition of T91 alloy.

Elements	C	S	N	O
PPM	58	260	525	208

The x-ray diffraction pattern of the specimen (T91 alloy under study) is shown in Fig. 1(a). The xrd pattern matches with the pattern reported in ICDD file no. 04–004-9069. The x-ray diffraction pattern of the oxidized product is also shown in Fig. 1 b. The peaks pertaining to oxides of iron (Fe_2O_3) ICDD file no. 04-015-9572, chromium (Cr_2O_3) ICDD file no. 00-006-0532 and molybdenum ICDD file no. 00-013-0345 indicates the formation of oxides of constituent elements present in T91 alloy.

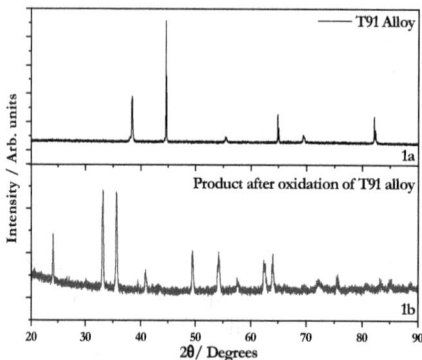

Fig. 1. XRD pattern of T91 alloy before and after oxidation

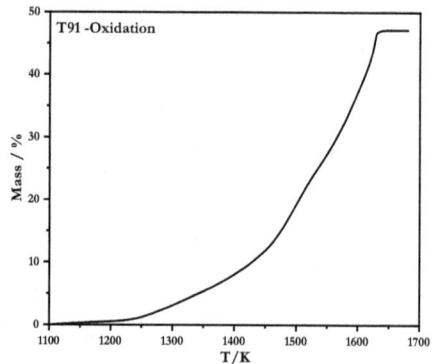

Fig. 2. Thermogram of T91 oxidation

3.2   Oxidation of T91 Alloy

It was observed in a separate experiment that the percent mass change during oxidation of T91 alloy from room temperature to 1100 K was insignificant indicating that the T91 alloy was resistant to oxidation up to 1100 K. In view of this for all non-isothermal experiments the start temperature for the oxidation of alloy was kept 1100 K. A typical thermogram for the oxidation of T91 alloy is shown in Fig. 2. The oxidation reaction started at ~ 1170 K and completed at 1650 K. The completion of the oxidation reaction is indicated by the line parallel to temperature axis, constancy of mass with temperature. The total mass gain (%) during the oxidation reaction was found to vary from ~ 45 to 36% (Table 2). The total mass gain (%) was found to be dependent on the heating rate and decrease in mass gain at higher heating rate was due to incomplete oxidation reaction.

Table 2. Mass (%) gain during non-isothermal experiments

Heating rate (K min^{-1})	1	3	5	7	10	15	20
Mass (%)	44.8	43.2	41.6	41.6	41.3	40.5	36.2

The change in the slope of the curve in temperature interval 1100 to 1680 K shows that the rate of percent increase in mass is varying with temperature during the oxidation reaction of T91 alloy.

For the purpose of kinetic analysis, the mass obtained after the completion of oxidation reaction was taken as the final mass 'm_f' and the extent of oxidation 'α', was evaluated by the following relations:

$$\alpha = \frac{m_T - m_i}{m_f - m_i} \qquad \text{for non-isothermal experiments} \qquad (1)$$

where m_i, m_f and m_T correspond to initial, final and intermediate masses at temperature 'T' respectively.

$$\alpha = \frac{m_t - m_i}{m_f - m_i} \qquad \text{for isothermal experiments} \qquad (2)$$

where m_i, m_f and m_t correspond to initial, final and intermediate masses at time 't' respectively.

3.3 Non Isothermal Kinetics by Using Isoconversional Method

The variation in extent of oxidation 'α' for T91 alloy with temperature at various heating rates is shown in Fig. 3. The extent of oxidation varies with heating rate at a fixed temperature. At a given temperature the extent of oxidation increases with decrease in heating rate. This is due to better thermal equilibration of sample at lower heating rates. As the reaction progresses, more and more alloy gets oxidized.

As per the isoconversional method [11] for a given conversion 'α' and a set of experiments performed under different linear heating rates 'β_i', the effective activation energy '$E\alpha$' for each 'α' value is determined by the following:

$$\text{In} = \left(\frac{\beta_i}{\Delta T_{\alpha,i}}\right) = \text{const.} - \frac{E_\alpha}{RT_{\alpha,i}} \qquad (3)$$

The slope of the plot of $\ln\left(\frac{\beta_i}{\Delta T_{\alpha,i}}\right)$ vs $1/T_{\alpha,i}$ gives the value of the effective activation energy 'E_α'. A plot of '$E\alpha$' as a function of 'α' for the oxidation of T91alloy is shown in Fig. 4. The 'E_α' is varying with 'α' and it changes rapidly as the reaction progresses. The increase / decrease in the value of 'E_α' is probably due to presence of multi elements in T91.

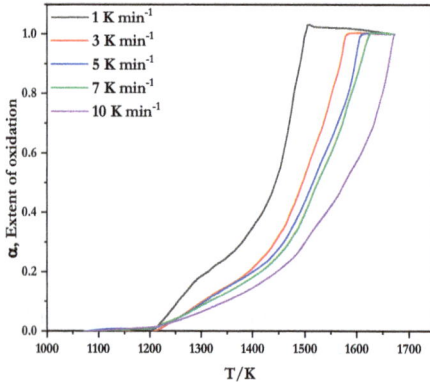

Fig. 3. Variation of extent of oxidation 'α' with temperature at various heating rates

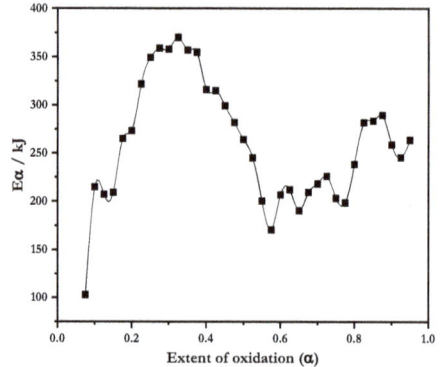

Fig. 4. A plot of '$E\alpha$' as a function of 'α' for T91 alloy (Non isothermal experiments)

Each element viz. Fe, Cr, Mo, Nb, V, Si and carbon having varying tendency to form their respective oxides. Thought the formation of oxide is thermodynamically feasible but the rate at which the individual element reacts with oxygen to form oxide varies. The formation of an oxide from metal is a function of temperature, time and thermodynamic stability of the oxide. Whether the presence of different humps in the plot corresponds to the formation of different oxides in sequence is the matter of further research.

3.4 Isothermal Kinetics

Isothermal experiments were performed in the temperature interval where the change in mass was found large (as observed from the non-isothermal data). Four temperature points viz. 1169, 1218, 1268 and 1318 K were chosen for the isothermal experiment and each experiment was performed for 6 h. No attempt was made to increase temperature and time further as chemical reaction between the T91 alloy and crucible material was observed at higher temperature and time duration. The mass gain for the isothermal experiment at different temperature is given in the Table 3. The maximum mass gain observed at 1318 K isothermal temperature is almost half the value observed for non-isothermal experiments. Lower mass gain at lower temperatures (1169 and 1218 K) is due to only surface oxidation of the sample. A plot of 'α' with 't' at various temperatures for T91 alloy (Isothermal experiments) is shown in Fig. 5.

The variation of extent of oxidation w.r.t. time is not similar for four isothermal temperatures. It is possible that one element may undergo oxidation at one particular isothermal temperature but the other element may show resistant towards oxidation at the same temperature. Hence, the extent of oxidation is a function of time and temperature.

Table 3. Mass (%) gain during isothermal experiments

T / K	1169 K	1218 K	1268 K	1318 K
Mass (%)	5.5	8.6	10.1	20.07

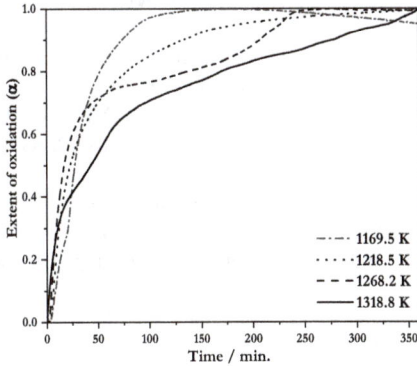

Fig. 5. A plot of 'α' with 't' at various temperatures for T91 alloy (Isothermal experiments)

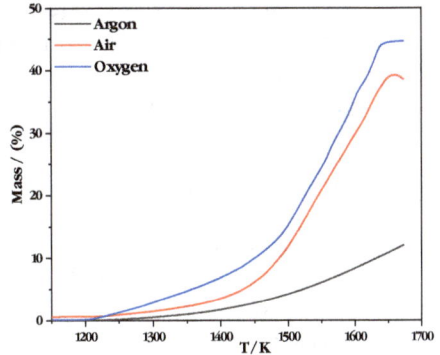

Fig. 6. A plot of percent increase in mass with 'T' under varying atmosphere

The sequential oxidation of constituent elements present in the alloy T91 was observed and this is also affected by the presence of other oxide.

3.5 Effect of Partial Pressure of Oxygen pO2 on the Oxidation of T91

pO_2 in the vicinity of the sample was varied by carrying out the experiment using three different reactive gases viz. Oxygen, air and UHP argon (2 ppm of O_2). The thermogravimetric curves obtained are shown in Fig. 6. It was observed that mass gain / extent of oxidation was a function of partial pressure of oxygen. Rate of oxidation could be controlled even at high temperature by suitable controlling the pO_2.

3.6 Conclusions

Oxidation of T91 alloy at high temperature was studied using thermogravimetry technique. Non isothermal and isothermal experiments were carried to find the extent of oxidation. The high temperature oxidation of T91 alloy was found to be a function of rate of heating temperature, time and partial pressure of oxygen. The effective activation energy 'E_α' was found to vary with extent of reaction and temperature. Further work to understand the kinetics and mechanism of oxidation is in progress.

References

1. Ampornrat, P., Was, G.S.: Oxidation of ferritic–martensitic alloys T91, HCM12A and HT-9 in supercritical water. J. Nucl. Mater. **371**, 1–17 (2007)

2. Gond, D., Chawla, V., Puri, D., Prakash, S.: Oxidation studies of T-91 and T-22 boiler steels in air at 900 °C. J. Miner. Mater. Charact. Eng. **9**(8), 749–761 (2010)
3. Das, S., Sanyal, Ṣ, Halder, P., Varma, A., Ravi Kumar, Y., Mandal, S.: Oxide scale characterization and study of oxidation kinetics in T91 steel exposed to dry air at high temperatures (873–1073 K). Met. Mater. Int. **28**, 1864–1880 (2022)
4. Ye, Z., Wang, P., Dong, H., Li, D., Zhang, Y., Li, Y.: Oxidation mechanism of T91 steel in liquid lead-bismuth eutectic: with consideration of internal oxidation. Sci. Rep. **6**, 35268 (2016). https://doi.org/10.1038/srep35268
5. Zhang, D., Xu, J., Zhao, G., Guan, Y., Li, M.: Oxidation characteristic of ferritic–martensitic steel T91 in water–vapour atmosphere. Chin. J. Mater. Res. **22**(6), 599–605 (2008)
6. Zhong, X., Xinqiang, W., Han, E.-H.: The characteristic of oxide scales on T91 tube after long-term service in an ultra-supercritical coal power plant. J. Supercrit. Fluids **72**, 68–77 (2012)
7. Fetni, S., Montero, D., Boubahri, C., Brouri, D., Briki, J.: Evolution mechanisms of T91 steel in subcritical conditions and role of an internal oxidation zone. Oxid. Met. **90**, 291–315 (2018)
8. Sayi, Y.S., Ramkumar, K.L., Venugoal, V.: Determination of non-metallic impurities in nuclear fuel. IANCAS Bull. **7**(3), 180–81 (2008)
9. Chetty, K.V., Radhakrishna, J., Sayi, Y.S., Balachander, N., Venkataramana, P., Natarajan, P.R.: Radiochem. Radioanal. Lett. **58**, 161–62 (1983)
10. Li, Y., Jinfeng, D., Li, L., Gao, K., Pang, X., Volinsky, A.A.: Mechanical properties and phase evolution in T91 steel during long-term high-temperature exposure. Eng. Fail. Anal. **111**, 104451 (2020)
11. Ortega, A.: A simple and precise linear integral method for isoconversional data. Thermochim. Acta **474**, 81–86 (2008)

Development of High Quality SWRH82A Grade Tyre Cord Steel Wire Rods

S. Monalisa Nayak, Abinash Dash, Gaurav Mehta, Sujit Kumar[✉], Vishwanathan N., and Surya Kumar Singh

Research & Development Department, Jindal Steel and Power, Raigarh 496001, India
{Sujit.kumar,vishwanathan.nagarajan}@jindalsteel.com

Abstract. The demand for high-quality tyre cords has increased significantly with the expansion of the automotive industry. High carbon steel wire rods are extensively used as drawn-wire materials for tyre cords and tyre beads, springs, suspension bridge cables, etc., due to their superior combination of strength and toughness. Tyre cord steel is typically drawn into thin wire sizes 0.15 to 0.38 mm dia. SWRH82A grade steel is widely used in tyre cord manufacturing due to its excellent mechanical properties and durability. However, the presence of grain boundary cementite and large hard non-deformable inclusions can lead to increased failures during wire drawing operations. This paper discusses the measures taken to reduce inclusions, segregation & grain boundary cementite to enhance performance of tyre cord by means of optimizing steelmaking, casting & rolling processes.

Keywords: High carbon steel · tyre cord · casting process parameters · laying head temperature · cooling rate · grain boundary cementite · tensile strength

1 Introduction

SWRH82A grade steel is a popular choice due to its high tensile strength and excellent fatigue resistance. However, the presence of grain boundary cementite and inclusions can negatively impact the drawability, necessitating the development of improved steelmaking and hot rolling manufacturing techniques to overcome these challenges (Fig. 1).

Wire breakages during the wire drawing operations can be a significant issue depending on the quality of steel rods. Two major reasons for wire breakages in SWRH82A Grade wires are i) carbon segregation and ii) steel cleanliness. Carbon segregation in SWRH82A grade could lead to formation of Grain Boundary Cementite (GBC) network at the centre, leading to reduction in the drawability of the steel rods. Furthermore, presence of any hard non-deformable inclusions leads to formation of voids with sharp stress raisers at the inclusion sites along with reduction in the overall load-bearing cross-sectional area and hence causing wire breakages during wire-drawing. The present paper details the measures taken at Jindal Steel & Power to reduce the formation of GBC and inclusions in SWRH82A tyre cord steel.

© The Author(s), under exclusive license to Springer Nature Singapore Pte Ltd. 2024
S. Patra et al. (Eds.): METCENT 2023, *Proceedings of the International Conference on Metallurgical Engineering and Centenary Celebration*, pp. 269–272, 2024.
https://doi.org/10.1007/978-981-99-6863-3_27

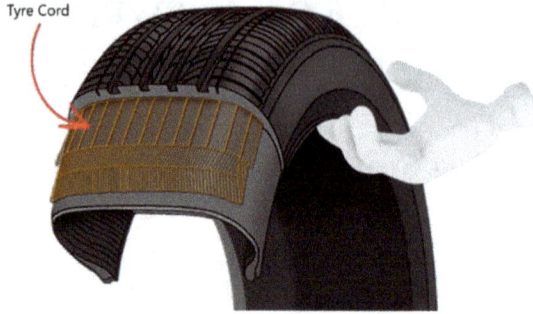

Fig. 1. Schematic of tyre cord assembled in heavy vehicle tyre [1]

2 Grain Boundary Cementite

The carbon content in SWRH82A is at the eutectoid point, and when there is carbon segregation in the billets, the central region will locally behave as a hypereutectoid steel giving rise to proeutectoid cementite [2] formation along the prior austenite grain boundaries during austenite to pearlite transformation upon cooling after hot rolling. A higher intensity of carbon segregation can lead to formation of a network of such grain boundary cementite which reduces the integrity of the wire at the central region significantly affecting the drawability of the wires, leading to wire breakages during wire drawing.

To reduce the GBC formation, firstly the carbon segregation in the billets was minimised, by optimising the casting parameters like superheat, casting speed, EMS current and frequency in the mould, volume of water flowing in the mould and in the secondary cooling region, etc. The result was a significant reduction in the carbon segregation index from a value of 1.07 (ratio of carbon content at the center to the average carbon content in the billet) to 1.01. The carbon contents in the billets were measured by LECO and the measured values in the billets produced before and after the above-mentioned casting parameter optimisation are shown in Table 1.

Table 1. Carbon segregation index measured in the billets before and after the casting process optimisation

Heat	Grade	Sample ID	%C		Carbon Segregation Index
			Avg.	Core	
A	SWRH82A	Before Improvement	0.842	0.90	1.07
B	SWRH82A	After Improvement	0.840	0.852	1.01

In addition to the improvement in the carbon segregation at the casting stage, the tendency for formation of GBC was further minimised by suppressing the proeutectoid cementite formation during the austenite to pearlite transformation after hot rolling by increasing the cooling rate, by 10 °C/s from the existing value, in the Stelmor conveyor.

This was achieved by increasing the blowing capacity of the existing blowers and addition of two new blowers in the Stelmor conveyor. Further to this, the laying head temperature was also increased by 40 °C, and the mill and conveyor speeds were optimised to achieve the required higher cooling rates in hot rolled wire rods in the Stelmor conveyor.

The above modifications resulted in a significant improvement in the GBC performance in the wire rods. The pass ratio for GBC has significantly improved from the previous value of 93.7% to the present values of 98.0% (Fig. 2).

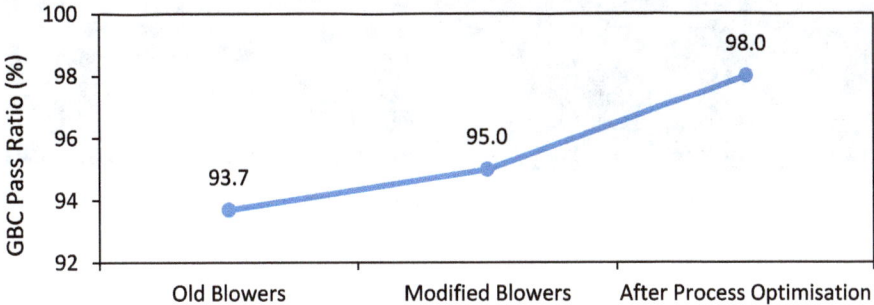

Fig. 2. Improvement in grain boundary cementite at different stages of modifications to the production process

3   Large Non-deformable Inclusions

The degree of cleanliness during steel casting is considered as a prime factor to achieve improved mechanical properties to meet the ever increasing demand of high-quality tyre cord steel. Presence of large non deformable inclusions result in breakage during wire drawing operation. In the deoxidation process during steelmaking, Aluminium killing is avoided due to their tendency for formation of non-deformable inclusions, and hence Silicon killing is preferred for such critical applications. Although the SWRH82A wire rods were Si-Mn killed, hard non-deformable inclusions of $Al_2O_3 \cdot MgO$ spinel and CaO-Al_2O_3-SiO_2-MgO type were also observed occasionally in the wire rods. The presence of Al_2O_3 makes the inclusions non deformable [3, 4] and hence detrimental to the wire drawing operation. Scanning electron micrograph (SEM) images along with their elemental mapping through EDS of specific types of inclusions observed are shown in Fig. 3.

The spinel inclusions observed were large, measuring up to 40 microns in some cases, and angular in shape with sharp corners. Several Ca related inclusions (calcium alumino silicates, calcium alumino magnesium silicates, etc.) were also observed, but were of globular shape and smaller in size (<6 microns). Hence, the spinel inclusions were the primary cause for wire breakage during the wire drawing operation.

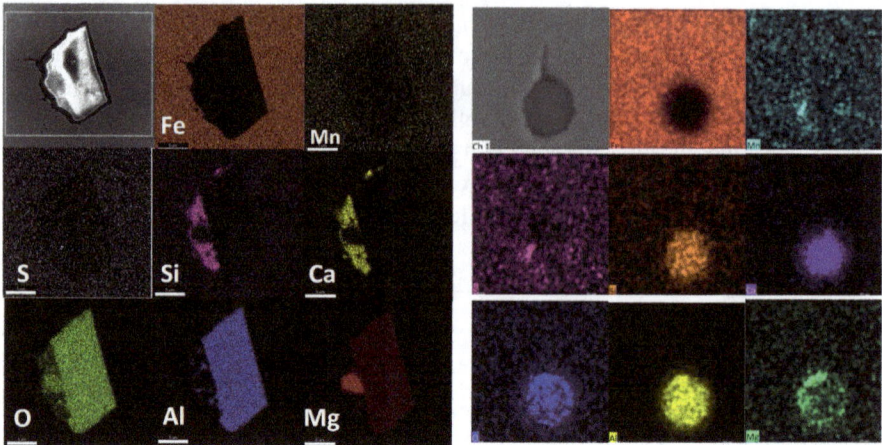

Fig. 3. SEM images and EDS mapping of few types of inclusion observed in the SWRH82Agrade Wire Rod Samples

To track formation of spinel inclusions, individual heats starting from tapping to casting were analysed. Slag samples and steel chemistry was analysed for each heat at various stages of steelmaking, starting from EAF tapping to up to tundish. Samples were also taken from the used ladle refractory wall to study the nature of dissolution of the MgO based refractory into the steel. Furthermore, several steelmaking parameters were optimized. The most important of them were 1) ladle slag basicity was maintained at 1.2 to 1.4, 2) Al_2O_3 content in the slag was reduced to lower than 10% to minimize Al_2O_3 entrapment from ladle slag and 3) increased rinsing time after degassing, to facilitate flotation of calcium based inclusions.

References

1. Newsroom Posco. https://newsroom.posco.com/en/tire-cord-a-hidden-ingredient-for-tire-saf ety/
2. de Ouro Preto, F.: Characterization of the proeutecoid cementite networks observed in the SAE 1092 wire rod steel grades by Fabio da Silva Borchardt Metallurgical engineer Universidade (1997)
3. Liao, J., Qing, G., Zhao, B.: Phase equilibria studies in the CaO-MgO-Al2O3-SiO2 system with Al2O3/SiO2 weight ratio of 0.4. Metals **13**(2), 224 (2023). https://doi.org/10.3390/met 13020224
4. Li, Y., Yang, W., Zhang, L.: Formation mechanism of MgO containing inclusions in the molten steel refined in MgO refractory crucibles. Metals **10**(4), 444 (2020). https://doi.org/10.3390/ met10040444

Processing and Structure-Property Correlation

Rotary Friction Welding of Cast Nickel Base Superalloy with Martensitic Stainless Steel for Aero Engine Application

Shweta Verma[✉], Vijay Petley, and A. Manjunath

Gas Turbine Research Establishment, DRDO, Bangalore, Karnataka, India
shweta.gtre@gov.in

Abstract. In the present investigation, rotary friction weld joints were prepared between cast Nickel base super alloy & Martensitic stainless steel which is used for disk and shaft assembly of aero engine respectively. The joints were obtained by rotary friction welding process (RFW), which combines the heat generated from friction between two surfaces and plastic deformation. Test Samples were prepared with different welding process parameters. These test samples were subjected to tensile tests, Vickers micro hardness, metallographic tests and energy dispersive spectrometry (EDS) analysis in scanning electron Microscope after clearance by radiography & fluorescence Penetrant inspection. The friction weld joints revealed radial plastic flow, axial plastic flow, thermo mechanically affected zone in the weld microstructure. EDS analysis at the interface of the junction showed that inter diffusion occurs between the main chemical components of the materials involved. At low friction force & friction time, proper fusion of the two materials was not observed, while micro cracks were observed at the joint when friction force & friction time are increased. An optimum combination of the two, revealed weld joint with proper fusion and no micro crack. These weld joints were evaluated for their mechanical properties by using conventional tensile tests at room temperature and recorded the maximum fracture strength of 1009 MPa. The strength of the joints varied with increasing friction time and the use of different pressure values. Finally, the results were concluded that most of the failure has been observed in the steel side thermo mechanical affected zone (TMAZ). The aim of this research works to emphasis the dissimilar weld joint characteristics for aerospace and defence application. The optimized weld parameter was used to realize the rotor & disk assembly for the aero engine application. The established welding parameters are friction force of 2 ton, upset force 3 Ton, friction time 3.42 s, rotational speed 1700 rpm.

Keywords: Axial plastic flow · Friction welding · Inter diffusion · Radial plastic flow · Thermo Mechanical Affected Zone

1 Introduction

Friction welding belongs solid state welding process wherein, a rotating part comes in contact with a stationary component. The heat generated due to friction force between the two parts helps in joining the parts. This metallic bond gets created at a temperature

© The Author(s), under exclusive license to Springer Nature Singapore Pte Ltd. 2024
S. Patra et al. (Eds.): METCENT 2023, *Proceedings of the International Conference on Metallurgical Engineering and Centenary Celebration*, pp. 275–285, 2024.
https://doi.org/10.1007/978-981-99-6863-3_28

lower than the melting point of the base metal & hence the joint has properties similar to the base metal properties [1, 2].

Inertia driven welding & continuous driven welding are two types of friction welding process. In the first welding process, one component is clamped in a spindle chuck with a attached fly wheel while the second is kept still to join them. A predetermined amount of energy is stored by rotating the fly wheel and chuck assembly at a certain speed(s). The drive to the fly wheel is declutched and two components are brought together under axial pressure for welding.

In continuous drive friction method, one of the components held stationary while the other rotated at a constant speed(s). The two components are brought together under axial pressure for a certain friction time. Then the clutch is separated from the drive, and the rotary component is brought to stop within the braking time while axial pressure on the stationary part is increased to a higher upset pressure for a predetermined upset time. This is also called as rotary friction welding (RFW) [2].

Efficient distribution & utilization of the welding heat is very important in order to obtain a sound weld joint, hence during the heating phase when the two parts are under pressure, the heated interface accumulates the heat and the temperature reaches up to the plasticity temperature of the base metal. Thus, a type of thermo mechanical treatment occurs in the welding region and this region has stable plastic structure [3].

RFW is the most common form of friction welding being used in industries because of relatively high overhead cost being balanced with higher production rate and low labour requirement. Most of the materials which cannot be welded by other welding techniques can be welded by friction welding. Its adjustable hardware and several dimensions make it useful for widespread application in the industry. Automotive industry, aviation and space industry machine production and spare part industry [3].

1.1 Material Property Requirement and Parameters of Friction Welding

Friction welding method requires good strength & ability to deform plastically under the applied force are the two important factors which results in a good quality weld joint. Axial pressure & torque applied during the process may deform the material excessively if the material does not have enough strength. A good response towards heat treatment is also another criterion for the material to result in defect free weld joint. The two materials which have similar properties results in asymmetrical deformations and in turn yields a higher welding strength. A good forgeable material can be friction welded even of it doesn't possess good friction properties. Alloy elements supplying dry oil prevent the joining section from reaching welding temperature [4].

There are various parameters which need to be controlled while friction welding. The rpm of the parts, friction time & pressure, forging time and pressure are the parameters needs to be considered while optimizing the welding process. Geometry of the part and material properties are also significant [3].

The low energy requirement and relatively less welding time are the main attributes which makes friction welding a very beneficial process. Friction welding is generally compared to electrical resistance welding. The power requirement of friction welding is about one tenth of electrical resistance welding [3].

2   Experimental

2.1   About Blisk Assembly

Present study shows the use of rotary friction welding for joining of steel shaft with Ni base super alloy blisk used in gas turbine industry. The low pressure turbine blisk assembly along with shaft rotates at speed higher than 30000 rpm and has diameter more than 200 mm & shaft length more than 500 mm. The welding trails were performed with simulated diameter of shaft & blisk on test specimens. Weld joint properties are established before welding the actual parts.

2.2   Material

The materials AE961W steel, a Martensitic stainless steel (MSS) in hardened & tempered condition and BZL-12Y BE Ni base super alloy in as cast condition were used in the present work. The elemental compositions for both the alloys are mentioned in Table 1 and Table 2.

Table 1. Elemental composition of AE961W steel [4]

% Element	C	Si	Mn	Ni	Cr	P	S	Mo	W	V	Cu
AE961W	0.1–0.16	0.6 max	0.6 max	1.5–1.8	10.5–12.0	0.03	0.025	0.35–0.5	1.6–2.0	0.18–0.30	0.25 max

Table 2. Elemental composition of BZL-12Y BE alloy [4]

% Element	C	Cr	Ti	Al	Mo	W	Co	Nb	V	B	Zr	Ce
BZL-12Y BE	0.14–0.20	8.5–10.5	4.2–4.7	5.0–5.7	2.7–3.4	1.0–1.8	12.0–15.0	0.5–1.0	0.5–1.0	0.015 max	0.02 max	0.025 max

2.3   Friction welding method

The test specimens of MSS & Ni base super alloy of dimension12 mm diameter & 29 mm diameter respectively were joined together using rotary friction machine (ETA make friction welding machine, TRW) (Fig. 1a and b) of capacity 20 ton using four different parameters as mentioned in Table 3.

Table 3. Welding Parameter

Sample	RPM	Friction force (T)	Burn off time (s)	Upset force (T)	Total burn off (mm)
1	1700	1	7.87	2	4.83
2	1700	0.6	8.82	2	5.03
3	1700	1.5	5.23	3	5.27
4	1700	2.0	3.42	3	5.27

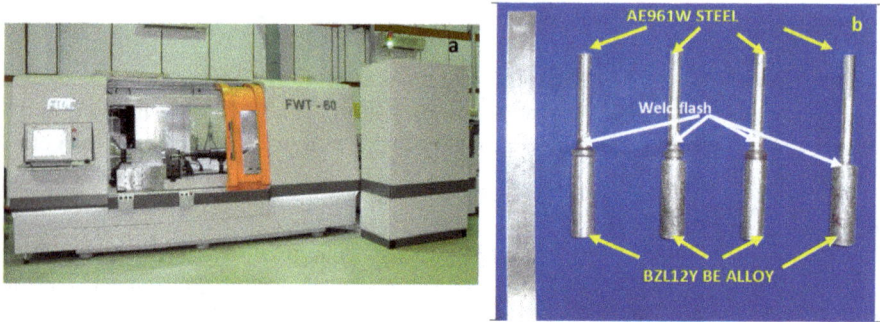

Fig. 1. (a) Rotary Friction welding machine (b) Friction welded sample

2.4   Mechanical Properties

The friction welded samples were EDM wire cut to fabricate the tensile specimen as per ASTM E8 Fig. 2. Room temperature tensile test were carried out using MTS make 50 T Universal testing machine. Samples were tested till fracture to record the ultimate strength of the joint.

2.5   Microstructure Characterization

Specimens welded with different weld parameter were prepared along the axial direction. Standard metallography techniques were employed for the microstructure preparation. The polished specimens were etched with Marble reagent to reveal microstructure. Microstructure images and micro hardness readings were taken to analyse the effect of parameters on the weld joint, so that the best parameter can be chosen for joining a dummy blisk assembly with no defects. Optical microscopy was carried out by Lecia Make Digital Video microscope system & Scanning Electron Microscopy was performed using Carl Zesis make SEM. Micro hardness across the weld joint at an interval of 0.25 mm is measured as per ASTM E 384 using Wilson micro hardness tester with 300 gmf load and dwell time of 10 s.

Fig. 2. Tensile test specimen

3 Results and Discussion

3.1 Tensile Test

Tensile tests were conducted for all four dissimilar weld joints and are compared with the individual tensile properties of the material. All the samples failed from the AE961W steel portion and exhibited intermediate strength as compared with the strength of the both the material Table 4 and Fig. 3. Many researchers also have found failure in steel at thermo-mechanically affected zone [5–10].

Table 4. Room temperature tensile properties of parent materials & samples

SL. No	0.2% Yield stress (MPa)	Peak Stress (MPa)	Elongation (%)
BZL12YBE [4]	716–755	883–915	5–10
AE961W [4]	961	1079	17–20
Sample 1	812	878	4
Sample 2	809	900	3.6
Sample 3	921	1009	2.7
Sample 4	864	907	1.2

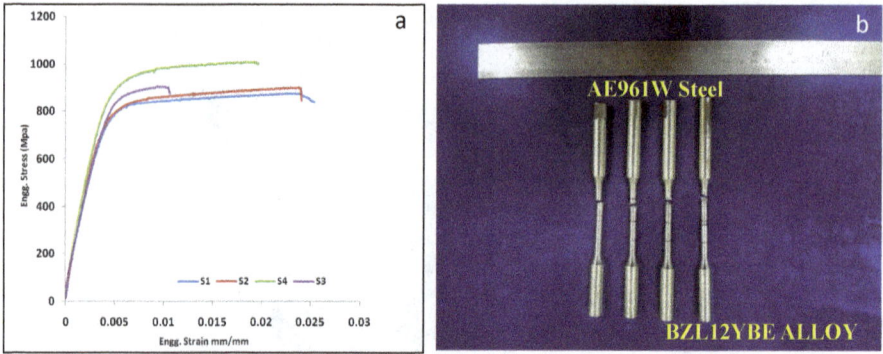

Fig. 3. (a) Tensile test results of FW samples (b) Tested samples

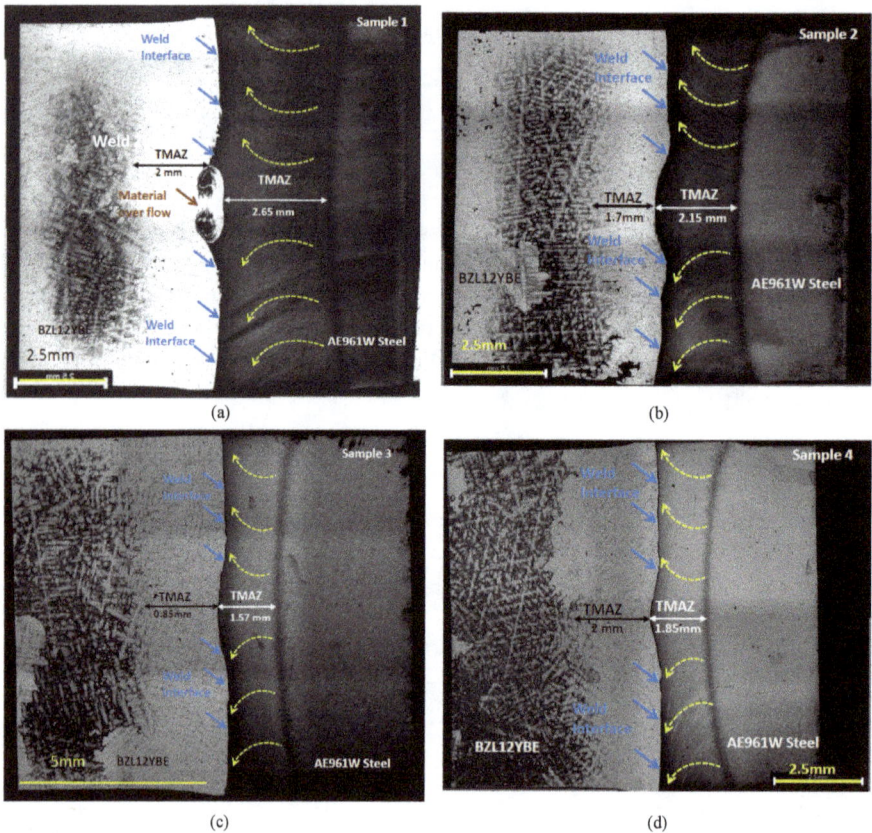

Fig. 4. Optical microstructure of (a) sample 1, (b) sample 2, (c) sample 3 & (d) sample 4

3.2 Microstructure Analysis

The samples are prepared metallographically to analyze under microscope. Figure 4 shows optical micrographs of all four samples and weld interface between AE961W steel and BZL12Y BE alloy can be seen. The plastic flow lines were observed for all the four test samples from AE961W steel. This is attributed to the more plastic flow behaviour of steel material at high temperature compare to the Ni base super alloys. Sample1 showed material overflow at the centre of the weld interface. Sample 2, 3 & 4 did not show any cracks under optical microscope and exhibited smooth weld interface between the two materials Fig. 4a, b, c & d. Further the samples were analysed under SEM. Sample 1, 2 & 3 revealed cracks when observed under SEM; while samples 4 exhibited smooth weld interface. With reference to sample 1 & 2, it is observed that for same upset force; increase in burn off time & decrease in friction force lead to cracking of the sample Fig. 5a & b. The decreased burn off time with same upset force & friction force yielded a smooth weld joint with no cracks in sample 4 when compared between sample 3 & sample 4 (Fig. 5c & d). Line scan is carried out across thermo mechanically affected zone (TMZ) to identify the elements diffusing during the welding process (Fig. 6).

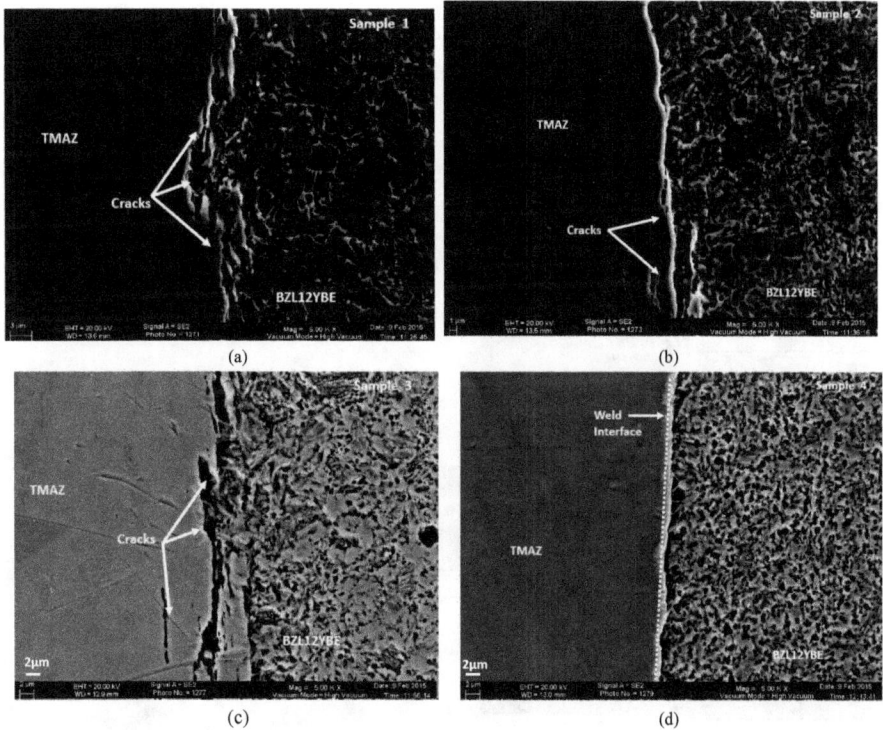

Fig. 5. Scanning electron microscope images of (a) sample 1, (b) sample 2 (c) sample 3 & (d) sample 4

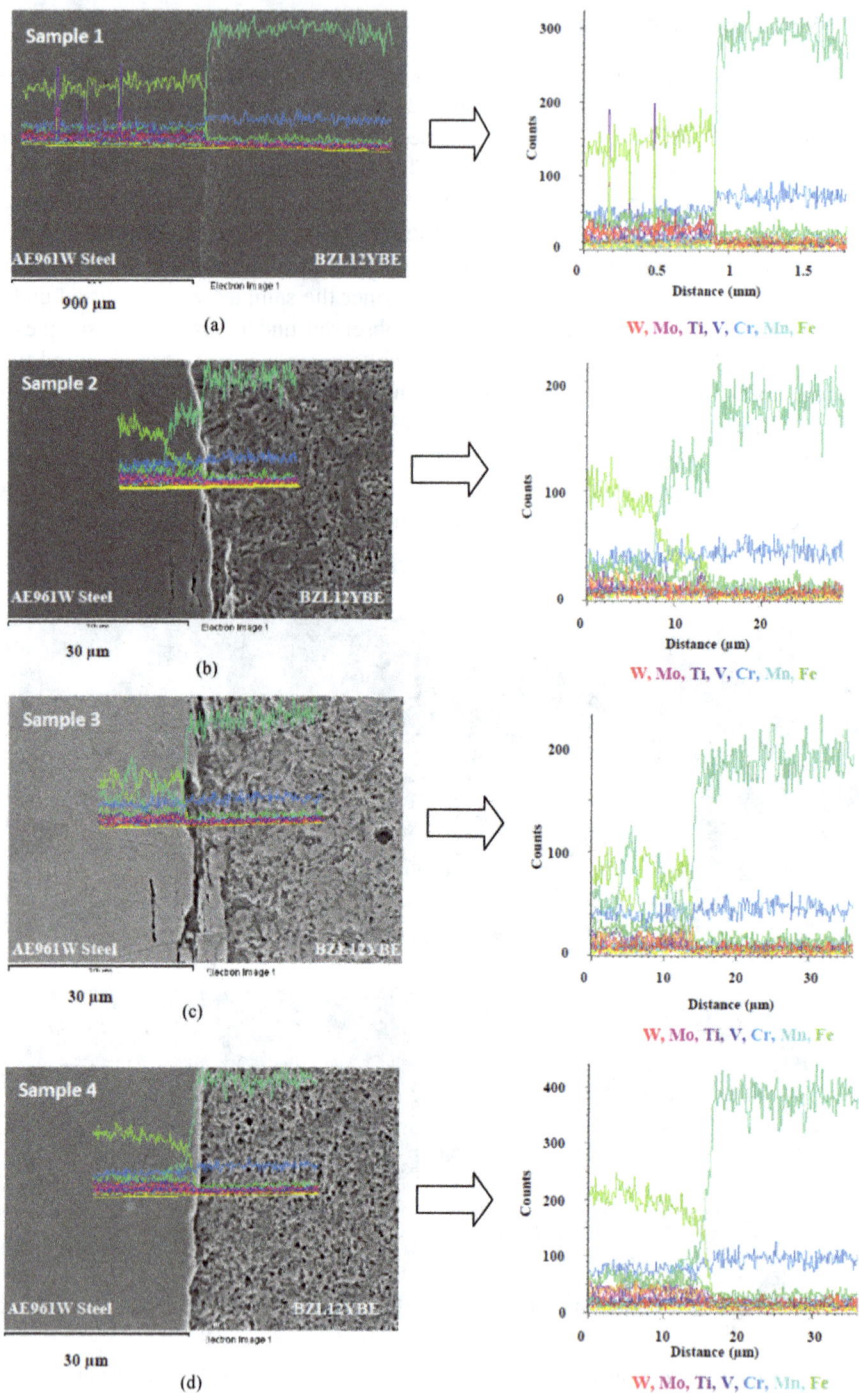

Fig. 6. EDS line scan of (a) sample 1 (b) sample 2, (c) sample 3 & (d) sample 4

3.3 Energy Dispersive Spectroscopy Analysis

Figure 6 shows the line scan performed across the weld joint for all the four samples. Major alloying elements W, Mo, Ti, V, Cr & Fe which are common in both the alloys have diffused across due to chemical composition gradient. Fe has diffused towards BZL12YBE from steel while V & Mn have diffused towards steel from BZL12YBE. This flow of elements is quiet evident by comparing amount of these elements in Table 1 & Table 2 and it helps in strengthening the weld joint.

3.4 Thermo Mechanically Affected Zone

Thermo mechanically affected zone (TMAZ) was measured for all the sample 1, 2, 3 & 4 (Table 5). It is observed that for the same upset force upset force, decreased friction force resulted in reduced TMAZ width. Analysis of TMAZ in all the four samples revealed that AE961 steel material has exhibited wider TMAZ in all four samples than that of BZL12 YBE alloy. This indicates that as the temperature increased during the welding process, more plastic deformation has occurred for steel material than BZL12 YBE material. The broken dendrites can be seen in all the samples in TMAZ location of BZL 12YBE alloy, similarly grains are refined in the TMAZ zone of AE961W steel compared to the parent location. The rise in temperature & cooling during the welding process has caused the grain refinement in TMAZ.

Table 5. Width (mm) of thermo mechanically affected zone

Material	Sample 1	Sample 2	Sample 3	Sample 4
BZL12YBE	2	1.7	0.85	2
AE961W	2.65	2.15	1.57	1.85

3.5 Microhardness Study

Microhardness profile across the weld interface is shown in Fig. 7. Increase in hardness in TMAZ for both the materials can be noticed. Mean hardness of various zones is recorded in Table 6. This is attributed to the grain refinement in TMAZ zone.

Highest hardness was observed at weld interface for all the samples. The TMAZ of BZL 12Y BE showed lesser hardness compared to the TMAZ of AE961W. This may be due to hard martensite phase present in the AE961W steel. J Alex Anandraj et al. [11] also observed less hardness for Ni base super alloy compared to the martensitic stainless steel.

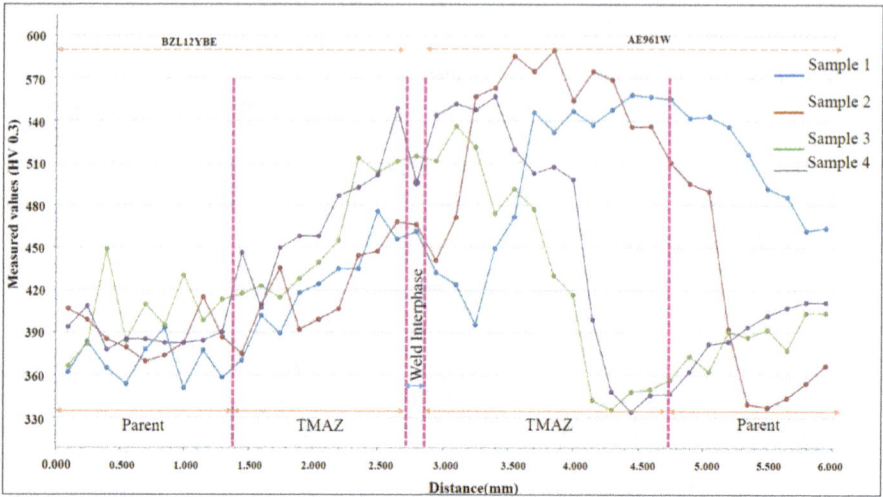

Fig. 7. Microhardness profile across the weld interface

Table 6. Mean Microhardness (Hv) in various zones

Location	Sample1	Sample 2	Sample 3	Sample 4
Parent -BZL12Y BE	370	390	402	388
TMAZ - BZL12Y BE	432	433	458	460
Weld Interface	470	471	528	497
Parent- AE961W	542	528	454	516
TMAZ – AE961W	476	348	371	379

4   Conclusions

1. Weld parameter of sample 4 has resulted in acceptable weld joint both from microstructural & mechanical property aspects of the weld joint. Hence, the weld parameter of sample four can be utilized for welding the low pressure turbine & shaft assembly.
2. Sample 3 & Sample 4 exhibited better strength values compared with sample 1, 2. This may be due to high upset force in sample 3 & 4 which forged more pressure to join the materials and hence resulted in joint with better strength. Sample 1 & 2 have shown comparable mechanical properties. Sample 4 exhibited lesser strength than sample 3 but better than sample 1 & 2.
3. Micro cracks were observed for all the samples except sample 4. Defect free weld joint was observed in sample 4.
4. Extensive grain refinement along with plastic flow lines were observed at the joint for all the samples. Wider TMAZ is observed in martensitic stainless steel.

5. The EDS analysis of the TMAZ for both the materials has shown elements of the base materials. The difference in the chemical composition of the two materials set up a concentration gradient across the mating interface causing species migration.

Acknowledgements. The authors thank Director, Gas Turbine Research Establishment (GTRE), DRDO for his encouragements in carrying out this work. Mr. Sai Venkata Kumar is acknowledged for his support in sample preparation & tensile testing.

References

1. Ali, M.: Study of heat affected zone in friction welding process. J. Mech. Eng. **1**(1), 11–17 (2012)
2. Mumin, S., Misirl, C.: Mechanical and metallurgical properties of friction welded Aluminum joints (2012). https://doi.org/10.5772/51130
3. Uzkut, M., Ünlü, B.S., Yilmaz, S.S., Akdağ, M.: Friction welding and its application in todays world. International Burch University (2020). https://www.scribd.com/document/335581086/issd2010-science-book-p710-p724-pdf
4. Gupta, B.: Aerospace Material, vol. 2, pp. 204–207 & vol. 3, pp.623–655. Certification Data from Mishra Dhatu Nigam Pvt. LTD., Hyderabad, India (1990)
5. Roger, C.: Reed: The Superalloys Fundamentals and Applications. Cambridge University Press, Cambridge (2006)
6. Donachie, M.J., Donachie, S.J.: Superalloys: a technical guide. In: ASM International, Materials Park, OH (2002)
7. Karadge, M., Grant, B., Baxter, P.J., Preuss, M.: Thermal relaxation of residual stresses in nickel-based superalloy inertia friction welds. Metall. Mater. Trans. A **42A**, 2301–2311 (2011)
8. Attallah, M.M., Preuss, M.: Welding and Joining of Aerospace Materials. Woodhead Publishing, Oxford (2012)
9. Luo, J., Li, L., Dong, Y., Xiaoling, X.: A new current hybrid inertia friction welding for nickel-based superalloy K418–alloy steel 42CrMo dissimilar metals. Int. J. Adv. Manuf. Technol. **70**(9–12), 1673–1681 (2013). https://doi.org/10.1007/s00170-013-5441-8
10. Anitha, P., Majumder, M.C., Saravanan, V., et al.: Microstructural characterization and mechanical properties of friction-welded IN718 and SS410 dissimilar joint. Metallogr. Microstruct. Anal. **7**, 277–287 (2018). https://doi.org/10.1007/s13632-018-0447-0
11. Anandaraj, J.A., Rajakumar, S., Balasubramanian, V., Petley, V.: Investigation on mechanical and metallurgical properties of rotary friction welded In718/SS410 dissimilar materials. Materials Today Proceedings **45**, 962–966 (2021)

Brittle Fracture Failure Analysis under Mixed-Mode Condition Using Asymmetric Edge Cracked Semicircular (AECS) Configuration

L. C. Shashidhara[1] ⬤, M. A. Umarfarooq[2(✉)] ⬤, N. R. Banapurmath[2(✉)],
Tabrej Khan[3(✉)], and Tamer A. Sebaey[4(✉)]

[1] Dayananda Sagar University, Bangalore 560068, Karnataka, India
[2] KLE Technological University, Hubballi 580031, Karnataka, India
umarfarooq.ma@gmail.com, nr_banapurmath@gmail.com
[3] Prince Sultan University, Riyadh 11586, Saudi Arabia
tkhan@psu.edu.sa
[4] Zagazig University, Zagazig, Sharkia, Egypt
tsebaey@psu.edu.sa

Abstract. The non-destructive assessment of structural elements enables the high-resolution identification of fractures and other failures. Fracture mechanics offers concepts for lifetime prediction of components when component geometry, material properties, and load intensity under steady state or impact loading are known. This is true when cracks are involved in mechanical failure events. Mixed-mode brittle fracture is one of the frequent forms of mechanical failure in components made up of quasi-brittle or brittle materials. Investigating the structural integrity of damaged components under mixed-mode loads is crucial.

For studying mixed mode fracture in brittle materials, an Asymmetric Edge Cracked Semi-circular (AECS) specimen exposed to asymmetric three-point bend load was recommended. The crack parameters were derived using finite element analysis for various crack lengths and varied loading point locations. It was demonstrated that entire mode mixities from pure mode I to pure mode II could be accomplished by choosing the proper places for the loading points. The proposed specimen was then used to perform a number of fracture tests on epoxy resin material. The experimental findings and those predicted by the maximum tangential stress criteria showed a very high degree of agreement.

Keywords: Mixed-mode Fracture · AECS · Stress Intensity Factor (SIF)

1 Introduction

In structures and components used in engineering, the occurrence of flaws is very often unavoidable. During the process of production or due to environmental causes or cyclic loading cracks can be created. The deformation modes like opening mode (mode-I) & sliding mode (mode-II) cause cracks in components exposed to in-plane load conditions.

© The Author(s), under exclusive license to Springer Nature Singapore Pte Ltd. 2024
S. Patra et al. (Eds.): METCENT 2023, *Proceedings of the International Conference on Metallurgical Engineering and Centenary Celebration*, pp. 286–296, 2024.
https://doi.org/10.1007/978-981-99-6863-3_29

In reality, cracked components and structures undergo a mixture of loads, often mixed-mode K_I & K_{II}. Brittle fracture in the mixed-mode state is the typical form in components and structures with cracks made up of brittle materials. Thus, under mixed-mode conditions, it is essential to study the reliability of structures and components with cracks [1–3].

Researchers have been recommended several experimental and theoretical approaches for discovering a mixed-mode form of brittle fracture. Laboratory specimens are preferred by the researchers to conduct their tests as the experimental studies on real machine components and structures are difficult and very expensive. However, to check the test results obtained from the test configurations with the fracturing effect on notched components under working load conditions, sufficient mixed-mode fracture parameters are also required. A series of tests has to be conducted on suitable test materials using proper test configurations to confirm a fracture criterion.

Fracture tests are conducted on PMMA by [4–6] for performing brittle fracture investigations. Similarly, researchers [7, 8] preferred epoxy material to investigate the fracture characteristics in brittle materials. PMMA and epoxy resin materials are used to conduct fracture tests as these are accepted as favored material models for performing brittle fracture investigations. Epoxy resin fractures at room temperature in a brittle way and are a reasonably isotropic and homogeneous material. Optical transparency of epoxy allows direct observation of the fracture path. In brittle fracture investigations, the ease of adding a sharp crack and machining to the desired shapes are other benefits of epoxy. Preparation of cracked specimens from epoxy is economical and comparatively easy.

Specimens like Cracked Brazilian Disk (CBD) and Semi- Circular Bend (SCB) are used by [9]. Semi-circular types of specimens with edge cracks are often used for investigating the fracture toughness in different materials like PMMA, rocks, asphalt, and epoxy in mixed-mode (I&II) condition by many researchers like [4–6, 10, 11]. This investigation mainly focuses on the outcomes of brittle fracture tests on epoxy resin from mode-I to mode-II in the full range achieved from the Asymmetric Edge Cracked Semi-circular bend (AECS) specimen. The crack growth path and the load at fracture are determined from the experimentations.

2 Mixed-Mode Fracture Toughness Testing

The AECS specimen is exposed to 3-point bending and many combinations of mode I, mixed-mode I/II, and mode II are achieved reliant on (a/R, S_1/R & S_2/R). The finite element method is used to determine its parameters near the crack tip for pure mode I, mixed-mode I&II, and pure mode II using several FE models.

Figures 1 shows the loading conditions and geometry of the new test configuration AECS specimen for pure opening mode (I) and pure sliding mode (II) respectively. The specimen configuration is an edge cracked semi-circular disk of radius R comprising of an edge crack of length 'a' and thickness 'B'. The specimen is placed on the bottom supports S_1 and S_2 and it is loaded at the center by the vertical load P.

In the AECS specimen, the mixed-mode condition can be achieved by adjusting the two bottom support positions (S_1 and S_2). If the lower supports are positioned symmetrically around the line of crack ($S_1 = S_2$), pure mode I loading can be achieved. Similarly, mixed-mode (I/II) loading can be accomplished by placing the two bottom supports (S_1 and S_2) asymmetrically along the line of crack ($S_1 \neq S_2$). When the lower support S_2 is closer to the crack line, pure mode II loading can be done relative to support S_1. Mode I and II influences are managed by selecting suitable locations (S_1 and S_2) for the bottom support. Different mode mixes can thus be obtained in the proposed sample. Earlier, this configuration was sometimes used but only to test pure mode I fracture strength using basic symmetric load conditions for various engineering materials, including polymers, asphalt, rocks, concrete, etc.

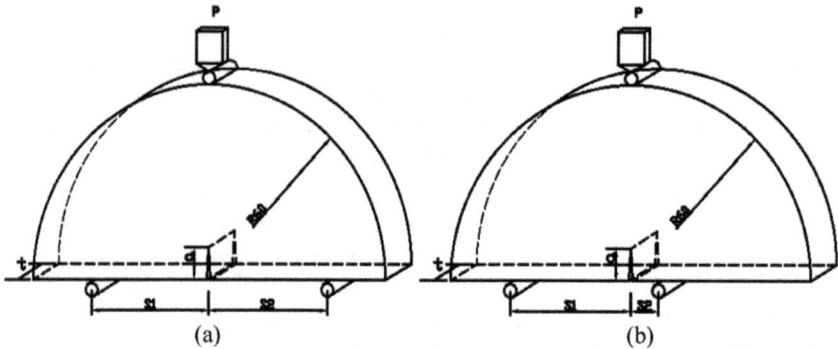

Fig. 1. AECS configuration (a) for pure mode I loading (vertical crack – symmetric loading), (b) for pure mode II loading (vertical crack – asymmetric loading)

The Perspex mold is used for the preparation of 10 mm thick sheets of epoxy resin. This mold is prepared by inserting three 10 mm thick aluminum spacers on three sides between sheets of perspex. For the final assembly, these spacers are equipped with bolts and nuts. The epoxy resin solution is then prepared by combining a 1:10 ratio of Lapox L-12 resin with a K-6 hardener. (i.e. 1 kg of resin was mixed with 100 gm of K6 hardener). The notch is prepared by using a fret saw of 0.6 mm thickness. The initial depth of the notch is made slightly less than 20 mm and later a razor blade is pressed along the notch to make a notch length of 20 mm, a sharp pre-crack was made. A total of 28 specimens were machined from a 10 mm thick epoxy resin sheet. Material properties of the epoxy resin are $v = 0.33$ and $E = 3$ GPa is considered for FE analysis [9]. Figure 2(a) & 2(b) shows the perspex mold used to prepare the epoxy sheets and fabricated specimen respectively.

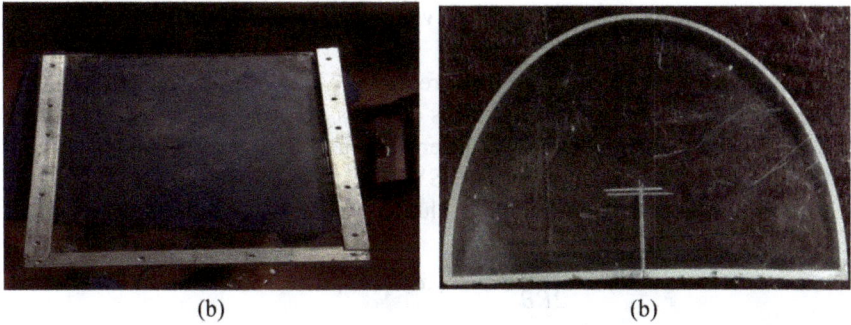

(b)   (b)

Fig. 2. Fabrication of specimens (a) Perspex Mold, (b) Fabricated AECS specimen

A UTM is used for the three-point bending test. Flexural load and flexural extension have been calculated for different mixed-mode combination. The load is applied on the specimen at the center using a three-point bend fixture as shown in Fig. 3.

Bottom support distance S_1 was kept constant (40 mm) for all tests and support distance S_2 value was varied for all range of mixed-mode I/II tests. For experiments, the values of S_2 in mm selected are $S_2 = 6$ (pure mode II), 7.5, 9, 12, 15, 23, and 40 (pure mode I). S_2 and S_1 values have to be arranged for each specimen respectively when it is loaded on the machine by the loading fixture.

Fig. 3. AECS specimen Loaded on 3-point bend fixture.

The test loading rate of 0.5 mm/min is maintained constantly for each test till the specimen fractures and the value of fracture load is recorded for the particular specimen. This machine is controlled completely by a servo system of the universal testing machine of 150 KN capacity.

Desired locations of the supports can be achieved by 2 lower anvils that move independently. So that we can obtain the desired locations of the supports by fixing one of the anvils at a constant distance (i.e. S_1) and moving the other anvil to the given location (i.e. S_2).

3 Stress Intensity Factor (SIF) Evaluation

The supports with their loading locations are S_1 and S_2 and the crack length is given as 'a'. These two parameters are the functions of stress intensity factors (SIF) namely K_I and K_{II} for the specimen. The SIF for different values of S_2 is calculated by using the post-processing command KCALC in ANSYS [12]. The geometric factors Y_I and Y_{II} are determined from the SIF K_I and K_{II} values using the following formulas.

$$K_I = \frac{P}{2RB}\sqrt{\pi a}\, Y_I(a/R, S1/R, S2/R) \tag{1}$$

$$K_{II} = \frac{P}{2RB}\sqrt{\pi a}\, Y_{II}(a/R, S1/R, S2/R) \tag{2}$$

where, P refers to applied load and Y_I and Y_{II} are corresponding geometry factors of mode-I and mode-II. Various finite element models of AECS specimen were evaluated using ANSYS to find Y_I and Y_{II}. Support position S_1 is fixed at 40 mm and S_2 varied between 6 and 40 mm to adjust the state of the mixed-mode. In the finite element models, the elastic material properties of epoxy resin given in Table 1 were considered.

Figure 4(a) shows the typical mesh pattern generated for evaluating mixed-mode SIF (K_I and K_{II}) in the AECS specimens. Which was finalized after trial and test methods for best results for the applied boundary conditions. Figure 4(b) shows the singularity generated at the crack tip.

Table 1. Specimen Dimensions & Material Properties

Specimen Dimensions	$L = 120$ mm, $S1 = 40$ mm, $S2 = 6.07$ mm to 40 mm, $a = 20$ mm, $B = 10$ mm, $R = 60$ mm
Material Properties of Epoxy Resin	$E = 3$ Gpa, $\mu = 0.33$

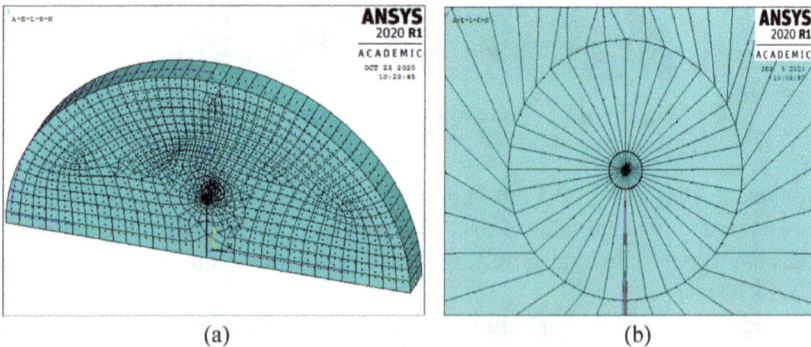

(a)   (b)

Fig. 4. (a) A typical finite element mesh used for simulating AECS specimen (b) Singularity generated at the crack tip.

4   Results and Discussion

For each iteration that has been carried out during this process, the contour plots obtained from the ANSYS code provide an overview of the stress intensity that has been generated. The plastic, elastic, and transitional zones are clearly defined, confirming the approach followed for the issue we are addressing. Figures 5(a) and 5(b) shows the stress plot of Von Mises for mode-I loading condition and contour plot respectively at the crack tip for pure mode-I loading. Similarly, at the crack tip for pure mode-II loading, Figs. 6(a) and 6(b) show mode-II loading state Von Mises stress plot and contour plot respectively.

Figure 7(a) and 7(b) show the variation of mode I and II geometry factors (Y_I and Y_{II}) against S_2/R respectively. The mode-I geometry factor Y_I decreases for all the a/R values as the support location S_2 moves closer to the crack plane (varies between $40 \geq S_2 \geq 6.07$). And increased values of geometry factor Y_I can be observed as the crack length to radius ratio a/R decreases from 0.5 to 0.25. Similarly, the mode-II geometry factor Y_{II} increases as the support location S_2 (varies between $40 \geq S_2 \geq 6.07$) moves closer to the crack plane for all the a/w values. And decreased values of geometry factor Y_{II} can be observed as the crack length to radius ratio a/R decreases from 0.5 to 0.25. Stress intensity factor solutions for the problem on hand obtained using Abaqus software are also reported in [5]. It is gratifying to note a close agreement between the two.

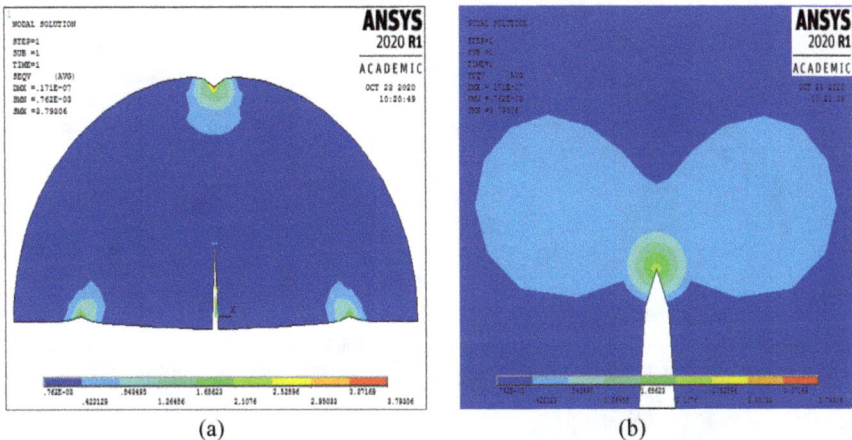

Fig. 5. (a) Mode-I loading condition-Von Mises stress plot (b) Pure mode-I Contour plot at the crack tip.

Similarly, the mode-II geometry factor Y_{II} increases as the support location S_2 (varies between $40 \geq S_2 \geq 6.07$) moves closer to the crack plane for all the a/w values. And decreased values of geometry factor Y_{II} can be observed as the crack length to radius ratio a/R decreases from 0.5 to 0.25. Stress intensity factor solutions for the problem on hand obtained using Abaqus software are also reported in [5]. It is gratifying to note a close agreement between the two.

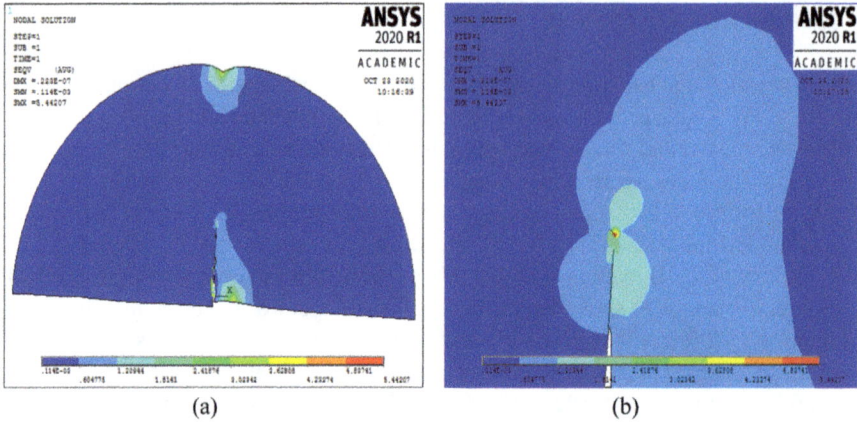

Fig. 6. (a) Mode-II loading condition-Von Mises stress plot (b) Pure mode-II Contour plot at the crack tip.

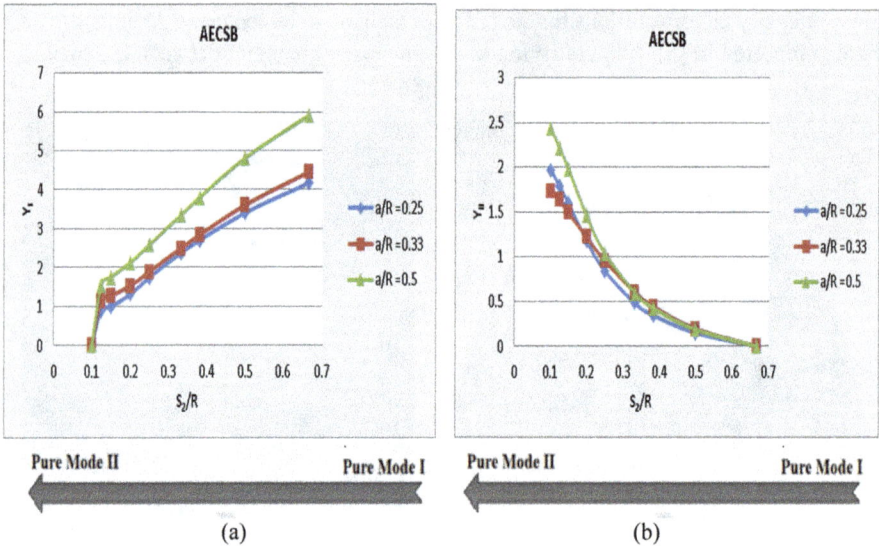

Fig. 7. (a) Variation of mode-I geometry factor (b) Variation of mode-II geometry factor

A series of mixed-mode fracture tests on the epoxy resin was performed to investigate the functional applicability of the AECS sample. Table 2 describes the findings attained from fracture experiments carried out on epoxy resin made specimens. Fracture toughness at mod-I decrease as the support location S_2 moves closer to the crack plane. And mode-II fracture toughness increases as the S_2 moves closer to the plane of crack.

Table 2. Summary of the mi xed-mode I/II fracture tests results

Sl No	Support span (S1-S2) in mm	Fracture Load, PCr in N	Fracture Toughne ss, KIc Kpa $\sqrt{mm}$	Fracture Toughne ss, KIIc Kpa $\sqrt{mm}$
1	40-40 (Pure-Mode I)	2440.9886	2.7231	0
2	40-40 (Pure-Mode I)	3152.4982	3.5169	0
3	40-40 (Pure-Mode I)	2170.6345	2.4215	0
4	40-40 (Pure-Mode I)	2763.9441	3.0834	0
5	40-23 (Mixed-mode)	4292.8241	1.8997	0.2949
6	40-23 (Mixed-mode)	4318.0801	1.9109	0.2966
7	40-23 (Mixed-mode)	5105.5472	2.2594	0.3507
8	40-23 (Mixed-mode)	3268.4255	1.4464	0.2245
9	40-15 (Mixed-mode)	3609.933	0.8941	0.4545
10	40-15 (Mixed-mode)	4986.7485	1.2352	0.6278
11	40-15 (Mixed-mode)	5106.0632	1.2647	0.6428
12	40-15 (Mixed-mode)	6389.1582	1.5825	0.8043
13	40-12 (Mixed-mode)	4777.1495	0.9813	0.8276
14	40-12 (Mixed-mode)	6051.2986	1.243	1.0484
15	40-12 (Mixed-mode)	5845.3139	1.2007	1.0127
16	40-12 (Mixed-mode)	3039.1065	0.6243	0.5265
17	40-09 (Mixed mode)	4724.3096	0.7801	0.9221
18	40-09 (Mixed-mode)	3519.0791	0.5811	0.6868
19	40-09 (Mixed mode)	6172.6995	1.0193	1.2048
20	40-09 (Mixed-mode)	5802.2585	0.9581	1.1325
21	40-7.5(Mixed-mode)	5140.0572	0.7197	1.0457
22	40-7.5(Mixed-mode)	3040.242	0.4257	0.6185
23	40-7.5(Mixed-mode)	7658.1456	1.0723	1.5579
24	40-7.5(Mixed-mode)	5339.2414	0.7476	1.0862
25	40-06 (Pure Mode-II)	4316.457	0	0.7829
26	40-06 (Pure Mode-II)	5983.0047	0	1.0852
27	40-06 (Pure Mode-II)	6851.7782	0	1.2427
28	40-06 (Pure Mode-II)	7398.8684	0	1.342

Figure 8 depicts the fracture origination angle (θo) direction of and the crack route in epoxy resin specimens evaluated with the AECSB configuration in various modes I and II combinations. Except for pure mode-I loadings, where the progress of the crack occurs alongside the original fracture line, mixed-mode I/II cracks enlarge in a non-coplanar means and bend off the initial line of crack. The MTS criteria may be utilized to look into the fracture beginning angles of the AECSB specimens that have been examined. 'M^e' is a parameter of mode mixity with the following definition

$$M^e = \frac{2}{\pi} \arctan \frac{K_I}{K_{II}} \qquad (3)$$

The fracture origination angle direction for the examined AECSB test specimens is also displayed in Fig. 9, along with a comparison to the traditional MTS criterion. Though there is a difference b/w the MTS curve and angles measured, the experimental findings and the MTS criteria curve are in good agreement with [5].

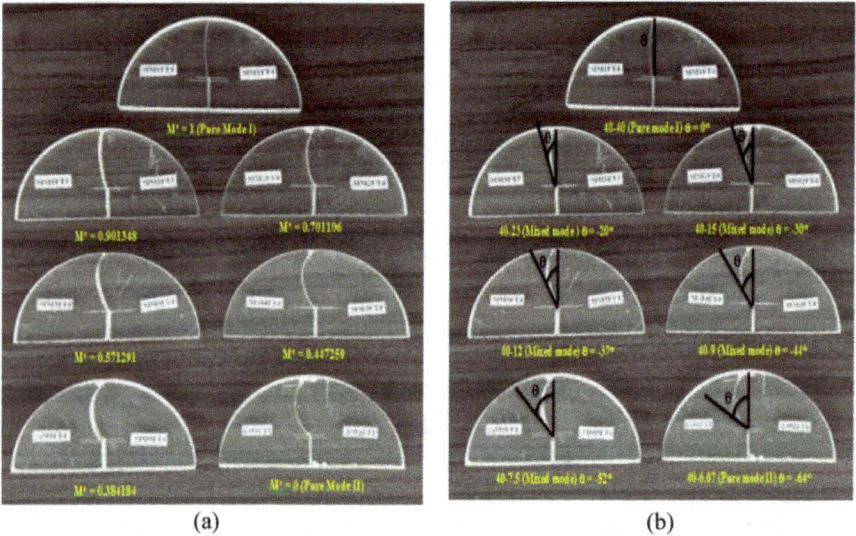

(a)   (b)

Fig. 8. (a) Patterns of fracture in the epoxy resin made AECSB test samples. (b) The direction of fracture origination angle in various combinations of modes I and II.

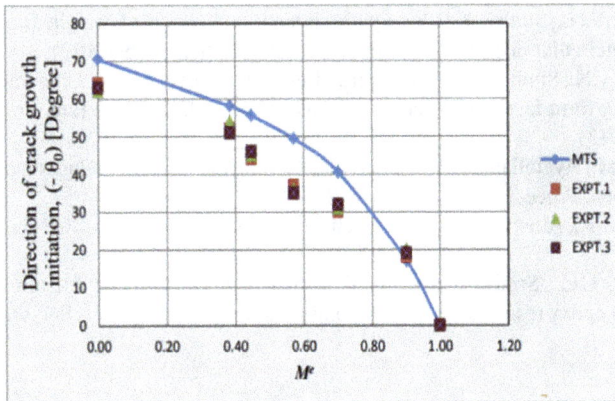

Fig. 9. The fracture origination angle direction in the verified AECSB specimens and MTS criterion comparison.

5 Conclusions

For performing mixed-mode (I/II) fracture experiments on brittle materials, the AECS test configuration is confirmed. The key advantages of the samples are a simple geometry, simple test configuration, and simple fixtures. Under mode-II, the load absorbed by the material before it fails is higher than that in mode-I. Fracture load is calculated from a series of experiments for a different combination of mode-I and II fracture. The difference between the experimental and theoretical outcomes was more substantial for mode II-controlled loading conditions. As mode-I fracture absorbs a very less amount of fracture load before it fails it will lead to catastrophic failures in the material. It is fair to determine that a good agreement occurs between the test results and prediction.

References

1. Umarfarooq, M.A., Gouda, P.S., Banapurmath, N.R., Edacherian, A.: Impact of process induced residual stresses on interlaminar fracture toughness in carbon epoxy composites. Compos. A Appl. Sci. Manuf. **27**, 105652 (2019)
2. Gouda, P.S., Sridhar, I., Umarfarooq, M.A.: Crack suppression in glass epoxy hybrid L-bend composites through drawdown coating technique using nano and micro fillers. Mater. Today Proc. **62**, 7292–7296 (2022)
3. Parveez, B., Kittur, M.I., Badruddin, I.A., Kamangar, S., Hussien, M., Umarfarooq, M.A.: Scientific advancements in composite materials for aircraft applications: a review. Polymers **714**(22), 5007 (2022)
4. Aliha, M.R.M., Bahmani, A., Akhondi, S.: Mixed mode fracture toughness testing of PMMA with different three-point bend type specimens. Eur. J. Mech. A/Solids **58**, 148–162 (2016)
5. Ayatollahi, M.R., Aliha, M.R.M., Saghafi, H.: An improved semi-circular bend specimen for investigating mixed-mode brittle fracture. Eng. Fract. Mech. **78**(1), 110–123 (2011)
6. Ayatollahi, M.R., Aliha, M.R.M., Hassani, M.M.: Mixed mode brittle fracture in PMMA - an experimental study using SCB specimens. Mater. Sci. Eng. A **417**(1–2), 348–356 (2006)
7. Surendra, K.V.N., Simha, K.R.Y.: Analysis of cracked and un-cracked semicircular rings under symmetric loading. Eng. Fract. Mech. **128**(C), 69–90 (2014)

8. Surendra, K.V.N., Simha, K.R.Y.: Synthesis and application of weight function for edge cracked semicircular disk (ECSD). Eng. Fract. Mech. **107**, 61–79 (2013)
9. Surendra, K.V.N., Simha, K.R.Y.: Design and analysis of novel compression fracture specimen with constant form factor: edge cracked semicircular disk (ECSD). Eng. Fract. Mech. **102**, 235–248 (2013)
10. Aliha, M.R.M., Ayatollahi, M.R.: Mixed mode I/II brittle fracture evaluation of marble using SCB specimen. Procedia Eng. **10**, 311–318 (2011)
11. Ayatollahi, M.R., Alborzi, M.J.: Rock fracture toughness testing using SCB specimen, pp. 1–7 (2013)
12. Shashidhara, L.C., Srinivas, H.K., Anil Kumar, C.: Mixed-mode fracture (KI and KII) evaluation in epoxy resin using ASENB configuration. J. Mines Metals Fuels **69**(12A) (2021)

Analyzing the Effect of Dynamic Impact in 6061 Al Alloy Using MATLAB as a Post Processing Tool

Ravi Kumar Singh and Nikhil Kumar[✉]

School of Materials Science and Technology, Indian Institute of Technology (BHU),
Varanasi 221005, India
nikhil.mst@itbhu.ac.in

Abstract. This study focuses on writing MATLAB code for the post processing of data received after investigating the behaviour of friction stir processed samples under impact conditions. For this study the experimental techniques was used which includes Electron Backscatter Diffraction (EBSD), data processing through MATLAB, and microstructural analysis to have a comprehensive understanding of the material's microstructure and mechanical property. The key findings and conclusions drawn from this work are that a MATLAB code was developed to process data obtained from Split Hopkinson Pressure Bar (SHPB) experiments. The raw data was processed to generate true stress versus true strain graph. Microstructural analysis through Electron Backscatter Diffraction (EBSD) showed significant grain refinement in the impacted samples compared to the base sample. Additionally, the impacted samples showed reduction in geometrically necessary dislocations (GNDs) and an increase in misorientation of high-angle grain boundaries (HAGB). The impacted samples showed ultimate tensile strength of 400 MPa and a ductility of approximately 20%. Also, the formation of adiabatic shear bands (ASBs) was observed in the impacted samples, demonstrating their significant influence on the plastic deformation behaviour of the material. Overall, this research study provides valuable insights into the impact behaviour of friction stir processed samples. The findings contribute to a deeper understanding of the material's mechanical response, microstructural changes, and the role of adiabatic shear bands. These insights have implications for optimizing materials for high-strain-rate applications. The code written for the SHPB data processing can be used for any material.

1 Introduction

Metals and alloys are extensively used in various industries because of their several beneficial properties like high strength, ductility, and good durability [1, 2]. Understanding the structure property relationship between the alloys and their application is critical for optimizing their overall performance in engineering applications. Dynamic impact loading conditions are encountered in a range of engineering applications which includes aerospace sector, automotive engineering, and for various structural design [3–6]. During such impact applied to the material, it experiences rapid and intense deformation

S. Patra et al. (Eds.): METCENT 2023, *Proceedings of the International Conference on Metallurgical Engineering and Centenary Celebration*, pp. 297–310, 2024.
https://doi.org/10.1007/978-981-99-6863-3_30

which further leads to complex changes in its mechanical behaviour. Material dynamic impact response is influenced by various factors, including microstructure, grain size, heat treatment, alloy composition, and processing conditions [7–10]. These metallurgical factors can significantly affect the material's strength, energy absorption capacity, and fracture resistance, ultimately impacting its overall performance.

In order to improve the mechanical response of the material many researchers used Friction Stir Processing (FSP) route [11–14]. FSP has potential to modify the grain structure as well as to induce super plasticity in the material [15, 16]. Recently Fujie Wang et al. got significant improvement in mechanical properties of 6061 Aluminum alloy and achieved desired microstructure modification [17]. In spite of using a strong processing route like FSP, dynamic testing is crucial to test. Since dynamic loading involved in almost all type of engineering applications e.g., automobile parts, in aerospace industry, bridges, train tracks, and turbines etc. [18, 19].

MATLAB, a powerful numerical computing environment, offers a wide range of tools and functions that facilitate data processing, analysis, visualization, and modelling [20–24]. Leveraging the capabilities of MATLAB streamlines the post-processing workflow and enables researchers to extract key parameters and metrics from the experimental data efficiently [25]. This significantly enhances the understanding of the material's dynamic response and aids in the development of predictive models. Furthermore, MATLAB enables researchers to perform advanced statistical analysis on the processed data [26]. Descriptive statistics, hypothesis testing, regression analysis, or machine learning algorithms can be applied to identify trends, correlations, or critical features related to the material's dynamic response. MATLAB's statistical toolbox and machine learning toolbox provide a rich set of functions and algorithms [27]. In this research work the MATLAB code was written which could to be used for the post processing of SHPB data for any material, also for Al 6061 sample the same code was used and data was analyzed.

2 Experimental

The present research utilized a tempered T6 commercial alloy with the following composition: Silicon (0.68%), Iron (0.49%), Copper (0.21%), Manganese (0.84%), Chromium (0.06%), Nickel (0.01%), Zinc (0.07%), Titanium (0.07%), and Aluminum (97.55%). Before the impact test, the alloy was processed under friction stir processing (Fig. 1) to improve its mechanical properties through grain refinement. Afterwards, the alloy samples were prepared in cylindrical shape with dimensions of 3 mm × 3 mm. To conduct dynamic impact tests, a Split Hopkinson Pressure Bar (SHPB) setup was employed. The SHPB system consisted of an incident bar, a transmitter bar, a pressure chamber, and a data acquisition system. The data acquisition system included a strain conditioner/amplifier and a 4-channel digital oscilloscope. For each dynamic impact trial, the sample was positioned between the incident and transmitter bars. The striker bar within the pressure chamber was adjusted to achieve a strain rate of 4200/s. Once initiated, the striker bar impacted the incident bar, generating elastic waves that propagated through the incident bar. A portion of these waves contributed to deforming the specimen, while the remaining portion propagated through the transmitter bar. The elastic wave reflected upon contact between the specimen and incident bar. Both the incident and transmitter bars were equipped with strain gauges to collect wave signals in the form of voltage-time data. These strain gauges captured the wave signals and transmitted them to the data acquisition system for further analysis.

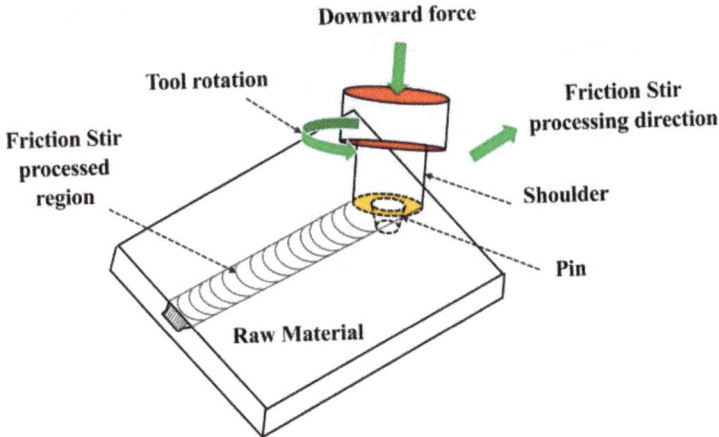

Fig. 1. Illustration of the FSP process

To facilitate EBSD (Electron Backscatter Diffraction) characterization, a sequential polishing procedure was implemented on the sample's surface. Initially, emery polishing was applied, followed by cloth polishing. Subsequently, the sample underwent electro-polishing using a mixture of 80% methanol and 20% perchloric acid to achieve a mirror-like finish. This electro-polishing process effectively eliminated any pre-existing strains present on the sample's surface. The resulting stress-free and pristine surface was deemed suitable for EBSD examination. For the EBSD scan, a step size of 0.05 μm was utilized. This step size determined the spatial resolution of the scan, ensuring that a precise and detailed analysis of the microstructure could be performed.

The processing of Split Hopkinson Pressure Bar (SHPB) experimental data in MATLAB involves several essential steps. Firstly, the raw data obtained from the pressure bar sensors, strain gauges, and displacement transducers is imported into MATLAB. Once imported, the data undergoes a series of operations including data cleaning, filtering, and calibration. In MATLAB, signal processing techniques are applied to generate data and extract crucial parameters from the SHPB data. Time-domain and frequency-domain analysis methods are commonly utilized to obtain signals related to stress, strain, strain rate, and displacement. MATLAB functions such as fast Fourier transform, wavelet, or custom algorithms can be employed to perform Fourier analysis, wavelet analysis, or other relevant signal processing techniques in order to extract the desired information from the data. The program developed in MATLAB serves to reduce the data obtained from a compression SHB experiment. By employing the mentioned signal processing techniques and custom algorithms, the program enables the extraction and analysis of pertinent information from the SHPB data.

2.1 Model and Procedure

After the experiment performed in SHPB, a folder containing the raw files received. The raw file includes two csv files containing the waveform results. After that MATLAB program was used to open compression_SHB_reduction.m file to run the code or After the implementation of this code, data in the waveform function generated (Fig. 2), from this output of the system could be calculated. The desired data would be received in excel file (command has been given in code). The actual code is provided in supplementary data.

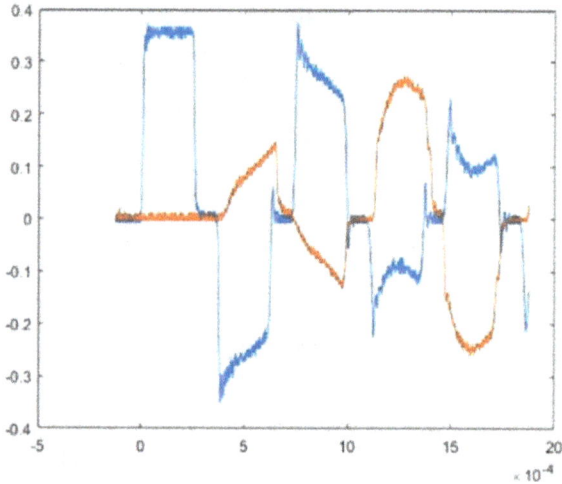

Fig. 2. Waveform (pulse) function with a pulse width 1 s as an input function, x(t), and the same exponential decay function as impulse response function, h(t). Using MATLAB to calculate the output of the system y(t).

3 Results and Discussion

In the present study, the microstructure of the tested sample subjected to a strain rate of 4200/s was compared with that of the base sample. The comparison revealed the presence of refined grains in the impacted samples, as depicted in Fig. 3 (a, b). The refined and fragmented grain structure indicated a significant alteration in the material's microstructural characteristics resulting from the impact testing.

Further analysis was conducted through grain boundary quick map images, which provided insights into the nature of grain boundaries before and after the impact testing. Prior to the impact testing, the base sample exhibited a relatively lower proportion of high-angle grain boundaries (HAGB) with a misorientation of approximately 61.9%. However, after the impact testing of the sample, the HAGB increased significantly to 85.7%. This observation shows a higher frequency of misoriented grain boundaries in the impacted samples. It suggests deformation in the grains and rearrangement of the grain boundaries under the higher dynamic stresses.

Additionally, the presence of Geometry Necessary dislocations (GNDs) was investigated which are associated with plastic deformation. GNDs were accumulated in the regions with significant plastic strain presence. It was ensuring strain compatibility across the material. These dislocations were maintaining the overall deformation compatibility within the material. The base sample before the impact testing, showed a relatively high fraction of GNDs present in the microstructure. Total fraction of GNDs present before the impact test was approximately 80% of the total. However, after the impact testing the fraction of GND significantly reduced to 27.9%. The reduction in the density of GNDs, implying a transformation in the deformation mechanism within the material due to the dynamic impact loading.

This study suggests that the impact testing at a strain rate of 4200/s leads to a significant microstructural change in the tested sample. The refinement of grains, the increase in high-angle grain boundaries, and the reduction in Geometry Necessary dislocations collectively indicate the occurrence of plastic deformation and grain rearrangement in response to the dynamic loading conditions. Results from these tests provide valuable insights for optimizing the material's performance in similar loading scenarios and contributing to a greater understanding of the material's response to high-strain-rate impacts.

The data obtained from the SHPB test was processed using the written MATLAB code and the true stress versus true strain curve was generated (Fig. 4). The analysis of the true stress versus true strain curve showed that the impacted sample exhibited an ultimate true stress of 400 MPa. The ductility of the impacted sample was determined to be approximately 20%. The observed ductility of 20% suggests that the impacted sample possessed a notable ability to undergo plastic deformation before reaching its ultimate failure point. The possible reason for such type of failure might be the formation of adiabatic shear bands (ASBs) in the impacted sample. This was might be attributed to the high strain rate experienced during the dynamic loading conditions. ASBs are localized regions of sudden plastic deformed zones that generally occurs at high strain rates and have unique microstructural features.

The processed Al sample was given high strain rate for the testing. During this process the material was not able to dissipate heat rapidly due to very high strain rate which ultimately leads to formation of adiabatic shear bands. Also, this fast deformation rate without allowing proper heat dissipation leads to the formation of localized deformation zones, which further becomes the point of crack initiation upon increasing load or strain rate.

The high strain rate present with dynamic impact loading condition induce rapid and severe deformation within the material as it is visible from Fig. 3b, causing localized shear stresses to exceed the critical threshold for ASB formation. These shear stresses initiate the nucleation and propagation of ASBs, which serve as pathways for the redistribution of stress and plastic deformation. The intense shear deformation within ASBs leads to significant microstructural changes as observed by Fig. 3.

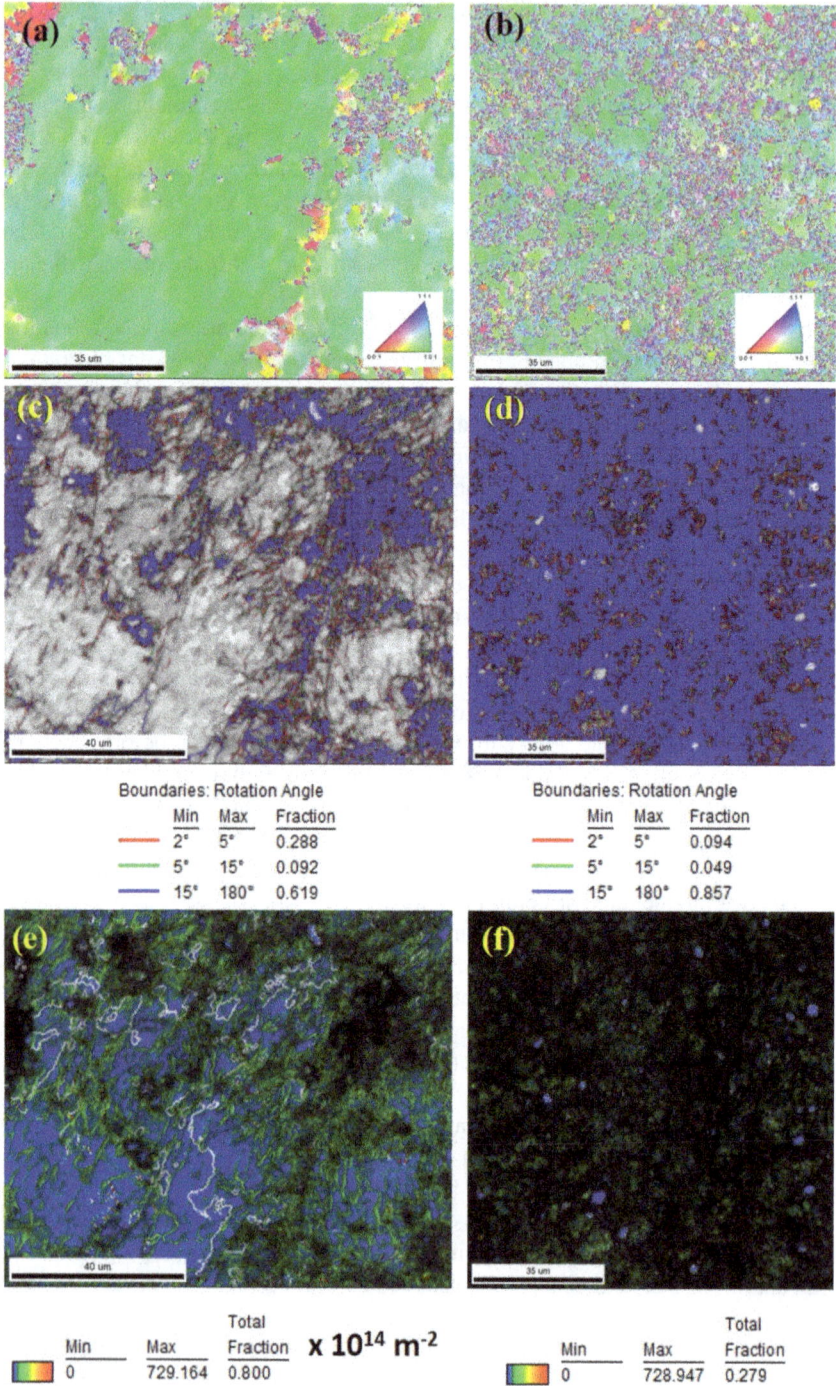

Boundaries: Rotation Angle

	Min	Max	Fraction
	2°	5°	0.288
	5°	15°	0.092
	15°	180°	0.619

Boundaries: Rotation Angle

	Min	Max	Fraction
	2°	5°	0.094
	5°	15°	0.049
	15°	180°	0.857

	Min	Max	Total Fraction	$\times 10^{14}$ m^{-2}
	0	729.164	0.800	

	Min	Max	Total Fraction
	0	728.947	0.279

Fig. 3. Inverse pole figure map for (a)initial sample and (b)Impacted sample, Grain boundary quick map for (c)initial sample and (d)Impacted sample and Geometry Necessary dislocations for (e)initial sample and (f)Impacted sample

Therefore, the high strain rates experienced during dynamic impact loading create conditions conducive to the formation of adiabatic shear bands. These shear bands play a crucial role in accommodating the deformation and redistributing stresses, contributing to the overall response and failure mechanisms of the material under such loading conditions.

The true stress versus true strain graph, provided further insights into the loading and unloading behaviour of the impacted sample. Clear distinctions between the loading and unloading phases were observed in the graph, indicating the reversible nature of the deformation process. These distinct phases of loading and unloading demonstrate the material's ability to withstand external forces and exhibit elastic behaviour during unloading [28].

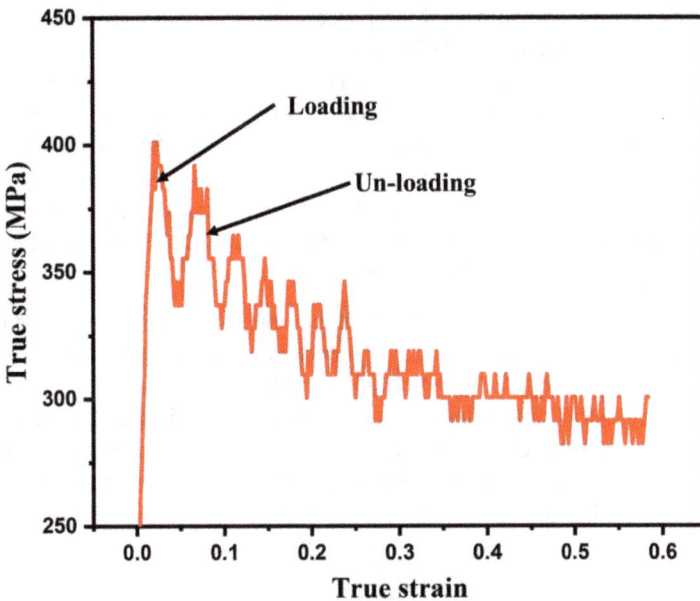

Fig. 4. True Stress Vs True strain for impacted sample.

4 Conclusion

In conclusion, this research study aimed to investigate the impact behavior of friction stir processed samples. Several key conclusions have been drawn based on the findings of this work:

1. A MATLAB code was successfully developed and utilized to process the data obtained from the Split Hopkinson Pressure Bar (SHPB) experiments. This code facilitated the generation of the true stress versus true strain graph, providing insights into the mechanical response of the impacted samples.
2. The EBSD characterization of the impacted samples revealed a significant grain refinement compared to the base sample. Additionally, the analysis indicated a reduction in the presence of geometrically necessary dislocations (GNDs) in the impacted samples. Conversely, the impacted samples exhibited an increased misorientation of high-angle grain boundaries (HAGB) compared to the base sample.
3. The ultimate tensile strength of the impacted sample was determined to be 400 MPa, signifying the maximum stress the material sustained before failure. Furthermore, the impacted sample exhibited a ductility of approximately 20%, demonstrating high ductility.
4. The formation of adiabatic shear bands (ASBs) was observed in the impacted sample. These localized regions of intense plastic deformation played a crucial role in the overall plastic deformation behaviour of the sample.

Overall, this research study sheds light on the impact behaviour of friction stir processed samples, along with that MATLAB code written in this paper could be used for other materials to study the same. The findings provide valuable insights into the mechanical response, microstructural changes, and the influence of adiabatic shear bands on the deformation mechanisms. This knowledge contributes to a better understanding of the material's performance under dynamic loading conditions, aiding in the development of optimized materials for high-strain-rate applications.

Supplementary Data

The MATLAB code used for this research work is written below:

```
% reduction follows Gray's "Classic Split-Hopkinson Pressure Bar Testing"
%
datafile='comp10a.xls';
outfile='comp10a_R.xls';
%
% Specimen Specific Information (often is customized to experiment)
%
Lg=0.118;              % Gage length (in)
d=0.118;               % Gage diameter (in)
Acs=pi*(d/2)^2;        % Cross-sectional area of the specimen
%
% Gage Start Times
%
GT1=-2.6E-6;           % Start of incident pulse
GT2=366.5E-6;          % Start of reflected pulse
GT3=422.5E-6;          % Start of transmitted pulse
GT4=767.0E-6;          % End of transmitted pulse
%
% Hopkinson Bar Properties (Ti-6Al-4V)
%
rho_bar=0.16*0.0310809501716; %slug/in^3;
E_bar=16500; %ksi
cb=sqrt((E_bar*1000*12)/rho_bar); %(in/s) Elastic Wave Speed
Acsb=pi*0.25^2; %in^2  Cross sectional area of Kolsky bar
%
% Import Data
%
Np=xlsread(datafile, 'Sheet1', 'B3');
time11=xlsread(datafile, 'Sheet1', 'D3:D15000');
gav=xlsread(datafile, 'Sheet1', 'E3:E15000');
gbv=xlsread(datafile, 'Sheet1', 'F3:F15000');
%
% Zero and Convert Data
%
%Calibration_11-28-2011
sfa=-9.5247*1000;     %scale factor for Gage A (lbf/V)
sfb=-9.5048*1000;     %scale factor for Gage B (lbf/V)
%
% find the average zero point (first 50 pts)
%
gav1=0.0;
gbv1=0.0;
for i=1:50
   gav1=gav(i)+gav1;
   gbv1=gbv(i)+gbv1;
end
zpa=(gav1/50);
zpb=(gbv1/50);
```

```
%
% zero and convert records gages A, B and C records to lbf
%
for i=1:Np
  time1(i)=time11(i);
  ga(i)=(gav(i)-zpa)*sfa;
  gb(i)=(gbv(i)-zpb)*sfb;
end
dt=time1(2)-time1(1);
%
% find indices where force components start
%
for i=1:Np
  if time1(i)>GT1
    int1=i;
    break
  end
end
for i=1:Np
  if time1(i)>GT2
    int2=i;
    break
  end
end
for i=1:Np
  if time1(i)>GT3
    int3=i;
    break
  end
end
for i=1:Np
  if time1(i)>GT4
    int4=i;
    break
  end
end
%
% Set up force records, calculate relative bar velocity, displacement
% strain rate, strain, force,one wave and two wave stresses
%
u(1)=0.0;
for i=1:int4-int3;
  time2(i)= dt*(i-1);
  F1(i)=ga(int1+i-1);              %incident pulse
  F2(i)=ga(int2+i-1);             %reflected pulse
  F3(i)=gb(int3+i-1);             %transmitted pulse.
  %calculate relative velocity and strain rate
  u_dot(i)=-((2*cb)/((E_bar*1000)*Acsb))*(F2(i));  %check units!
  eps_dot(i)=(1/Lg)*u_dot(i);
  if i>=2
    u(i)=u(i-1)+(dt/2)*(u_dot(i-1)+u_dot(i)); %displacement
  end
  eps(i)=(1/Lg)*u(i); %engineering strain
  sigma_1w(i)=(1/Acs)*F3(i); %1-wave engineering stress
  sigma_2w(i)=(1/Acs)*(F1(i)+F2(i)); %2-wave engineering stress
  eps_t(i)=log(1+eps(i)); %true strain
```

```
   sigma_t(i)=sigma_1w(i)*(1+eps(i)); %true stress
end
%
% Plotting
%
figure(1)
plot(time1,ga,'b')
hold on
plot(time1,gb,'r')
xlabel('Time (sec)','FontSize',12)
ylabel('Force (lbf)','FontSize',12)
title('Wave Data','FontSize',14)
legend('Gage A','Gage B')
%
figure(2)
plot(time2,eps_dot,'b')
xlabel('Time (sec)','FontSize',12)
ylabel('Strain Rate(1/s)','FontSize',12)
title('Strain Rate','FontSize',14)
%
figure(3)
plot(u,F3,'b')
xlabel('Displacement (in)','FontSize',12)
ylabel('Force (lbf)','FontSize',12)
title('Force vs. Displacement','FontSize',14)
%
figure(4)
plot(eps,sigma_1w,'b')
hold on
plot(eps,sigma_2w,'r')
plot(eps_t,sigma_t,'g')
xlabel('Strain','FontSize',12)
ylabel('Stress (psi)','FontSize',12)
title('Stress vs. Strain','FontSize',14)
legend('1-Wave','2-Wave','True')
%
% Write Data to Excel Sheet
%
Test Information
heading5={'Lg (in)','Lg;'d (in)',d;'Acs (in^2)',Acs;'GT1 (s)',GT1;'GT2 (s)',GT2;'GT3 (s)',GT3};
s = xlswrite(outfile, heading5, 'Test_Info', 'A1')
%Headings
heading1={'Time','Gage A','Gage B','Gage A','Gage B'};
s = xlswrite(outfile, heading1, 'Data', 'A1')
heading2={'Time','Velocity','Displacement','Force','Strain    Rate','Eng.    Strain','1W    Stress','2W    Stress','True
Strain','True Stress'};
s = xlswrite(outfile, heading2, 'Data', 'G1')
heading3={'(s)','(V)','(V)','(lbf)','(lbf)'};
s = xlswrite(outfile, heading3, 'Data', 'A2')
heading4={'(s)','(in/s)','(in)','(lbf)','(1/s)',' ','(ksi)','(ksi)',' ','(ksi)'};
s = xlswrite(outfile, heading4, 'Data', 'G2')
%Raw Data
xlswrite(outfile, time1, 'Data', 'A4') %Time1
xlswrite(outfile, gav, 'Data', 'B4') %Gage A (V)
xlswrite(outfile, gbv, 'Data', 'C4') %Gage B (V)
xlswrite(outfile, ga', 'Data', 'D4') %Gage A (lbf)
```

```
xlswrite(outfile, gb', 'Data', 'E4') %Gage B (lbf)

%Reduced Data
xlswrite(outfile, time2', 'Data', 'G4')          %Time2
xlswrite(outfile, u_dot', 'Data', 'H4')          %Velocity (in/s)
xlswrite(outfile, u', 'Data', 'I4')              %Displacement (in)
xlswrite(outfile, F3', 'Data', 'J4')             %Force (lbf)
xlswrite(outfile, eps_dot', 'Data', 'K4')        %Strain Rate (1/s)
xlswrite(outfile, eps', 'Data', 'L4')            %Eng. Strain
xlswrite(outfile, (1/1000)*sigma_1w', 'Data', 'M4')   %1W Stress (ksi)
xlswrite(outfile, (1/1000)*sigma_2w', 'Data', 'N4')   %2W Stress (ksi)
xlswrite(outfile, eps_t', 'Data', 'O4')          %True Strain
xlswrite(outfile, (1/1000)*sigma_t', 'Data', 'P4')    %True Stress (ksi)


After this give the command,
>> plot(time11, gav)

>> hold on

>> plot(time11, gbv)
```

References

1. Renganathan, G., Tanneru, N., Madurai, S.L.: Orthopedical and biomedical applications of titanium and zirconium metals. In: Fundamental Biomaterials: Metals, pp. 211–241 (2018). https://doi.org/10.1016/B978-0-08-102205-4.00010-6
2. Nasiri, Z., Ghaemifar, S., Naghizadeh, M., Mirzadeh, H.: Thermal mechanisms of grain refinement in steels: a review. Met. Mater. Int. **27**, 2078–2094 (2021). https://doi.org/10.1007/s12540-020-00700-1
3. Nečemer, B., Vuherer, T., Glodež, S., Kramberger, J.: Fatigue behaviour of re-entrant auxetic structures made of the aluminium alloy AA7075-T651. Thin-Walled Struct. **180**, 109917 (2022). https://doi.org/10.1016/J.TWS.2022.109917
4. Qin, G., et al.: Influence of single or multi-factor coupling of temperature, humidity and load on the aging failure of adhesively bonded CFRP/aluminum alloy composite joints for automobile applications. Int. J. Adhes. Adhes. **123**, 103345 (2023). https://doi.org/10.1016/J.IJADHADH.2023.103345
5. Ganilova, O.A., Cartmell, M.P., Kiley, A.: Application of a dynamic thermoelastic coupled model for an aerospace aluminium composite panel. Compos. Struct. **288**, 115423 (2022). https://doi.org/10.1016/J.COMPSTRUCT.2022.115423
6. Singh, R.K., et al.: Influence of wire rolling on Zircalloy-2: tensile behaviour and microstructural investigation. J. Market. Res. **25**, 2001–2013 (2023). https://doi.org/10.1016/J.JMRT.2023.06.052
7. Hallberg, H.: Influence of process parameters on grain refinement in AA1050 aluminum during cold rolling. Int. J. Mech. Sci. **66**, 260–272 (2013). https://doi.org/10.1016/j.ijmecsci.2012.11.016
8. Hongfu, Y., et al.: Effect of rolling deformation and passes on microstructure and mechanical properties of 7075 aluminum alloy. Ceram Int. **49**, 1165–1177 (2023). https://doi.org/10.1016/j.ceramint.2022.09.093
9. McQueen, H.J.: Development of dynamic recrystallization theory. Mater. Sci. Eng., A **387–389**, 203–208 (2004). https://doi.org/10.1016/j.msea.2004.01.064
10. Acharya, S., et al.: High strain rate dynamic compressive behaviour of Al6061-T6 alloys. Mater. Charact. **127**, 185–197 (2017). https://doi.org/10.1016/j.matchar.2017.03.005

11. Yadav, D., Bauri, R.: Effect of friction stir processing on microstructure and mechanical properties of aluminium. Mater. Sci. Eng. A **539**, 85–92 (2012). https://doi.org/10.1016/j.msea.2012.01.055

12. Chaudhary, A., Kumar Dev, A., Goel, A., Butola, R.: The mechanical properties of different alloys in friction stir processing: a review (2018). www.sciencedirect.comwww.materials today.com/proceedings

13. Karthikeyan, L., Senthilkumar, V.S., Balasubramanian, V., Natarajan, S.: Mechanical property and microstructural changes during friction stir processing of cast aluminum 2285 alloy. Mater. Des. **30**, 2237–2242 (2009). https://doi.org/10.1016/j.matdes.2008.09.006

14. Singh, R.K., Guraja, S.S.S., Ajide, O.O., Owolabi, G.M., Kumar, N.: Investigation of initial metallurgical factors on the dynamic impact response and adiabatic shear bands formation of the 6061 Al alloy. Mater. Sci. Eng. A **865**, 144636 (2023). https://doi.org/10.1016/J.MSEA. 2023.144636

15. Cavaliere, P., Squillace, A.: High temperature deformation of friction stir processed 7075 aluminium alloy. Mater. Charact. **55**, 136–142 (2005). https://doi.org/10.1016/J.MATCHAR. 2005.04.007

16. Feng, A.H., Ma, Z.Y.: Enhanced mechanical properties of Mg–Al–Zn cast alloy via friction stir processing. Scr. Mater. **56**, 397–400 (2007). https://doi.org/10.1016/J.SCRIPTAMAT. 2006.10.035

17. Wang, F., Wei, J., Wu, G., Qie, M., He, C.: Microstructural modification and enhanced mechanical properties of wire-arc additive manufactured 6061 aluminum alloy via interlayer friction stir processing. Mater. Lett. **342**, 134312 (2023). https://doi.org/10.1016/J.MATLET.2023. 134312

18. Lan, C., et al.: Experimental study on wayside monitoring method of train dynamic load based on strain of ballastless track slab. Constr. Build. Mater. **394**, 132084 (2023). https://doi.org/ 10.1016/J.CONBUILDMAT.2023.132084

19. Cai, W., Hu, Y., Fang, F., Yao, L., Liu, J.: Wind farm power production and fatigue load optimization based on dynamic partitioning and wake redirection of wind turbines. Appl. Energy **339**, 121000 (2023). https://doi.org/10.1016/J.APENERGY.2023.121000

20. Volk, M.W.R., Fu, R.R., Trubko, R., Kehayias, P., Glenn, D.R., Lima, E.A.: QDMlab: a MATLAB toolbox for analyzing quantum diamond microscope (QDM) magnetic field maps. Comput. Geosci. **167**, 105198 (2022). https://doi.org/10.1016/J.CAGEO.2022.105198

21. Charpentier, I., Sarocchi, D., Rodriguez Sedano, L.A.: Particle shape analysis of volcanic clast samples with the Matlab tool MORPHEO. Comput. Geosci. **51**, 172–181 (2013). https://doi. org/10.1016/J.CAGEO.2012.07.015

22. Jiao, S., et al.: KSSOLV 2.0: an efficient MATLAB toolbox for solving the Kohn-Sham equations with plane-wave basis set. Comput. Phys. Commun. **279**, 108424 (2022). https:// doi.org/10.1016/J.CPC.2022.108424

23. Tikhomirov, D., Amiri, N.M., Ivy-Ochs, S., Alfimov, V., Vockenhuber, C., Akçar, N.: Fault scarp dating tool - a MATLAB code for fault scarp dating using in-situ chlorine-36 supplemented with datasets of Yavansu and Kalafat faults. Data Brief **26**, 104476 (2019). https:// doi.org/10.1016/J.DIB.2019.104476

24. Yang, T., et al.: AFDeter: a MATLAB-based tool for simple and rapid determination of the structural parameters and the airflow-related properties of fibrous materials. SoftwareX **20**, 101213 (2022). https://doi.org/10.1016/J.SOFTX.2022.101213

25. Gustafson, S.E., Pagan, D.C., Sanborn, B., Sangid, M.D.: Grain scale residual stress response after quasi-static and high strain rate loading in SS316L. Mater. Charact. **197**, 112692 (2023). https://doi.org/10.1016/J.MATCHAR.2023.112692

26. Roa, J.J., Mateo, A.M., Llanes, L.: Implementation of massive nanoindentation coupled with statistical analysis to evaluate complex heterogeneous microstructures in materials manufactured following powder metallurgy processing routes. In: Caballero, F.G. (ed.) Encyclopedia

of Materials: Metals and Alloys, pp. 465–470. Elsevier, Oxford (2022). https://doi.org/10.1016/B978-0-12-819726-4.00097-1

27. Terparia, S., Mir, R., Tsang, Y., Clark, C.H., Patel, R.: Automatic evaluation of contours in radiotherapy planning utilising conformity indices and machine learning. Phys. Imaging Radiat. Oncol. **16**, 149–155 (2020). https://doi.org/10.1016/j.phro.2020.10.008

28. Wu, X., Li, L., Liu, W., Li, S., Zhang, L., He, H.: Development of adiabatic shearing bands in 7003-T4 aluminum alloy under high strain rate impacting. Mater. Sci. Eng. A **732**, 91–98 (2018). https://doi.org/10.1016/j.msea.2018.06.087

Ratcheting Fatigue Behaviour of Advanced Structural Materials

Prerna Mishra[1]($\boxtimes$), N. C. Santhi Srinivas[2], G. V. S. Sastry[2], and Vakil Singh[2]

[1] Department of Mechanical Engineering, MKSSS Cummins College of Engineering, Pune 411052, India
prerna.mishra@cumminscollege.in

[2] Department of Metallurgical Engineering, Indian Institute of Technology (Banaras Hindu University), Varanasi 221005, India
prernam.rs.met16@iitbhu.ac.in

Abstract. The advanced power generating industries requires high operating steam temperatures and pressures, to attain higher efficiency. This led to development of alloys with superior properties at elevated temperatures. The two alloys under investigation are Modified 9Cr-1Mo steel and Inconel 617 alloy. From the application point of view, the two alloys are used as piping and tubing materials in various components such as steam generator, super heater, re heater and heat exchanger etc. The present investigation deals with the comparative study of the two alloys under asymmetrical stress controlled cyclic loading, with the help of normalized mean stress and stress amplitude with constant stress rate, at ambient and homologous temperatures. Fractographic studies of the failed specimens were also observed, at room and homologous temperatures with the aid of scanning electron microscope. The difference in fractographic features and deformation mechanism of the alloys has been elaborated. The key results of this investigation points that there is accumulation of plastic strain under asymmetrical cyclic loading. Modified 9 Cr-1Mo steel exhibited unique fracture behaviour under ratcheting which was not observed in case of IN-617 superalloy. On comparing the ratcheting of the alloys at ambient temperature and homologous temperatures it was concluded that uniform strain is a crucial parameter to consider when selecting materials for piping and tubing components that are subjected to ratcheting. Inconel 617 alloy showed remarkable strength and sustained very high strain values compared to modified steel.

Keywords: mean stress · stress amplitude · homologous temperature · ratcheting

1 Introduction

Failure occurring under cyclic loading is termed as fatigue. The loading condition may change from symmetric to asymmetric. When there is symmetrical loading (R = −1), plastic strain does not accumulate; however, when there is asymmetrical loading, hysteresis loops do not shut, and plastic strain increases due to the application of positive

© The Author(s), under exclusive license to Springer Nature Singapore Pte Ltd. 2024
S. Patra et al. (Eds.): METCENT 2023, *Proceedings of the International Conference on Metallurgical Engineering and Centenary Celebration*, pp. 311–322, 2024.
https://doi.org/10.1007/978-981-99-6863-3_31

mean stress; this process is known as ratcheting or cyclic creep. In comparison to low cycle fatigue, cyclic life is dramatically reduced due to steady increase in tensile permanent deformation. Because of fluctuations in internal pressure, temperature, and seismic occurrences, components such as pipes and tubes in many industries and power plants ratchet throughout service; thus, ratcheting fatigue has gained significant importance [1, 2].

From 1990 to 2020, the area of ratcheting fatigue has attracted the attention of numerous investigators. First investigation in the area of ratcheting was reported by Bairstow [3] in 1911. Accumulation of plastic strain was observed in a steel specimen under uniaxial cyclic stressing with tensile mean stress. Numerous other researchers also reported progression of axial strain under uniaxial asymmetric cyclic loading at elevated, ambient and low temperatures [4–6]. Researchers mainly focussed on various experimental aspects of ratcheting including influence of mean stress, stress amplitude, stress rate, path of loading, planar anisotropy, temperature, previous loading history. Ratcheting in engineering and true stress modes and their comparison has also been addressed. Ratcheting- tensile, ratcheting- LCF, ratcheting-creep and their life assessment has also been discussed [7–9]. During the last three decades, much of the research related to ratcheting fatigue is focussed on microstructural characterization and constitutive modelling related to ratcheting fatigue in various structural materials such as different grades of steels, alloys of aluminium, magnesium, zirconium and copper. In most of these studies, the effect of various parameters influencing ratcheting fatigue behaviour, modes of conducting ratcheting fatigue experiments and studies on ratcheting from specimen to component level are mentioned. Also the correlation of ratcheting fatigue with microstructural modifications, effect of ratcheting fatigue on other phenomenon such as creep, tensile and low cycle fatigue, comparison of ratcheting fatigue behaviour of materials with different crystal structures are discussed [10–12].

The advanced new generation power plants and industries use high temperature, pressure and aggressive environments to achieve higher efficiency and to reduce CO_2 emissions. Therefore, there is requirement of alloys of better performance to sustain such conditions. The alloy should possess superior mechanical properties, microstructural stability and resistance against oxidation and corrosion at elevated temperatures for long duration. Modified 9Cr-1Mo steel and Inconel 617 alloys are used as piping and tubing material in various engineering components in power generating, chemical and petroleum industries. These alloys undergo different types of cyclic loading during their service and fail due to fatigue [13, 14]. This necessitates the study of ratcheting fatigue in these two alloys therefore the present investigation deals with the comparative study of these two alloys at ambient and homologous temperature.

2   Materials and Methods

Modified steel in normalized (1060 °C−1 h) and tempered (780 °C−1 h) condition was supplied by IGCAR, Kalpakkam, India, in form of 25 mm thick plate. Superalloy Inconel 617 was procured from Bharat Aerospace Metals in the form of rod of 14 mm diameter. The rods are given solution treatment at 1175 °C for 40 min and quenched in water. Chemical compositions of alloys has been addressed in our earlier publications

[1, 14]. Tensile tests were conducted at strain rate of 10^{-3} s^{-1} at ambient and elevated temperatures, using 100 kN Instron (Model 5982). Uniaxial asymmetrical ratcheting fatigue tests at room and high temperatures were conducted using servo hydraulic test system (MTS model 810). The schematic geometry of tensile and fatigue specimens is depicted in Fig. 1(a and b). MTS extensometer (Model: 632.53E) was mounted on gauge section of specimen for strain measurement. Since a comparison has to be made between the alloys at ambient and homologous temperatures hence the stress rate ($\sigma^{\cdot}$) was kept constant at 150 MPa/s, and identical normalized mean stress and stress amplitude was used for both the alloys. The values of the mean stress and stress amplitude were normalized with respect to ultimate tensile strength of the respective material so that except the material all other parameters are constant. The service temperature of both the alloys are different thus the best way to compare the mechanical properties of different materials at elevated temperatures is in terms of the ratio of the test temperature to the melting point, expressed in degrees Kelvin. This ratio is often referred to as the homologous temperature.

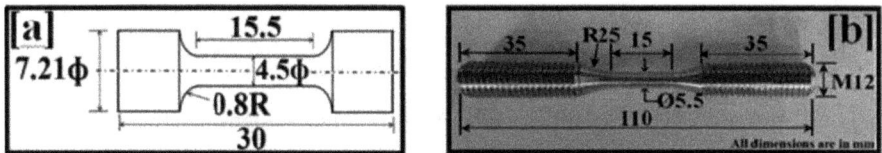

Fig. 1. Schematic geometry of cylindrical (a) tensile specimen (b) fatigue specimen.

3 Theory and Calculation

3.1 Calculation of Homologous Temperature

Mathematically, homologous temperature is expressed as:

$$T_H = \frac{T}{T_m} \tag{1}$$

where; T_H, T, and T_m are homologous, service, and melting temperature in Kelvin, respectively.

The service temperature of the modified steel is lower than that of the IN 617 alloy; therefore, for calculation of the homologous temperature, service temperature of modified steel (873 K) was taken into consideration.

(T_m) modified steel = 2073 K

(T_m) IN-617 = 1623 K

(T) modified steel = 873 K

(T) IN-617 = 1123 K

Thus, in case of modified steel:

$$T_H = \frac{873}{2073}$$

$$T_H = 0.42$$

Hence, for comparing the two alloys at elevated temperature, the homologous temperature is kept identical for both the materials and the test temperature corresponding to homologous temperature of 0.42, for the IN 617 alloy is calculated as:

$$T_H = \frac{T}{T_m}$$

$$0.42 = \frac{T}{1623}$$

$$T = 682K$$

Thus, corresponding to service temperature of 600 °C for the modified steel and the homologous temperature 0.42, the corresponding test temperature for the IN 617 alloy works out to be 409 °C.

3.2 Normalizing of Mean Stress and Amplitude of Stress

There was difference in the yield strength (YS) and ultimate tensile strength (UTS) of the two materials; therefore, maximum stress for the two materials will be variation in the mean stress and the amplitude of stress; hence, comparison of the two materials will not be justifiable. Therefore, mean stress and amplitude were normalized with UTS of the two materials, and the stress rate was kept constant at 150 MPa/s.

$$\text{Normalised mean stress } \sigma_{nm} = \frac{\sigma_m}{\sigma_{UTS}} \tag{2}$$

$$\text{Normalised stress amplitude } \sigma_{na} = \frac{\sigma_a}{\sigma_{UTS}} \tag{3}$$

4 Results

4.1 Microstructure

In line with earlier investigations the microstructure of modified steel in normalized and tempered state exhibits lath martensitic structure (Fig. 2a) with average grain size of 20 μm. The second phase particles of $M_{23}C_6$, M_6C and MX type are distributed within laths and along boundaries. Similarly, microstructure of solutionized superalloy IN-617, reveals single phase matrix of austenite (γ) with average grain size of 100 μm and second phase particles distributed along grain boundaries (Fig. 2b) [1, 2, 13, 14].

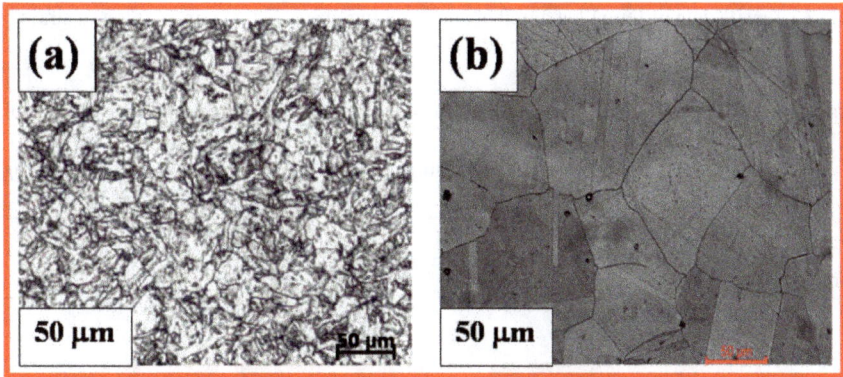

Fig. 2. Optical micrographs of heat-treated (a) Modified 9Cr-1Mo steel showing lath martensitic structure (b) Inconel 617 alloy showing single phase γ matrix.

4.2 Tensile Properties

For finalizing the asymmetrical stress controlled loading experiments and assessment of various tensile parameters monotonic tensile tests were performed at a strain rate of 10^{-3} s^{-1} to evaluate various tensile parameters. All the asymmetrical stress controlled experiments were performed above the yield strength of the materials. High temperature tests were conducted using the same machine with the help of a single zone resistance heating furnace. Table 1 and Table 2 shows the tensile properties of modified steel and superalloy Inconel 617 at ambient and homologous temperatures. Figure 3 shows the engineering stress- strain curves for both the alloys at ambient temperature.

Table 1. Tensile data of Modified 9Cr-1Mo steel at ambient and homologous (0.42T_m) temperatures at strain rate of 10^{-3} s^{-1}.

Temperature (°C)	Yield strength (MPa)	Ultimate Tensile Strength (MPa)	Uniform Plastic Elongation (%)	Elongation (%)	Reduction in Area (%)	Degree of work hardening (S_{UTS}/S_{YS})
Room	575	713	8	26	74	1.23
Homologous	305	380	2.2	34.7	63	1.24

Table 2. Tensile data of IN-617 at ambient and homologous ($0.42T_m$) temperatures at strain rate of 10^{-3} s^{-1}.

Temperature (°C)	Yield strength (MPa)	Ultimate Tensile Strength (MPa)	Uniform Plastic Elongation (%)	Elongation (%)	Reduction in Area (%)	Degree of work hardening
Room	347	835	54	66	68	2.45
Homologous	235	665	60	72	63	2.82

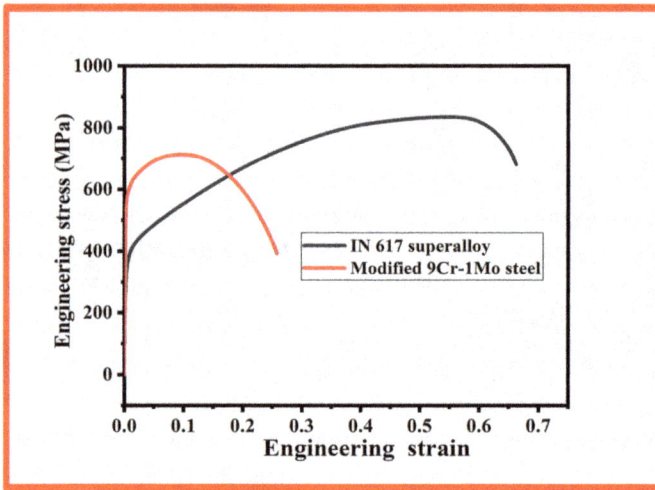

Fig. 3. Engineering stress-strain curves of the alloys tested at ambient temperature at the strain rate of 10^{-3} s^{-1}.

4.3 Uniaxial Ratcheting Behaviour of the Alloys at Ambient and Homologous ($0.42T_m$) Temperatures

Asymmetrical stress controlled fatigue tests were conducted at ambient and homologous temperatures for both the alloys with the help of normalized mean stress and stress amplitude at a constant stress rate of 150 MPa/s to observe the difference in ratcheting behaviour of the alloys. The reason behind the normalized values is the difference in the yield and ultimate tensile strength of the alloys. Due to the above mentioned difference, the maximum stress will differ and thus the values of mean stress and stress amplitude will differ for the alloys. Therefore, for the purpose of comparison all the values should be same except the temperatures hence the parameters are normalized with respect to tensile strength of the respective alloys at that particular temperature. The test matrix for the alloys at ambient and homologous temperatures is shown in Table 3.

Table 3. Ratcheting fatigue test matrix at ambient and homologous ($0.42T_m$) temperatures.

AMBIENT TEMPERATURE

Material	Minimum stress σ_{min} (MPa)	Maximum stress σ_{max} (MPa)	Stress Amplitude σ_a (MPa)	Mean stress σ_m (MPa)	Normalized stress amplitude (σ_a/σ_{UTS})	Normalized mean stress (σ_m/σ_{UTS})	No. of cycles to failure (N_f)
Modified 9Cr-1Mo steel	−200	620	410	210	410/713 = 0.58	210/713 = 0.29	1506
Inconel 617 alloy	−240	730	485	245	500/835 = 0.58	250/835 = 0.29	4206

HOMOLOGOUS TEMPERATURE ($0.42T_m$)

Material	Maximum stress σ_{max} (MPa)	Minimum stress σ_{min} (MPa)	Stress Amplitude σ_a (MPa)	Mean Stress σ_m (MPa)	Normalized stress amplitude (σ_a/σ_{UTS})	Normalized mean stress (σ_m/σ_{UTS})	No. of cycles to failure (N_f)
Modified 9Cr-1Mo steel	365	−175	270	95	270/380 = 0.71	95/380 = 0.25	2894
Inconel 617 alloy	640	−310	473	165	473/665 = 0.71	165/665 = 0.25	2252

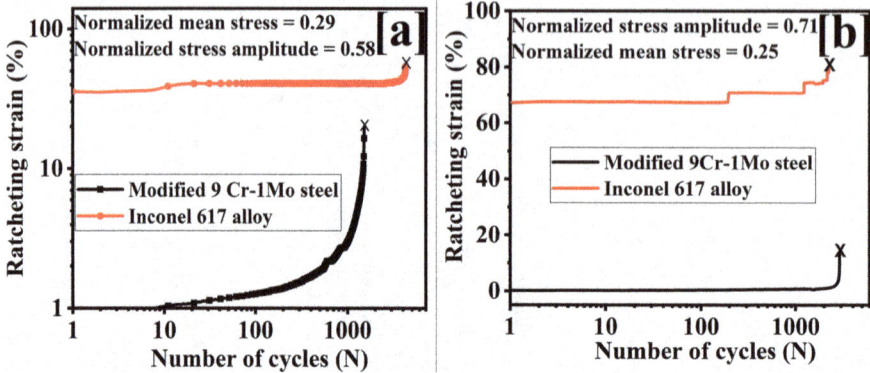

Fig. 4. Ratcheting strain plots for modified 9Cr-1Mo steel at (a) ambient temperature (b) homologous temperature $(0.42T_m)$.

It was observed that at ambient temperature ratcheting strain accumulation (Fig. 4a) in IN-617 alloy (52%) is significantly higher than that in modified 9Cr-1Mo steel (20%) and the alloy also exhibited higher fatigue life. At homologous temperature $(0.42T_m)$ ratcheting strain accumulation is very high in IN-617 alloy (65%) as compared to modified 9Cr-1Mo steel (15%) and there is very slight difference between the fatigue lives of the two alloys. Modified 9Cr-1Mo steel exhibited very stable ratcheting behaviour whereas IN-617 alloy showed stepped increase in ratcheting strain (Fig. 4b).

4.4 Fracture Behaviour of the Alloys at Room and Homologous $(0.42T_m)$ Temperatures

On critical examination of the fracture surfaces it was observed that modified steel exhibited tapered annular region, surrounding central region of tensile fracture in Fig. 5(a), which clearly shows occurrence of necking in modified 9Cr-1Mo steel; in sharp contrast, there is no necking in Inconel 617 alloy (Fig. 5b). Dimples on fracture surface of the modified 9Cr-1Mo steel (Fig. 5c) occurred essentially due to necking that led to tensile fracture. On the other hand, in IN-617 alloy failure occurred by usual process of fatigue, through crack initiation and propagation as depicted in Figs. 5 (b and d).

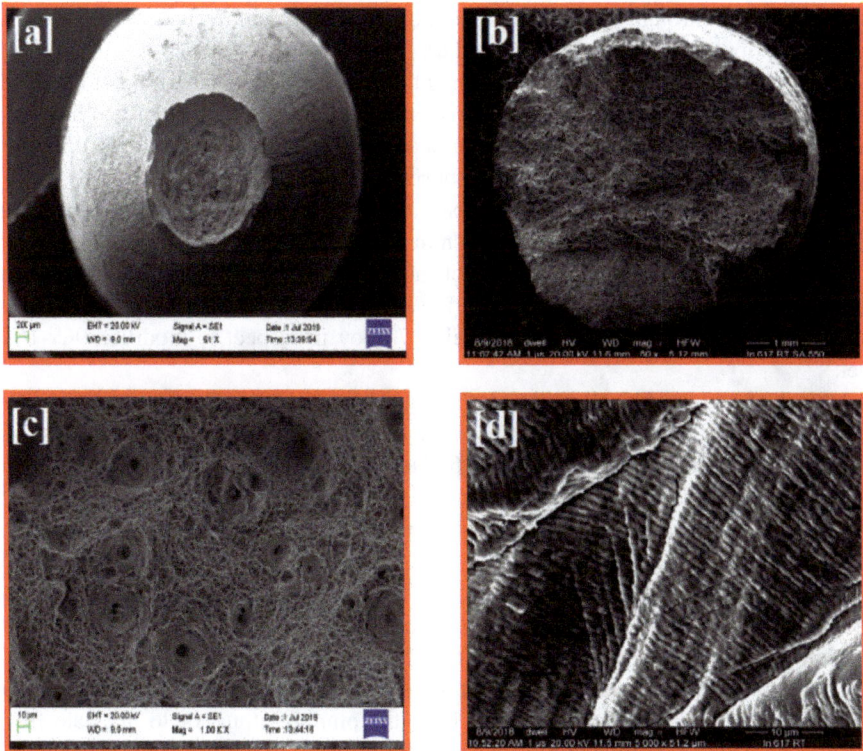

Fig. 5. Fractographs of both the alloys at room temperature tested under ratcheting: (a and c) modified 9 Cr-1Mo steel, (b and d) IN-617 alloy.

5 Discussion

Ratcheting strain is much higher in IN-617 alloy, still it exhibits higher fatigue life as compared to the life depicted by modified 9Cr-1Mo steel. The reason for observed behaviour is that in IN-617 alloy, the degree of work hardening is much higher, in comparison with that of modified 9Cr-1Mo steel.

Table 1 and Table 2 shows that uniform plastic strain in superalloy is much higher than that of modified steel (nearly by seven times). Figure 4(a) clearly reveals that cumulative ratcheting strain in the superalloy IN-617 alloy is 52% and that in modified steel is 20%, even then also, the cyclic fatigue life of modified steel (1506) is much smaller than of superalloy IN-617 alloy (4206). The reason behind this may be attributed to the fact that the value of uniform plastic strain in the modified steel was only 8%; therefore, once the value of cumulative ratcheting strain exceeded 8% (the uniform plastic strain), there was commencement of necking, which quickly reduced the material's load bearing capacity and resulted in rapid fracture in the tensile mode (Fig. 5a). However, even though the cumulative ratcheting strain for the Inconel 617 alloy was much higher (52%), it did not exceed the uniform plastic strain of 54%. As a result, necking did not happen, and the failure happened through the typical process of fatigue, which involves crack initiation and propagation, as shown in Fig. 5(b and d).

Excessive plastic deformation, exceeding the degree of work hardening of the material, would increase the tendency for fatigue crack initiation and propagation, however, the degree of work hardening would accommodate the accumulating plastic strain resulting from ratcheting and would not allow onset of necking, and the fatigue life will be enhanced. Low uniform plastic strain, resulting from low degree of work hardening, would cause necking and thereby reduction in fatigue life.

On comparing the ratcheting strain plot for both the alloys at homologous temperature, it can be very well observed that during initial cycles of stressing, there is abrupt rise in ratcheting strain in case of Inconel 617 alloy in respect to modified 9Cr-1Mo steel.

In case of Inconel 617 alloy, the ratcheting strain was 70% whereas modified 9Cr-1Mo steel showed very stable cyclic behaviour. The fatigue life of modified 9Cr-1Mo steel is higher as compared to that of IN-617 alloy. There is stepped increase observed in ratcheting strain for IN-617 alloy after certain number of cycles. The 0.42 Tm homologous temperature of IN-617 alloy is also the temperature at which the alloy exhibits dynamic strain aging behaviour.

Dynamic strain aging (DSA) is a process of aging that takes place during plastic deformation over a particular strain rate and temperature range, where the interactions between solute atoms and dislocations, result in a strong pinning of dislocations. A material exhibiting DSA requires higher stresses to produce further straining of the material, either to pull dislocations free from the pinning atoms or to nucleate fresh dislocations. Thus in the ratcheting strain plots, there is observation of steps like increase in the strain that must be due to unpinning effect and due to which there is abrupt increase in the most stabilized region of the ratcheting curve, hence there is reduction in cyclic life of the alloy.

Inconel 617 alloy showed tremendous capability of sustaining such high amount of strain under asymmetrical loading (almost 70%) at normalized mean stress of 0.25 and normalized stress amplitude of 0.71 as compared to modified 9Cr-1Mo steel. Therefore, such materials are required to avoid catastrophic failures. Uniform elongation and total elongation are very important parameters for designing of piping and tubing components.

6   Conclusions

The major findings are summarised below:

1. On comparing the ratcheting behaviour of the alloys at ambient temperature, Inconel 617 alloy showed higher fatigue life in comparison with that of the modified 9Cr-1Mo steel. The reason behind this is higher uniform elongation value (54%) for Inconel 617 alloy. Thus, uniform strain can be considered as an important tensile parameter for selection of materials for piping and tubing components, experiencing ratcheting fatigue.
2. On comparing the alloys at 0.42 Tm, modified 9Cr-1Mo steel exhibited more significant and stable result with higher fatigue life in comparison with the Inconel 617 alloy.

3. Inconel 617 alloy showed very high stain accumulation of approximately 70% in comparison to modified 9Cr-1Mo steel and the difference in fatigue life was also of about 600 cycles; thus it can be attributed that under the same parameters though cyclic life is less for Inconel 617 alloy but it can sustain very high strain values.
4. Inconel 617 alloy showed abrupt increase in plastic strain accumulation at 0.42 Tm. The reason behind this may be due to dynamic strain aging of the alloy resulting from unpinning effect. Plastic strain accumulation resulting from pinning of dislocations causes hardening but release of dislocations resulting from unpinning of dislocations causes softening. Strain is localized.
5. Modified 9Cr-1Mo at 0.42 Tm under the same parameters exhibited very stable behaviour and higher cyclic life as compared to Inconel 617alloy. The capability to withstand such high strain values in case of Inconel 617 alloy is very high as compared to Modified 9Cr-1Mo steel. Thus it can be concluded that uniform strain plays a very important role at room as well as at elevated temperature for ratcheting fatigue.
6. In ratcheting fatigue, with tensile mean stress, there is progressive increment in the overall length of the specimen; thus, it is obvious that plastic strain in the specimen does not remain localized in the region of slip bands, like that in fatigue under symmetrical loading. Thus, consideration of uniform plastic strain in analysing ratcheting fatigue is quite relevant and important and tensile ductility of material can be considered an important parameter for LCF resistance.

References

1. Mishra, P., Rajpurohit, R.S., Srinivas, N.S., Sastry, G.V.S., Singh, V.: Ratcheting fatigue behavior of modified 9Cr–1Mo steel at room temperature. Metals Mater. Int. **27**(12), 4922–4936 (2021)
2. Mishra, P., Srinivas, N.C., Singh, V.: Ratcheting fatigue of Superalloy IN-617 under tensile mean stress at RT. Trans. Indian Inst. Metals, 1–12 (2022)
3. Bairstow, L.: The elastic limits of iron and steel under cyclical variations of stress. Philos. Trans. Roy. Soc. London Ser. A Containing Papers Math. Phys. Charac. **210**(459–470), 35–55 (1911)
4. Kennedy, A.J.: Effect of fatigue stresses on the recovery properties of metals. Nature **178**(4537), 810–811 (1956)
5. Benham, P.P., Ford, H.: Low endurance fatigue of a mild steel and an aluminium alloy. J. Mech. Eng. Sci. **3**(2), 119–132 (1961)
6. Moyar, G.J.: A mechanics analysis of rolling element failures. University of Illinois at Urbana-Champaign (1960)
7. Dutta, K., Sivaprasad, S., Tarafder, S., Ray, K.K.: Influence of asymmetric cyclic loading on substructure formation and ratcheting fatigue behaviour of AISI 304LN stainless steel. Mater. Sci. Eng., A **527**(29–30), 7571–7579 (2010)
8. Mishra, P., Santhi Srinivas, N.C., Sastry, G.V.S., Singh, V.: Influence of pre-ratcheting fatigue on tensile behavior of modified 9Cr-1Mo steel at ambient temperature. Trans. Indian Inst. Metals **74**(4), 937–948 (2021)
9. Paul, S.K., Sivaprasad, S., Dhar, S., Tarafder, S.: True stress control asymmetric cyclic plastic behavior in SA333 C-Mn steel. Int. J. Press. Vessels Pip. **87**(8), 440–446 (2010)
10. Lin, Y.C., Liu, Z.-H., Chen, X.-M., Chen, J.: Uniaxial ratcheting and fatigue failure behaviors of hot-rolled AZ31B magnesium alloy under asymmetrical cyclic stress-controlled loadings. Mater. Sci. Eng. A **573**, 234–244 (2013)

11. Wen, M., Li, H., Yu, D., Chen, G., Chen, X.: Uniaxial ratcheting behavior of Zircaloy-4 tubes at room temperature. Mater. Des. **46**, 426–434 (2013)
12. Kang, G., Liu, Y., Dong, Y., Gao, Q.: Uniaxial ratcheting behaviors of metals with different crystal structures or values of fault energy: macroscopic experiments. J. Mater. Sci. Technol. **27**(5), 453–459 (2011)
13. Verma, P., et al.: Dynamic strain ageing, deformation, and fracture behavior of modified 9Cr–1Mo steel. Mater. Sci. Eng. A **621**, 39–51 (2015)
14. Rao, C.V., Srinivas, N.C.S., Sastry, G.V.S., Singh, V.: Dynamic strain aging, deformation and fracture behaviour of the nickel base superalloy Inconel 617. Mater. Sci. Eng. A **742**, 44–60 (2019)

Study of Microstructure and Mechanical Properties of Similar and Dissimilar FSW of Al 7075 and Mg AZ31 Alloys Without and With Zn Interlayer

Satya Kumar Dewangan, Manwendra Kumar Tripathi$^{(\boxtimes)}$, Pragya Nandan Banjare, Abhijeet Bhowmik, and Manoranjan Kumar Manoj

Department of Metallurgical and Materials Engineering, National Institute of Technology Raipur, Raipur, India
mktripathi.mme@nitrr.ac.in

Abstract. The similar and dissimilar joining of Al and Mg alloys are important in aerospace and automobile industries. The Al 7075 and Mg AZ31 alloys are mostly used in the industries due to their light weight and high strength properties. In present work four FSW of similar Al 7075, similar Mg AZ31, dissimilar Al 7075 and AZ31 without interlayer and dissimilar Al 7075 and AZ31 with Zn interlayer has been studied at 1300 rpm rotational speed and 20 mm/min traverse speed, corresponding stir zones (SZ) have also been investigated. It was found that, the dissimilar FSW of Al 7075 and Mg AZ31 alloys without interlayer has formed defects (such as cracks, void, tunneling and kissing bonds) due to the formation of brittle phases *viz.* Al_3Mg_2 and $Al_{12}Mg_{17}$. Therefore, the Zn interlayer was induced in between the faying surface of Al and Mg alloys FSW which led to defect free joint. During stirring the Zn has reacted with Mg and Al, and formed different phases such as Al-Mg-Zn and Mg-Zn, which inhibited the formation of Al_3Mg_2 and $Al_{12}Mg_{17}$ IMCs layers. The joint efficiency of similar FSW of Al 7075 attained 55% (285 MPa) and Mg AZ31 attained 91% (225 MPa). The dissimilar FSW without interlayer has formed immediate cracking and no strength has been achieved. Whereas, the dissimilar FSW with Zn interlayer attained 34% joint efficiency of AZ31 (85 MPa).

Keywords: Friction stir welding · Al 7075 · Mg AZ31 · Intermetallic compounds · Zn interlayer

1 Introduction

The light weight and high strength of Al and Mg alloys are important for reducing the fuel consumption as well as environmental considerations. The similar and dissimilar joining of Al and Mg alloys are difficult by the fusion based joining process due to the formation of hot cracks and oxide formation [1]. Therefore, the solid-state joining technique is used for the joining of these Al and Mg alloys. Friction stir welding is a one of

S. Patra et al. (Eds.): METCENT 2023, *Proceedings of the International Conference on Metallurgical Engineering and Centenary Celebration*, pp. 323–334, 2024.
https://doi.org/10.1007/978-981-99-6863-3_32

the solid-state joining techniques in which the frictional heat is generated in between the frictional contact surface of tool and workpiece. The workpiece materials are plastically deformed to produce coalescence. Many researchers have worked to improve the joint strength by optimization of process parameter, tool design along with modification in the FSW techniques [2–4]. Tool rotational speed, transverse speed and plunge force are responsible parameters for the frictional heat generation between the tool and workpiece. The pin offset of the centre line towards Mg/Al alloy side has been used in the dissimilar FSW of Al and Mg alloys [5].

Al and Mg alloys are mixed during welding and form brittle intermetallic compounds such as $Al_{12}Mg_{17}$ and Al_3Mg_2 [6]. The liquation during the welding process is responsible for the formation of $Al_{12}Mg_{17}$ and Al_3Mg_2 [7, 8]. The continuous layer of these IMCs are detrimental to the joint strength. Several researchers have investigated formation of the IMCs to reduce the thickness by using SiC particle [9], Zn [4, 10, 11], Sn [1], Ni [12] and Cd [13] *etc*. Effect of Zn interlayer on the butt configuration of Al and Mg dissimilar FSW has been studied and formation of IMCs of Al-Mg-Zn has been reported [11]. The very limited work reported on comparative study of similar FSW of Al 7075 and Mg AZ31 alloy, and dissimilar FSW of Al 7075 and Mg AZ31 alloy without and with Zn interlayer.

The present work investigates the similar FSW of Al 7075 and Mg AZ31 alloys. The dissimilar FSW of Al 7075 and AZ31 alloys without and with Zn interlayer has also been investigated. The Zn interlayer has been incorporated between the faying surface of workpieces. The material flow behavior of similar and dissimilar (Al 7075 and Mg AZ31 alloy) have also been discussed. SEM/EDS and XRD have been done for the investigation of IMCs formed during the FSW of Al 7075 and AZ31 alloys without and with interlayer. Hardness and tensile test have been performed for mechanical characterization.

2 Experimental Details

Similar and dissimilar friction stir welding of Al 7075 and Mg AZ31 alloy has been performed at rotational speed 1300 rpm and transverse speed 20 mm/min by the cylindrical pin type tool [14, 15]. Figure 1(a) and (b) show the schematic diagram of the FSW arrangement and cylindrical pin tool, respectively. The dissimilar FSW without interlayer and with Zn interlayer has been performed. A ribbon of zinc having thickness of 0.3 mm was used as interlayer between Al and Mg alloy shown in Fig. 1(a). The tool pin offset of 0.3 mm was used towards the Al 7075 alloy from the center of the weld in FSW without interlayer and 0.45 mm offset was used in FSW with Zn interlayer. The cylindrical tool (H13 steel) having shoulder diameter of 16 mm, pin diameter 3 mm and pin length 2.8 mm was used. The tool tilt angle 3° was used for all the welding conditions.

After welding, image of the top surface has been taken using high resolution camera. Then the weld plates are cut along the perpendicular to the welding direction for the various tests and characterizations by using CNC wire cut EDM. The tensile sample (ASTM E8 standard) and metallography sample has been taken from the perpendicular to the welding direction shown in Fig. 1(a). Macrostructure and microstructure studied after the samples were polished by using different grade SiC papers followed by cloth polishing with alumina powder of 2-micron size. The Keller's reagent was used for etching of aluminum alloy side. Whereas, Picrol etchant was used for etching of magnesium alloy side. SEM-EDS was used for estimation of chemical composition. Hardness tests were performed on microhardness tester at 100 gm load. Tensile tests were conducted using tensile testing machine at cross head speed of 1 mm/min. Figure 1(a) shows the experimental setup of FSW, where the dotted line marked for the metallography and tensile sample. Figure 1(b) shows the cylindrical tool with flat shoulder surface before the experiment.

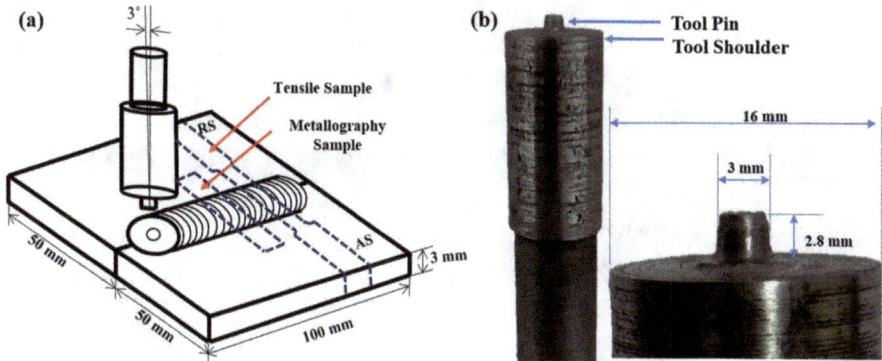

Fig. 1. (a) Schematics of FSW (b) Cylindrical tool for FSW

3 Results and Discussion

The joining of Al 7075 and Mg AZ31 alloy leads defects (cracks, void and tunneling) which is responsible for poor joint strength [1]. Present work studies the mechanical properties of similar FSW of Al 7075 and Mg AZ31 alloy and comparative studies of dissimilar FSW of Al 7075 and Mg AZ31 alloy with and without Zn interlayer at rotational speed 1300 rpm and travel speed 20 mm/min. All FSW samples such as similar Al 7075, similar Mg AZ31, dissimilar Al 7075 and AZ31 alloy without interlayer and dissimilar Al 7075 and AZ31 alloys have been named as *FSW-7075, FSW-AZ31, FSW-7075-AZ31 and FSW-7075-AZ31-Zn*, respectively. The FSW without interlayer (*FSW-7075-AZ31*) has formed defect. Therefore, attempt has been taken to investigate the effect of Zn interlayer on dissimilar FSW of Al 7075 and Mg AZ31 alloy at travel speed 20 mm/min. Defect free joint has been observed in the FSW with Zn interlayer. The optical microscopy, SEM, XRD, hardness and tensile test have been performed after defect free joint formation of FSW samples.

Figure 2(a), (b), (c) and (d) show the friction stir welding sample of FSW-7075, FSW-AZ31, FSW-7075-AZ31 and FSW-7075-AZ31-Zn, respectively. Flash has formed in all the samples and external crack has formed only in the sample FSW-7075-AZ31. Figure 2(e), (f), (g) and (h) represent the macrostructure of the cross-section of FSW samples of Fig. 2(a), (b), (c) and (d), respectively. Macrostructure represents the defect free joint formation in the samples FSW-7075, FSW-AZ31 and FSW-7075-AZ31-Zn. The external defect through thickness of the plate has been formed in sample FSW-7075-AZ31.

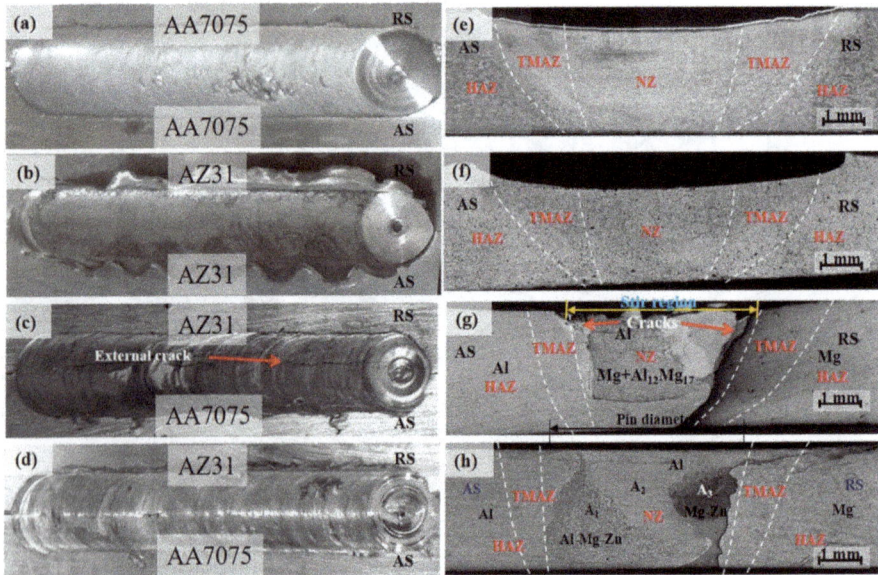

Fig. 2. (a), (b) (c) and (d) FSW samples FSW-7075, FSW-AZ31, FSW-7075-AZ31 and FSW-7075-AZ31-Zn, respectively (e), (f) (g) and (h) Macrostructure of FSW samples

Macrostructural observation in dissimilar FSW samples shows that the materials have mixed and flowed in both the welds in a similar manner, however, formation of new phases and their distribution is quite different. It has been observed that during stirring, material is displaced from the top surface of Mg base plate (RS) and transferred to Al base plate (AS) by the back portion of tool shoulder. Further, pin of the tool transfers to the mixed material from upper surface of the plates towards the bottom side in the Al base plate (AS). Detailed investigation of similar and dissimilar FSW has been discussed in the following sub sections.

3.1 Similar FSW of Al7075 Alloys (FSW-7075)

Figure 3 shows the microstructure of Nugget Zone (NZ), Thermo-Mechanical Affected Zone (TMAZ), Heat Affected Zone (HAZ) and Base Material (BM) in the sample FSW-7075. Figure 3(a) shows the elongated microstructure of the base material of Al 7075. Figure 3(b) shows the microstructure of the HAZ, which is slightly coarser than the base material. Figure 3(c) shows the macrostructure of FSW-7075, where all zones are separated by dotted line. Figure 3(d) represents the non-uniform microstructure of the TMAZ. Figure 3(e) shows the NZ, where the fine grains are formed due to the dynamic recrystallization of materials.

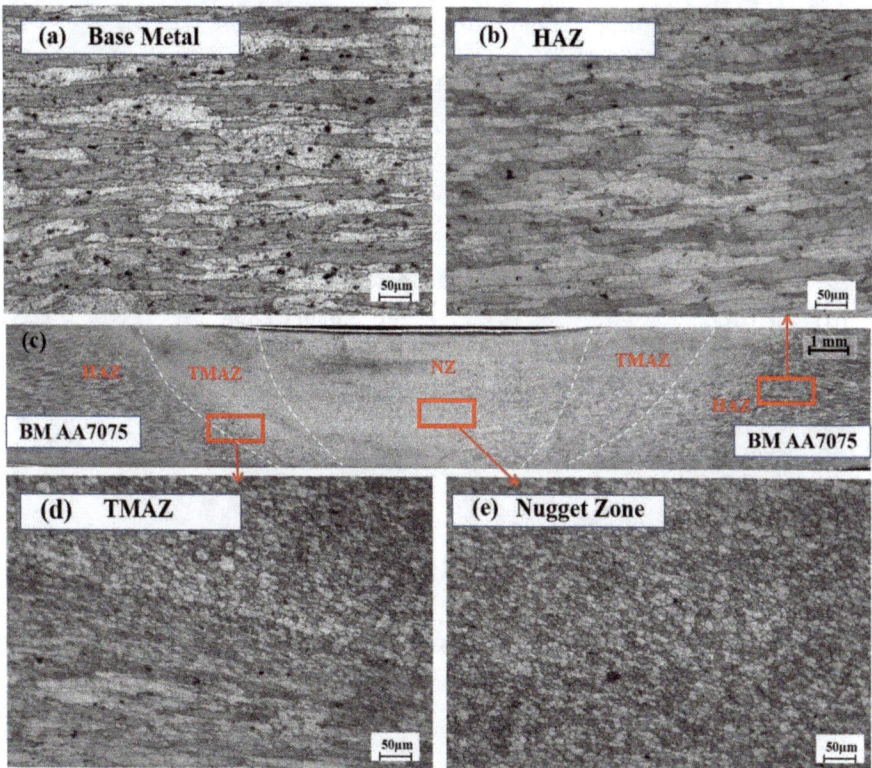

Fig. 3. (a) microstructure of base material (b) HAZ of FSW-7075 (c) macrostructure of FSW-7075 (d) TMAZ of FSW-7075 (e) NZ of FSW-7075

3.2 Similar FSW of AZ31 Alloys (FSW-AZ31)

Figure 4 shows the microstructure of different zones in sample FSW-AZ31. The microstructure of the base material AZ31 is shown in Fig. 4(a). Figure 4(b) shows the microstructure of the HAZ, where the grains are slightly coarser than the base material. Figure 4(c) shows the macrostructure of FSW-7075, where all zones are marked by dotted line. Figure 4(d) represents the TMAZ, where the non-uniform grains have been observed. Figure 4(e) shows the NZ, where the recrystallized fine grains are observed.

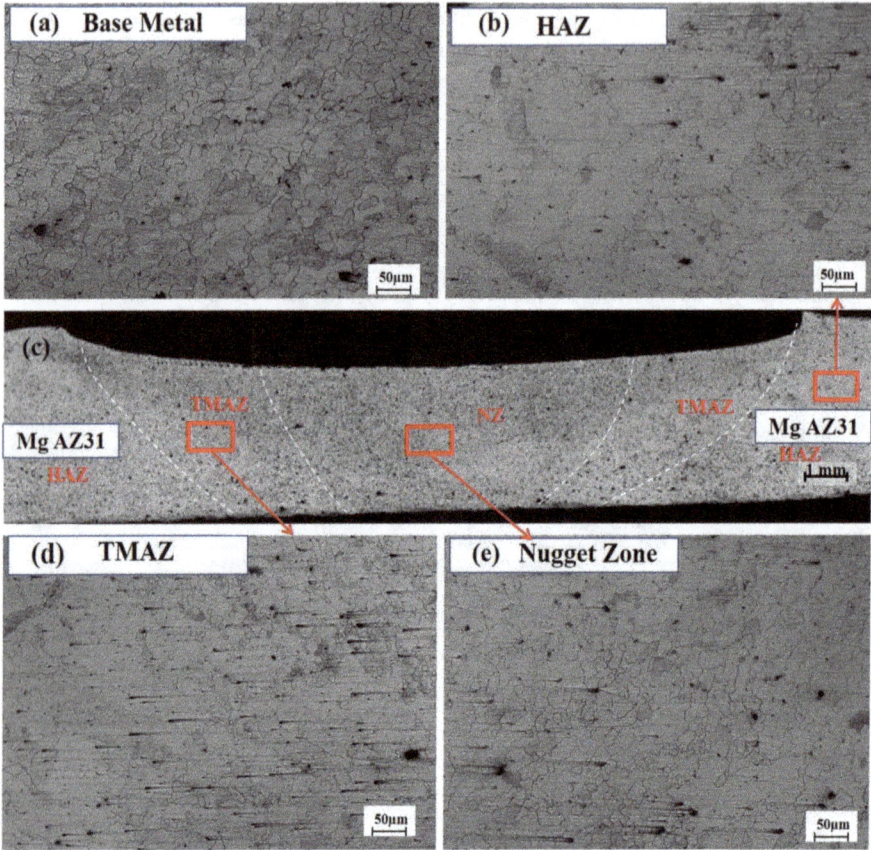

Fig. 4. (a) Microstructure of base material AZ31 (b) HAZ of FSW-AZ31 (c) macrostructure of FSW-AZ31 (d) TMAZ of FSW-AZ31 (e) NZ of FSW-AZ31

3.3 Dissimilar FSW of Al 7075 and Mg AZ31 Alloys Without Interlayer (FSW-7075-AZ31)

Figure 5(a) shows the microstructure of FSW-7075-AZ31, where the cracks and different morphology of mixed region have been formed in NZ due to the stirring action of the tool. Figure 5(b) and (c) show the magnified images of the rectangular regions marked in Fig. 5(a), where the cracks are represented by arrows. The EDS has been done for the chemical composition analysis on the different locations L_1 to L_5 as shown in Table 1. The location L_1 shows the α-Al to the shoulder affected region. Location L_2 and L_3 shows the Al-Mg mixed region (δ-Mg+Al$_{12}$Mg$_{17}$) in the NZ which is formed by the eutectic reactions L=δ-Mg+Al$_{12}$Mg$_{17}$ (at 437 °C) [16, 17]. Location L$_4$ is the α-Al in the NZ. The location L$_5$ shows the δ-Mg, which is also the shoulder affected region.

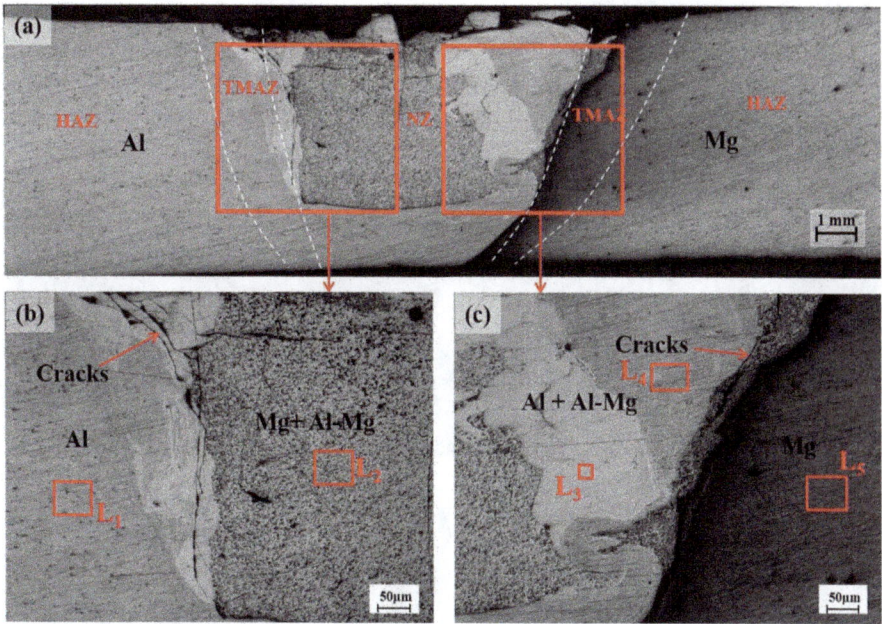

Fig. 5. (a) Macrostructure of FSW-7075-AZ31 shows rectangular region (b) and (c) Magnified image of rectangular region in Fig. 5(a)

Table 1. EDS analysis for the chemical composition (atomic %) in different location in Fig. 5

S.N	Mg	Al	Mn	Cu	Zn	Probable phase
L$_1$	3.05	93.90	–	0.56	2.49	α-Al
L$_2$	73.75	25.40	0.16	–	0.69	δ-Mg+Al$_{12}$Mg$_{17}$
L$_3$	62.57	35.87	0.10	0.18	1.02	Al$_{12}$Mg$_{17}$
L$_4$	2.83	94.08	–	0.60	2.37	α-Al
L$_5$	97.47	2.12	0.11	–	0.29	δ-Mg

The macrostructure of sample FSW-7075-AZ31 indicates that the defect free joint of Al 7075 and AZ31 alloy is not possible. Therefore, the additional Zn interlayer has been incorporated in the faying surface of the weld plates for obtaining the defect free joint, as discussed in the next section.

3.4   Dissimilar FSW of Al 7075 and Mg AZ31 Alloys With Zn Interlayer (FSW-7075-AZ31-Zn)

Figure 6(a) shows the macrostructure of cross section friction stir welding of Al 7075 and AZ31 alloy with Zn interlayer using cylindrical tool at travel speed 20 mm/min. Here, two different regions such as Al-Mg-Zn mixed (A_1) and Mg-Zn mixed (A_2) region have been observed in the NZ. Here also, it is noted that larger flashes formed in FSW-7075-AZ31-Zn. There is no external defect observed on the top surface as well as across the thickness direction of the weld.

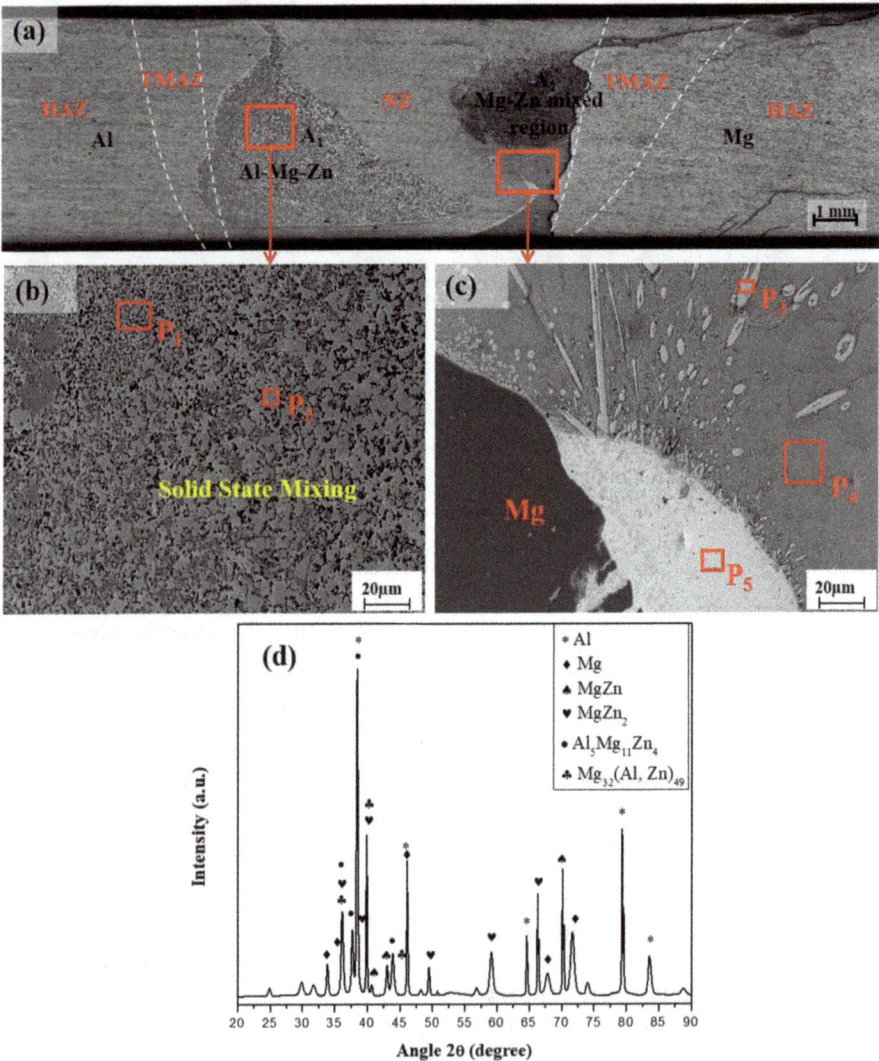

Fig. 6. (a) Macrostructure of FSW-7075-AZ31-Zn show two different regions of Al-Mg-Zn and Mg-Zn mixed region (b) Magnified image of Al-Mg-Zn mixed region in Fig. 6(a) (c) Magnified image of Mg-Zn mixed region in Fig. 6(a)

The region A_1 consist of Al-Mg-Zn mixed region, which is formed due to the tool stirring and mixing of Al and Mg with use of Zn interlayer. The variation of structural morphology such as finer and coarse structure have been observed in the region A_1 as shown in Fig. 6(b).

Table 2. EDS analysis for the chemical composition (atomic %) in different location in Fig. 6

S.N	Mg	Al	Zn	Si	Probable phases
P_1	61.35	20.18	17.84	0.63	δ-Mg+$Al_5Mg_{11}Zn_4$
P_2	57.57	24.82	17.62	–	$Al_5Mg_{11}Zn_4$
P_3	53.9	7.1	38.8	–	$Mg_{32}(Al, Zn)_{49}$
P_4	67.3	4.9	27.6	–	δ-Mg+MgZn
P_5	34.7	4.1	61.1	–	$MgZn_2$

The EDS analysis shows (Table 2) the chemical composition in location P_1 and P_2, which represents the phase formation of δ-Mg+$Al_5Mg_{11}Zn_4$ and $Al_5Mg_{11}Zn_4$, respectively as shown in Fig. 6(b). Region A_2 is the mixed region of Mg-Zn and small quantity of Al in NZ, their morphology is different than the region A_1 due to the composition variations. The EDS analysis on the location P_3, P_4 and P_5 shows the probable phases $Mg_{32}(Al, Zn)_{49}$, Mg+MgZn and $MgZn_2$, respectively in Fig. 6(c). These probable phases $Mg_{32}(Al, Zn)_{49}$, Mg+MgZn and $MgZn_2$ formed in the NZ is responsible for eliminating the formation of hard phases of Al_3Mg_2 and $Al_{12}Mg_{17}$ in sample FSW-7075-AZ31-Zn. Figure 6(d) is the XRD analysis of sample FSW-7075-AZ31-Zn also confirm the phase formation of MgZn, $MgZn_2$, $Al_5Mg_{11}Zn_4$ and $Mg_{32}(Al, Zn)_{49}$ as shown in Fig. 6(d).

3.5 Hardness and Tensile Behaviour

The variation in hardness is shown in Fig. 7. The hardness of base alloys Al 7075 and Mg AZ31 are 153HV and 58HV, respectively. The hardness test of the FSW samples have been performed on transverse cross section such that it provides variation in hardness of different zones/regions of the FSW joint. Maximum hardness has been observed on the NZ in all samples approximately 240 HV (for FSW without interlayer) and 320 HV (for FSW with Zn interlayer) due to the formation of intermetallic compounds whereas the lowest hardness is in the HAZ of Mg alloy having approx. Value of 50HV in all welding conditions.

Figure 8(a) and (b) shows the engineering stress vs. engineering strain curve of FSW samples. The tensile strength of similar FSW of Al 7075 was attained 285 MPa and for Mg AZ31, it was attained 225 MPa. The dissimilar FSW without interlayer has formed immediate cracking therefore no strength has been achieved. Whereas, the dissimilar FSW with Zn interlayer attained up to 85 MPa tensile strength. The joint efficiency has been achieved 55% in samples FSW-7075, 91% in FSW-AZ31 and 34% in FSW-7075-AZ31-Zn.

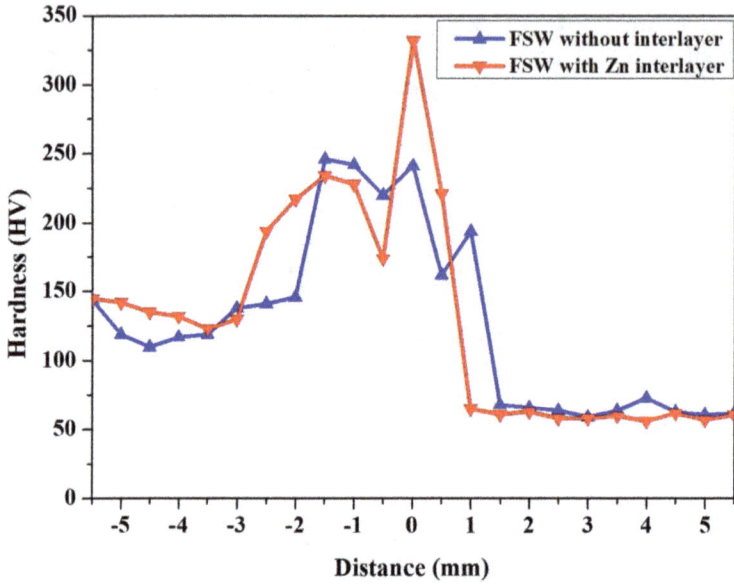

Fig. 7. Hardness profile of samples FSW-7075-AZ31 and FSW-7075-AZ31-Zn

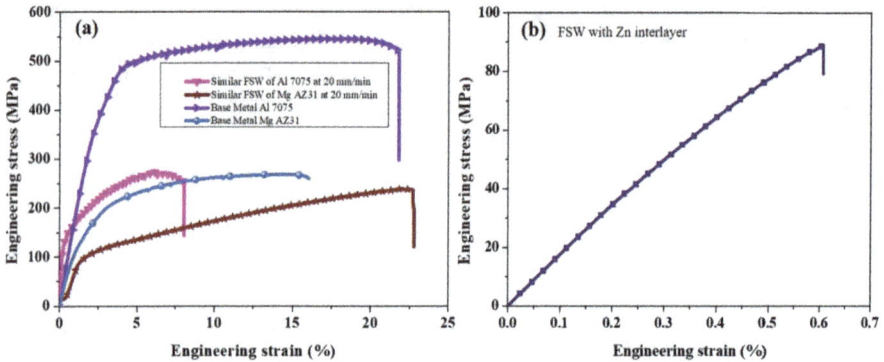

Fig. 8. Engineering stress *vs* strain curve of (a) Base material and similar FSW (b) Dissimilar FSW with Zn interlayer

4   Conclusion

The similar and dissimilar FSW of Al 7075 and Mg AZ31 alloys have been performed at 1300 rpm rotational speed and 20 mm/min traverse speed. The microstructure and mechanical properties of the FSW samples have been investigated:

- Defect free joint has been produced in Similar FSW of Al 7075 and Mg AZ31 alloys at rotational speed 1300 rpm and travel speed 20 mm/min.

- The external cracks have formed in dissimilar FSW of Al 7075 and Mg AZ31 alloys without interlayer at traverse speed 20 mm/min due to formation of intermetallic compound such as $Al_{12}Mg_{17}$ and Al_3Mg_2.
- The Zn interlayer was introduced in between the faying surface of Al 7075 and Mg AZ31 alloys FSW and the defect free joint has been observed at transverse speed 20 mm/min. Only the kissing bond has been observed at the interface of different phases. During stirring, Zn has reacted with Mg and Al and formed different phases such as $MgZn$, $MgZn_2$, $Mg_{32}(Al, Zn)_{49}$ and $Al_5Mg_{11}Zn_4$, which inhibited the formation of Al_3Mg_2 and $Al_{12}Mg_{17}$ layers.
- The tensile strength of similar FSW of Al 7075 attained 285 MPa (joint efficiency 55% of Al 7075) and Mg AZ31 attained 225 MPa (joint efficiency 91% of AZ31). The dissimilar FSW without interlayer has formed immediate cracking and no strength has been achieved. Whereas, the dissimilar FSW with Zn interlayer attained up to 85 MPa tensile strength (joint efficiency 34% of AZ31).

Acknowledgment. The authors would like to acknowledge and thanks to the Director, NIT Raipur, India for providing necessary facilities and constant encouragement for publication of the research work.

References

1. Karimi-dermani, O., Abbasi, A., Azimi, G., Javad, M.: Dissimilar friction stir lap welding of AA7075 to AZ31B in the presence of Sn interlayer. J. Manuf. Process. **68**, 616–631 (2021). https://doi.org/10.1016/j.jmapro.2021.05.068
2. Banjare, P.N., Manoj, M.K.: Effect of tool RPM and tool travers speed on mechanical properties of friction stir welded joints of dissimilar aluminium alloys. Int. J. Mech. Prod. Eng. Res. Dev. **10**, 215–222 (2020)
3. Banjare, P.N., Dewangan, S.K., Manoj, M.K.: Study of material flow and mechanical properties of friction stir welded AA2024 with AA7075 dissimilar alloys using top-half-threaded pin tool. Trends. Sci. **18**, 28–31 (2021)
4. Dewangan, S.K., Banjare, P.N., Tripathi, M.K., Manoranjan, M.K.: Effect of vertical and horizontal zinc interlayer on material flow, microstructure, and mechanical properties of dissimilar FSW of Al 7075 and Mg AZ31 alloys. Int. J. Adv. Manuf. Technol. (2023). https://doi.org/10.1007/s00170-023-11348-7
5. Liang, Z., Chen, K.E., Wang, X., et al.: Effect of tool offset and tool rotational speed on enhancing mechanical property of Al/Mg dissimilar FSW joints. Metall. Mater. Trans. A Phys. Metall. Mater. Sci. **44**, 3721–3731 (2013). https://doi.org/10.1007/s11661-013-1700-4
6. Kostka, A., Coelho, R.S., dos Santos, J., Pyzalla, A.R.: Microstructure of friction stir welding of aluminium alloy to magnesium alloy. Scr. Mater. **60**, 953–956 (2009). https://doi.org/10.1016/j.scriptamat.2009.02.020
7. McLean, A.A., Powell, G.L.F., Brown, I.H., Linton, V.M.: Friction stir welding of magnesium alloy AZ31B to aluminium alloy 5083. Sci. Technol. Weld. Join. **8**, 462–464 (2003). https://doi.org/10.1179/136217103225009134
8. Sato, Y.S., Park, S.H.C., Michiuchi, M., Kokawa, H.: Constitutional liquation during dissimilar friction stir welding of Al and Mg alloys. Scr. Mater. **50**, 1233–1236 (2004). https://doi.org/10.1016/j.scriptamat.2004.02.002

9. Farahani, M.T.M.: Dissimilar friction stir welding of 7075 aluminum alloy to AZ31 magnesium alloy using SiC nanoparticles. Int. J. Adv. Manuf. Technol. **86**, 705–715 (2016). https://doi.org/10.1007/s00170-015-8211-y

10. Zhong, X., Zhao, Y., Pu, J., et al.: Microstructure characterization and mechanical properties of Mg/Al dissimilar joints by friction stir welding with Zn interlayer. Phys. Met. Metallogr. **121**, 1309–1318 (2020). https://doi.org/10.1134/S0031918X20130190

11. Abdollahzadeh, A., Shokuhfar, A., Cabrera, J.M., et al.: The effect of changing chemical composition on dissimilar Mg/Al friction stir welded butt joints using zinc interlayer. J. Manuf. Process. **34**, 18–30 (2018). https://doi.org/10.1016/j.jmapro.2018.05.029

12. Dong, S., Lin, S., Zhu, H., et al.: Effect of Ni interlayer on microstructure and mechanical properties of Al/Mg dissimilar friction stir welding joints. Sci. Technol. Weld. Join. (2022). https://doi.org/10.1080/13621718.2021.2014742

13. Dewangan, S.K., Tripathi, M.K., Manoj, M.K.: Material flow behavior and mechanical properties of dissimilar friction stir welded Al 7075 and Mg AZ31 alloys using Cd interlayer. Met. Mater. Int. **28**, 1169–1183 (2022). https://doi.org/10.1007/s12540-021-00980-1

14. Zheng, B., Zhao, L., Lv, Q., et al.: Effect of Sn interlayer on mechanical properties and microstructure in Al/Mg friction stir lap welding with different rotational speeds. Mater. Res. Express **7**, 076504 (2010). https://doi.org/10.1088/2053-1591/ab9fbb

15. Xiaoqing, J., Yongyong, L., Tao, Y., et al.: Enhanced mechanical properties of dissimilar Al and Mg alloys fabricated by pulse current assisted friction stir welding. J. Manuf. Process. **76**, 123–137 (2022). https://doi.org/10.1016/j.jmapro.2022.02.007

16. Ranjole, C., Pratap, V., Basil, S.: Numerical prediction and experimental investigation of temperature, residual stress and mechanical properties of dissimilar friction-stir welded AA5083 and AZ31 alloys. Arab. J. Sci. Eng. **47**, 16103–16115 (2022). https://doi.org/10.1007/s13369-022-06808-3

17. Firouzdor, V., Kou, S., Society, M.: Formation of liquid and intermetallics in Al-to-Mg friction stir welding. Metall. Mater. Trans. A Phys. Metall. Mater. Sci. **41**, 3238–3251 (2010). https://doi.org/10.1007/s11661-010-0366-4

Mechanical and Corrosion Study of Dissimilar Friction Stir Welding of AZ31Mg Alloy and Cu-8Zn Alloy

Abhijeet Bhowmik$^{(\boxtimes)}$, Satya Kumar Dewangan, Pragya Nandan Banjare, and Manoranjan Kumar Manoj

Department of Metallurgical and Materials Engineering, National Institute of Technology, Raipur 492010, India
`abhowmik.phd2021.mme@nitrr.ac.in`

Abstract. Fusion welding techniques are not very viable for joining dissimilar metals like copper (Cu) and Magnesium (Mg) alloys, which finds applications in electronics and mobility industries. Basically, numerous researchers have explored Mg alloy and pure Cu joints. But very limited literature is available regarding joining of Mg alloy and Cu alloys. Further, corrosion behaviour of these joints needs to be explored. The present work focuses on friction stir welding (solid state welding) of Cu-8Zn alloy and AZ31 Mg alloy. The main process parameters taken for the work are tool rpm of 1200 and 20 mm/min as travel speed. The resultant weld specimens were examined for microstructural characterization, and for mechanical and corrosion behaviour. The optical microscopy and SEM-EDS analysis revealed that stir zone is made up of complex microstructure. The XRD analysis confirms that the stir zone consists of inter-metallic compounds (IMCs) like Mg_2Cu and $MgCu_2$. The potentiodynamic polarization test were conducted and it was found that welded samples have less corrosion resistance than the two parent metals. The micro-hardness of the stir zone reached a maximum value of 429.95 HV because of the existence of IMCs and fine grains. The maximum tensile strength achieved was 68 MPa, with fracture occurred in brittle manner.

Keywords: FSW · Mg-Cu Welding · Tensile Properties · Microstructural Characterization · Corrosion Behaviour

1 Introduction

Joints made of different materials viz. Magnesium and copper alloys are frequently employed in engineering constructions, electronics, and automotive industries due to higher thermal conductivity as well as high strength to weight ratio [1–3]. Currently, the most common techniques for joining magnesium and copper alloys include TIG Welding [4, 5], Diffusion Bonding [6–8] Cold Metal Transfer [9], Ultrasonic Spot Welding [10], Resistance Spot Welding [11], Ultrasound induced transient liquid phase bonding [3, 12], and Laser Welding [1, 2, 13]. A high-quality weld is difficult to produce since

© The Author(s), under exclusive license to Springer Nature Singapore Pte Ltd. 2024
S. Patra et al. (Eds.): METCENT 2023, *Proceedings of the International Conference on Metallurgical Engineering and Centenary Celebration*, pp. 335–344, 2024.
https://doi.org/10.1007/978-981-99-6863-3_33

it is obvious to find flaws like oxidation, problems related to welding like shrinkages and distortions, and coarse grains during the conventional welding and fusion welding process [14]. For welding such materials, friction stir welding (FSW) offers a far better choice. FSW, which is a type of solid-state welding was introduced in 1991 [15]. It uses material mixing, frictional heat, and severe plastic deformation to form the bond between two different materials. The FSW joints have fine grains because of severe plastic deformation, and defects that are present during traditional and fusion welding processes are also eliminated [14]. Friction stir welding was utilized in this investigation to join AZ31Mg alloy and Cu-8Zn alloy. The impacts on the welding seam mechanical and microstructure characteristics were explored. And also, the corrosion behaviour of FSW joint of Mg/Cu alloy has been explored.

2   Materials and Methodology

2.1   Parent Materials (PM)

Commercially available Cu-8Zn and Mg AZ31 alloy sheets with dimensions as length 100 mm, width 50 mm, and height of 3 mm were selected for the FSW technique. Mechanical properties and chemical composition of both parent materials were determined and reported in Table 1. Both parent metals were cleaned using emery paper and acetone before FSW to remove the oxide layer.

2.2   FSW Process

Friction stir welding was performed through the use of modified vertical milling machine. Dissimilar FSW process used is shown schematically in Fig. 1(a). To carry out the welding, a cylindrical tool fabricated with H13 steel with dimensions of 16 mm for shoulder, 3 mm for pin, and 2.8 mm for length was used and presented in Fig. 1(b). For the FSW technique, the following parameters are used: tool rpm as 1200, 20 mm/min as travel speed, tilt angle 3°, pin offset 0.5 mm towards Cu-8Zn from the weld centre. Cu-8Zn alloy plate was set up on the advancing side (AS) of the FSW process and AZ31 Mg alloy on the retreating side (RS). The welded sample is shown in Fig. 2.

2.3   Characterization

By using optical (ZEISS), stereo (Leica), and scanning electron (ZEISS) microscopes and XRD, the welded joint composition and microstructure were studied. The hardness and strength was examined by Vickers Microhardness Tester with a 100 g load, and MTS Universal Testing Machine with strain rate of 1 mm/min, respectively. Fractured surfaces were analyzed using a scanning electron microscope (SEM). Through the use of potentiostat (VersaSTAT), corrosion analysis was carried out in the potential range of -1.8 to 1 V with a scan rate of 1 mV/s in a medium containing 3.5% NaCl solution.

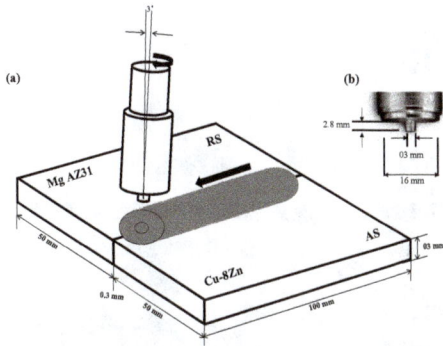

Fig. 1. (a) Diagram of the friction stir welding procedure, (b) H13 FSW tool

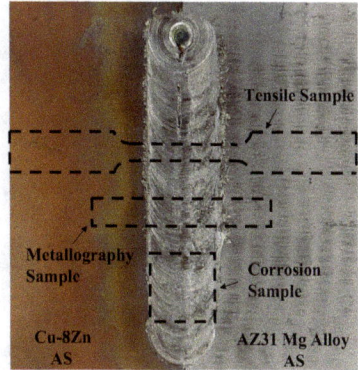

Fig. 2. FSW welded sample

Table 1. Chemical composition and mechanical properties of the parent materials

Parent Materials	Chemical Composition					Mechanical Properties			
	Elements (wt.%)								
	Mg	Al	Zn	Mn	Cu	YS (MPa)	UTS (MPa)	% Elongation	Hardness (HV)
AZ31Mg alloy	96.71	2.28	0.77	0.25	–	170	241	18.5	62
Cu-8Zn	–	–	7.93	–	92.07	231	280	30	115

3 Findings and Analysis

3.1 Optical Characterization

The FSW of the Cu-8Zn and AZ31 alloys' macro- and microstructures are depicted in Fig. 3. The macrostructure of the FSW is used to identify the several microstructural zones produced during FSW, which are known as the stir zone (SZ), the thermomechanically affected zone (TMAZ), and the heat affected zone (HAZ). Which are depicted in Fig. 3(a). The microstructure in Fig. 3(d), which is found at position 'D' in Fig. 3(a), indicates the SZ. The grain size at this site is very fine compared to other zones. This is because SZ experiences severe plastic deformation and dynamic recrystallization [15]. At point 'C' in Fig. 3(a), the microstructure appears to be of TMAZ on advancing side (AS) which is shown in Fig. 3(c) The grains in this region are not homogeneous due to plastic deformation and as well as heat exposure. The HAZ microstructure in advancing side is depicted in Fig. 3(b), for the point "B" in Fig. 3(a). Figure 3(e) and (f) depict the microstructure at locations "E" and "F" on the retreating side (RS) of Fig. 3(a), which correspond to TMAZ and HAZ.

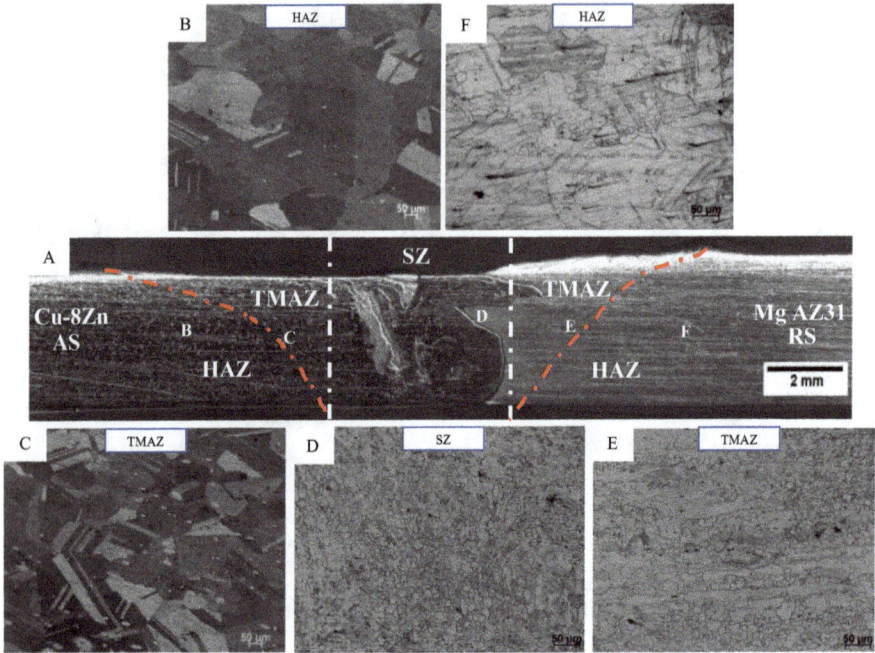

Fig. 3. (a) Macrograph of FSW Sample, (b) Microstructure of HAZ of AS of FSW sample, (c) Microstructure of TMAZ of AS of FSW sample, (d) Microstructure of SZ of RS of FSW sample, (e) Microstructure of TMAZ of RS of FSW sample, and (f) Microstructure of HAZ of RS of FSW

3.2   SEM and Phase Characterization

The stir zone macrostructure for FSW is depicted in Fig. 4(b). It has been observed that the Mg alloy had been penetrated by the Cu alloy, which was on the retreating side. As the Cu extruded into Mg from the bottom, the Mg is observed to extrude into Cu from the top. But as Mg have low strength and hardness in comparison to Cu, its penetration was restricted, and it reacted with Cu to form a thick reaction layer. Also, voids at different locations in the stir zone was observed.

B1, B2 and B3 are the three distinct locations that have been selected for subsequent SEM-EDS investigation. In Figs. 4(a), (c) and (e), these locations are further enlarged, respectively. Figure 4(d), and (f) provide enlarged images of the regions B11and B21,which are depicted in Fig. 4(a) and (c), respectively. The centre location of the stir zone is depicted in Fig. 4(d). The reaction layer contains δ-Mg and Mg_2Cu as revealed from EDS at location P2 and P5 and the findings were tabulated in Table 2. Within the layer, the bright colored fragments are α-Cu as EDS of location P1 revealed. The gray colored patch surrounding Cu is Mg_2Cu as identified at location P3 and P4. A small crack is visible around the Cu fragment which can be due to the presence of intermetallic [16–18]. Adjacent to the reaction layer towards the right is a dark colored patch which is basically δ-Mg as per EDS finding of location P6 of Fig. 4(e). Figure 4(f) shows the right side of the stir zone. A complicated intercalated structure is evident in this region. Cu has reacted here with Mg and formed river like structure of $MgCu_2$ which is revealed by

EDS analysis at point P9 and P10 This river like structure is flowing through α-Cu (EDS at location P11) and δ-Mg (EDS at location P8 which is δ-Mg and Mg_2Cu). Figure 5 shows XRD peaks confirming the existence of α-Cu, δ-Mg, Mg_2Cu and $MgCu_2$.

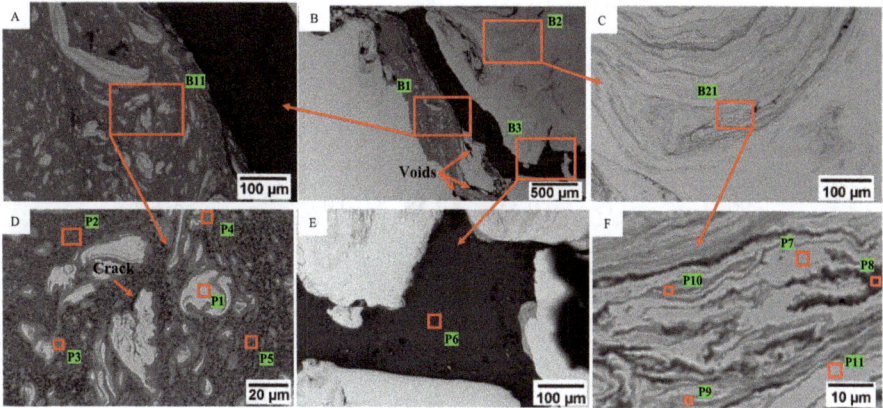

Fig. 4. (a) An enlarged SEM image of region B1 as shown in Fig. 4(b), (b) SEM image of stir zone of FSW, (c) An enlarged SEM image of B2 region as shown in Fig. Figure 4(b), (d) An enlarged SEM image of region B11 as shown in Fig. 4(a), (e) An enlarged SEM image of region B3 as shown in Fig. 4(b), and (f) An enlarged SEM image of region B21 as shown in Fig. 4(b)

3.3 Corrosion Analysis

During electrochemical corrosion experiments, the parent materials (PM) AZ31 and Cu-8Zn and the friction stir welded sample were placed in a container containing 3.5% NaCl solution. The test samples were cut into square shape and mirror polished before corrosion testing. To evaluate the corrosion behaviour of parent metals and FSW joint, a potentiodynamic polarization (direct current based) corrosion test was conducted on them. The platinum (Pt) plate serves as the counter electrode and Ag/AgCl electrode serves as the reference electrode in the potentiostat three-electrode test setup. The test sample was attached to the working electrode, which has an exposed surface area of 1 cm^2. A potentio-dynamic polarisation test was carried out in potential range of -1.8 to 1 V at a scan rate of 1 mV/s. Figure 6 displays the Tafel (potentiodynamic polarization) curves, produced from the corrosion test. With the help of Tafel curves corresponding corrosion current density (Icorr), (corrosion potential (Ecorr), Tafel slopes for anodic and cathodic curve (βa and βc) [19, 20] are evaluated through extrapolation method and the results were reported in Table 3.

Table 2. EDS analysis of different locations as shown in Fig. 4

Location	Chemical configuration (at%)						Probable phase
	Mg	Cu	Zn	Al	Mn	Si	
P_1	1.33	90.80	7.87	-	–	–	α-Cu
P_2	70.08	24.01	2.55	3.37	–	–	$Mg+Mg_2Cu$
P_3	64.19	31.57	2.72	1.52	–	–	Mg_2Cu
P_4	65.31	30.82	2.40	1.47	–	–	Mg_2Cu
P_5	72.78	21.78	2.43	3.02	–	–	$Mg+Mg_2Cu$
P_6	96.81	0.19	0.33	2.58	0.09	–	δ-Mg
P_7	2.21	89.89	7.91	–	–	–	α-Cu
P_8	49.97	44.77	2.65	2.60	–	–	$Mg+MgCu_2$
P_9	37.13	56.88	4.81	1.17	–	–	$MgCu_2$
P_{10}	36.93	54.96	4.49	3.62	–	–	$MgCu_2$
P_{11}	0.46	91.58	7.96	–	–	–	α-Cu

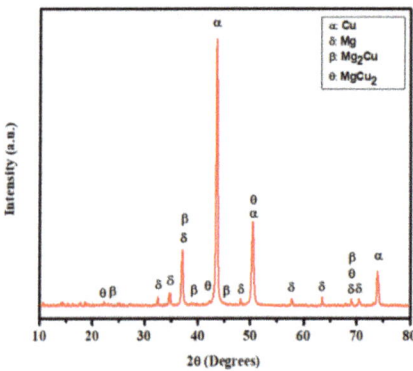

Fig. 5. XRD analysis of Cu/Mg dissimilar FSW

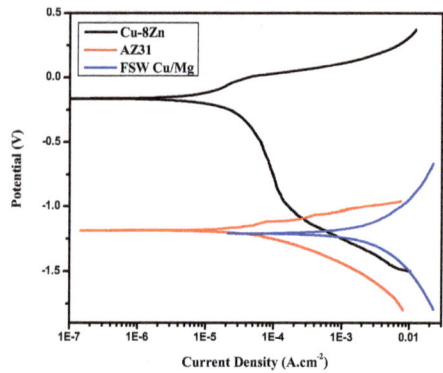

Fig. 6. Tafel curve for Cu-8Zn, AZ31 and dissimilar FSW Cu/Mg joint

E_{corr} which is corrosion potential is a measure of corrosion resistance offered by the material. The material which have low corrosion resistance shows high negative value of E_{corr} [21]. During the test, E_{corr} of parent materials Cu-8Zn and AZ31 found to be -0.164 V and -1.186 V, respectively. This demonstrated that Cu-8Zn has greater corrosion resistance than AZ31. Whereas E_{corr} of FSW sample was found to be -1.211 V which is much high negative value with respect to AZ31. This suggests that the FSW sample's corrosion resistance is the lowest compared to the parent materials. This reduction of corrosion resistance in FSW sample can be attribute to galvanic corrosion due to the presence of dissimilar material in the joint [22]. Measurement of polarisation resistance (R_p) was calculated to evaluate corrosion resistance of materials. Lower corrosion current means higher polarisation resistance. Polarisation resistance is the term for the specimen resistance to oxidation when an external voltage is applied. The polarisation resistance and corrosion rate are inversely proportional [20, 23]. The Polarisation resistance (R_p) has been calculated and tabulated in Table 3 with the help of Stern-Geary Eq. (1) [24] which is given by

$$Rp = \frac{\beta a * \beta c}{2.3 * Icorr*(\beta a + \beta c)} \tag{1}$$

The calculated R_p and obtained I_{corr} from Tafel curve of parent materials and FSW sample are compared.

It has been observed that Cu-8Zn PM shows lowest I_{corr} and highest R_p value in comparison to AZ31 PM and FSW sample indicating that the corrosion rate to be lowest and having high corrosion resistance. Whereas FSW sample shows highest value of I_{corr} and lowest R_p value indicating that it has highest corrosion rate and lowest corrosion resistance.

Table 3. Corrosion Analysis data for parent materials and dissimilar FSW joint

S.N	Sample	I_{corr}	E_{corr}	β_a	β_c	R_p
1	Cu-8Zn PM	14.078 $\mu A.cm{-2}$	−0.164 V	142.208 mV	392.561 mV	3220 mΩ.cm2
2	AZ31 PM	55.686 $\mu A.cm^{-2}$	−1.186 V	123.455 mV	177.53 mV	569 mΩ.cm^2
3	FSW Cu/Mg	2326 $\mu A.cm^{-2}$	−1.211 V	383.857 mV	399.541 mV	36 mΩ.cm^2

3.4 Hardness Variation

Microhardness values for the parent materials Cu-8Zn alloy and AZ31Mg alloy are 115 HV and 62 HV, respectively. According to the hardness profile for the FSW sample, the maximum hardness of 429.95 HV is reached in stir region as shown in Fig. 7. Hardness increases as distance from welding seam decreases. The presence of fine grains and thick reaction zone made up of the intermetallic compounds Mg_2Cu and $MgCu_2$, the stir zone witness an increase in hardness.

3.5   Tensile and Fractography

The engineering stress-strain graph of FSW Cu-8Zn/AZ31Mg alloy joint is shown in Fig. 8. The joint strength of 68 MPa was attained, which is 28% of the AZ31 Mg Alloy (241 MPa). Figure 9 show SEM images of fractured surface of the tensile sample. Figure 9 (a) shows macrostructure of fractured surface. In Fig. 9(b), region B from Fig. 9(a) is magnified. The region C in Fig. 9 (b) is depicted in an enlarged SEM image in Fig. 9 (c). The crack region is visible and cleavage pattern is observed. These fracture surfaces reveal that the fracture is brittle in nature (Fig. 9).

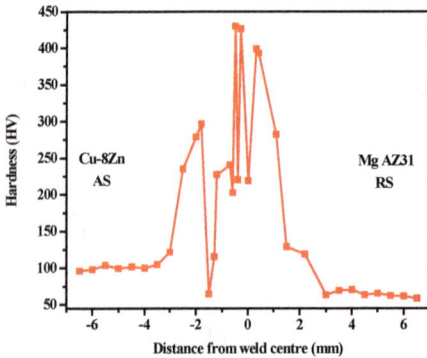

Fig. 7. Hardness variation of Cu-8Zn/AZ31Mg alloy FSW joint

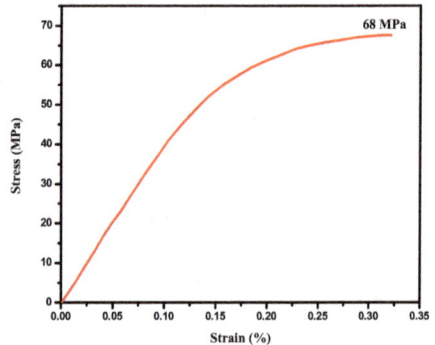

Fig. 8. Engineering Stress-strain curve of FSW sample of AZ31Mg /Cu-8Zn

Fig. 9. (a) Macrostructure fracture surface of FSW (b) Magnified image of region B in Fig. 9(a), and (c) Magnified image of region C in Fig. 9 (b),

4   Conclusions

At 1200 rpm and 20 mm/min travel speed, the dissimilar FSW of AZ31 Mg alloy and Cu-8Zn was carried out. FSW of AZ31 Mg alloy and Cu-8Zn resulted in the production of IMCs Mg_2Cu and $MgCu_2$. The stir zone had a thick reaction layer with IMC. Cracks and voids was observed in the samples. The corrosion analysis was done for FSW Cu/Mg joints, and it was found that corrosion resistance decreases, in order, Cu parent material (Cu-8Zn PM) < Mg parent material (AZ31) < FSW sample. The FSW sample's hardness reached a maximum value of 429.95 HV in the stir zone. 68 MPa was identified as the highest tensile strength. The fracture surface of welded tensile samples shows that brittle failure was caused by cracks and cleavage patterns.

References

1. Dai, J., Yu, B., Ruan, Q., Chu, P.K.: Improvement of the laser-welded lap joint of dissimilar mg alloy and Cu by incorporation of a Zn interlayer. Materials (Basel) **13**(9), 2–11 (2020). https://doi.org/10.3390/MA13092053
2. Tan, C., He, W., Gong, X., Li, L., Feng, J.: Influence of laser power on microstructure and mechanical properties of fiber laser-tungsten inert gas hybrid welded Mg/Cu dissimilar joints. Mater. Des. **78**, 51–62 (2015). https://doi.org/10.1016/j.matdes.2015.04.022
3. Mao, Z., et al.: Ultrasound-induced transient liquid-phase bonding of Mg/Cu dissimilar metals with Zn interlayer in air. Mater. Lett. **268**, 127483 (2020). https://doi.org/10.1016/j.matlet.2020.127483
4. Liu, L.M., Wang, S.X., Zhu, M.L.: Study on TIG welding of dissimilar Mg alloy and Cu with Fe as interlayer. Sci. Technol. Weld. Join. **11**(5), 523–525 (2006). https://doi.org/10.1179/174329306X122794
5. Liming, L., Shengxi, W., Limin, Z.: Study on the dissimilar magnesium alloy and copper lap joint by TIG welding. Mater. Sci. Eng. A **476**(1–2), 206–209 (2008). https://doi.org/10.1016/j.msea.2007.04.089
6. Mahendran, G., Balasubramanian, V., Senthilvelan, T.: Developing diffusion bonding windows for joining AZ31B magnesium and copper alloys. Int. J. Adv. Manuf. Technol. **42**(7–8), 689–695 (2009). https://doi.org/10.1007/s00170-008-1645-8
7. Mahendran, G., Balasubramanian, V., Senthilvelan, T.: Influences of diffusion bonding process parameters on bond characteristics of Mg-Cu dissimilar joints. Trans. Nonferrous Met. Soc. China (English Ed.) **20**(6), 997–1005 (2010). https://doi.org/10.1016/S1003-6326(09)60248-X
8. Du, S.-M., Hu, J.: Diffusion bonding of magnesium alloy (AZ31B) to Cu by using Sn foil interlayer. In: Proceedings of the 2015 International Conference on Material Science and Applications, vol. 3, no. Icmsa, pp. 186–189 (2015). https://doi.org/10.2991/icmsa-15.2015.35
9. Cao, R., Jing, M., Feng, Z., Chen, J.H.: Cold metal transfer welding-brazing of magnesium to pure copper. Sci. Technol. Weld. Join. **19**(6), 451–460 (2014). https://doi.org/10.1179/1362171814Y.0000000204
10. Macwan, A., Chen, D.L.: Microstructure and mechanical properties of ultrasonic spot welded copper-to-magnesium alloy joints. Mater. Des. **84**, 261–269 (2015). https://doi.org/10.1016/j.matdes.2015.06.104
11. Shan, H., Zhang, Y., Li, Y., Luo, Z.: Dissimilar joining of AZ31B magnesium alloy and pure copper via thermo-compensated resistance spot welding. J. Manuf. Process. **30**, 570–581 (2017). https://doi.org/10.1016/j.jmapro.2017.10.022
12. Lang, Q., Wang, Q., Han, J., Zhang, W., Zhang, Y., Yan, J.: Microstructure evolution of Mg/Cu dissimilar metal jointed by ultrasonic-induced transient liquid phase bonding with Zn interlayer. Mater. Charact. **157**, 109897 (2019). https://doi.org/10.1016/j.matchar.2019.109897
13. Zhao, X., et al.: Fiber laser welding-brazing characteristics of dissimilar metals AZ31B Mg alloys to copper with Mg-based filler. J. Mater. Eng. Perform. **27**(3), 1427–1439 (2018). https://doi.org/10.1007/s11665-018-3166-4

14. Sharma, N., Khan, Z.A., Siddiquee, A.N.: Friction stir welding of aluminum to copper—an overview. Trans. Nonferrous Met. Soc. Chin. (English Ed.) **27**(10), 2113–2136 (2017). https://doi.org/10.1016/S1003-6326(17)60238-3

15. Mishra, R.S., Ma, Z.Y.: Friction stir welding and processing. Mater. Sci. Eng. R Rep. **50**(1–2), 1–78 (2005). https://doi.org/10.1016/j.mser.2005.07.001

16. Sato, Y.S., Park, S.H.C., Michiuchi, M., Kokawa, H.: Constitutional liquation during dissimilar friction stir welding of Al and Mg alloys. Scr. Mater. **50**(9), 1233–1236 (2004). https://doi.org/10.1016/j.scriptamat.2004.02.002

17. Mohammadi, J., Behnamian, Y., Mostafaei, A., Gerlich, A.P.: Tool geometry, rotation and travel speeds effects on the properties of dissimilar magnesium/aluminum friction stir welded lap joints. Mater. Des. **75**, 95–112 (2015). https://doi.org/10.1016/j.matdes.2015.03.017

18. Hou, W., Shah, L.H.A., Huang, G., Shen, Y., Gerlich, A.: The role of tool offset on the microstructure and mechanical properties of Al/Cu friction stir welded joints. J. Alloys Compd. **825**, 154045 (2020). https://doi.org/10.1016/j.jallcom.2020.154045

19. Bade, G.E., Caraban, A., Sebesan, M., Dzitac, S., Cret, P., Stel, A.: Polarisation measurements used for corrosion rates determination. J. Sustainable Energy **1**(1), 2067–5538 (2010). http://www.energy-cie.ro/archives/2010/n1-1-14.pdf

20. Ropital, F.: Environmental degradation in hydrocarbon fuel processing plant: Issues and mitigation. Woodhead Publishing Limited (2011). https://doi.org/10.1533/9780857093783.5.437

21. Sinhmar, S., Dwivedi, D.K.: Investigation of mechanical and corrosion behavior of friction stir weld joint of aluminium alloy. Mater. Today Proc. **18**, 4542–4548 (2019). https://doi.org/10.1016/j.matpr.2019.07.426

22. Seo, B., Song, K.H., Park, K.: Corrosion properties of dissimilar friction stir welded 6061 aluminum and HT590 steel. Met. Mater. Int. **24**(6), 1232–1240 (2018). https://doi.org/10.1007/s12540-018-0135-2

23. Toshev, Y., et al.: Protective coating of zinc and zinc alloys for industrial applications. Woodhead Publishing Limited (2006). https://doi.org/10.1016/b978-008045263-0/50073-8

24. Stern, M., Geary, A.L.: Electrochemical polarization: I. A theoretical analysis of the shape of polarization curves. J. Electrochem. Soc. **104**(1), 56 (1957)

Review Papers

Enhancing Technologies to Improve Metallurgical Processes

Maneesh C. Srivastava[1]([⊠]) and Balachandran P. Kamath[2]

[1] Amity School of Engineering and Technology, Amity University Uttar Pradesh,
Lucknow 226028, India
s_maneesh@yahoo.com
[2] Freelance Metallurgical Process Consultant, Mumbai 400607, India

Abstract. Advancements in basic sciences, materials, engineering and instrumentation have opened up exciting possibilities in extractive metallurgy. The metallurgical processes generally are energy and water guzzlers, with extreme operating conditions, producing large amounts of effluents and waste products. Over the last few years, there has been an effort to innovate and integrate some of these developments into existing metallurgical processes, thus making them more energy efficient, resource efficient, and environment friendly. With the above intent, certain enhancing technologies, such as - microwave treatment, ultrasonic processing, electrostatic separation, and biological treatment methods have been discussed in the paper. These technologies could be integrated with several unit operations such as mineral beneficiation, sintering, smelting of minerals, leaching, metal precipitation, hot metal mixing, effluent treatment etc. There are many more technologies like ore sorting, chloride metallurgy, bioleaching but are left out to keep this paper brief.

The discussed technologies are futuristic and promising in nature. Some of them are in the initial stages of laboratory experimentation or pilot plants. In fact, many research laboratories round the globe are working on these technologies to develop energy efficient and resource efficient solutions to the conventional methods. This paper attempts to survey some of the above technologies to highlight their potential in the minerals and metals industry, their current status, and possible use in our context.

Keywords: Microwave · Ultrasonic · Electrostatic · Biological treatments · Minerals

1 Introduction

The potential of microwave treatment could be utilized in reducing the energy consumption in comminution process. The treatment weakens the bond strength of the ore particles, thus reducing energy requirement during grinding process. The technique also assists in increasing the recovery by processing highly complex ores having interlocked grains besides making them amenable for flotation [1]. The ultrasonic waves could be

S. Patra et al. (Eds.): METCENT 2023, *Proceedings of the International Conference on Metallurgical Engineering and Centenary Celebration*, pp. 347–357, 2024.
https://doi.org/10.1007/978-981-99-6863-3_34

used in various unit operations ranging from hydrometallurgy to cast house operations. The technique influences the formation of microbubbles during froth floatation process, leaching kinetics, solidification composition and microstructure, resulting in improving the separation, and recovery in hydrometallurgical operations, and customizing the cast house products [2]. The electrostatic separation technique may be used in mineral dressing to segregate conductive particles, a possible alternative to water gobbling froth floatation for sulphides, thus reducing the use of chemicals, water, and surfactants, besides reducing environmental damage [3]. Bioprocessing technologies of heavy metal bearing sulphate effluent could be developed using appropriate type of sulphate reducing bacteria [4].

2 Microwave Treatment

In mineral industry, chunks of rocks are crushed and ground (comminution) to make the ore to micron size for further processing or liberate the minerals entrapped in the gangue material. This crushing and grinding operation involve a huge amount of energy and the operational efficiency is also very poor. Only 1% of the energy input to the rock breakage process is used to generate new surfaces. The process is also marred with high erosion rate of crushing and grinding tools.

Microwave irradiation is a technique that can aid the mineral processing industry in improving the energy efficiency of the crushing process i.e. conversion of electrical energy into particle breakage, and can also contribute to improved recovery of the minerals.

Microwave is a part of electromagnetic radiation spectrum. Electromagnetic (EM) radiation may be characterized as waves propagating by the interactions of mutually perpendicular electric and magnetic field. The frequency of microwave ranges from 10^7 to 10^{10} Hz. The energy of the EM radiation increases with increasing frequency of the radiation. The microwave treatment of the ores leads to the differential heating of the ores based on the dielectric properties of the material [5].

Dielectric heating occurs through two mechanisms:

- Motion of dipolar species similar to rotation in the presence of electromagnetic field (dielectric relaxation)
- Drift of ionic species under the electromagnetic field resulting in resistive losses (electric conductivity)

2.1 Interaction of Microwave with the Material

Different minerals absorb microwave to different levels depending on the composition. Microwave can heat minerals selectively. Some minerals are transparent to microwave irradiation, while some minerals heat well [1, 5].

The selective absorption of microwave leads to differential heating within the material, leading to thermal gradient. The thermo-mechanical stresses developed within the material aids in reducing the strength of the rocks/ ores, thus making them weaker and requiring lesser energy for comminution. The heating in microwave irradiation process is driven from inside of the material to the surface of the material, whereas in conventional heating, the heat is transferred from the outer surface to the core.

On microwave irradiation, most of the arsenides, sulphides, sulphoarsenides and sulphosalts gets heated strongly, while most of the carbonates, silicates, and sulphates are transparent to the microwave irradiation. With change in temperature, some minerals undergo phase transformation, that is accompanied with volumetric change, that further leads to the development of thermomechanical forces within the material, and reduction in strength. At lower frequencies, the dipoles within the dielectric have time to reorient with the applied electric field. At higher frequencies, during external field reversal, the dipoles get less time to reorient, and they start to lose with regard to polarization, and less quantity of energy is stored. A further rise in frequency leads to further failure of polarization to keep pace with changing applied field. With changing applied field, the dielectric is no longer able to store the energy proportional to the applied electric field, and the loss in energy increases [1, 5, 6]. This loss in energy is dissipated and shows up as a rise in temperature of the minerals. Figure 1 (a) shows SEM image of raw ore and Fig. 1 (b) shows microwave treated ore with transgranular and intergranular microfractures. [7].

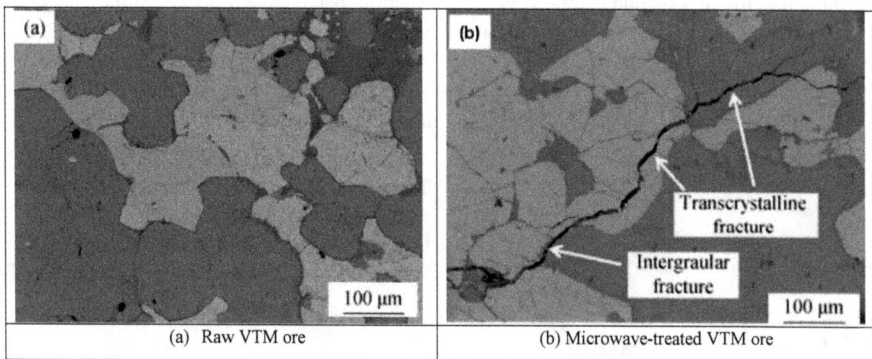

(a) Raw VTM ore   (b) Microwave-treated VTM ore

Fig. 1. SEM image of (a) raw ore and (b) microwave treated ore (showing transgranular and intergranular microfractures) [7].

The observation of microwave irradiation on different ores such as gold ore, copper ore, ilmenite ore, carbonatite ore, iron ore, vanadium titano-magnetite ore is summarized in Table 1.

The pre-treated microwave irradiated ore reduces the grinding energy and is dependent on the mineralogy, type of ore, size of the particles, location of microwave system in the mineral processing unit, microwave power level, and exposure time. The net effect on the grindability would be the result of complex interactions of the above factors.

Table 1. Observation and results of microwave irradiation on mineral beneficiation.

Sample and condition	Observation
10 g of gold ore containing magnetite, hematite, silicate and quartz pre-treated with microwave for 5 min at microwave power of 700W	Magnetite attained maximum temperature (~500 °C), hematite (~150 °C), and aluminosilicate samples (~100 °C). The differential heating of the phases leads to the thermal stress induced cracking, resulting in the reduction of crushing strength by 31.2%. [8]
40 g ilmenite ore, 1 kW and 2.45 GHz microwave, with exposure time up to 2 min	In about 1 min, the ilmenite ore attained 350 °C. The microwave treatment improves the efficiency of ilmenite ore. [9]
Four types of ore, ilmenite, sulphide ore, gold ore, and carbonatite ore. A sample of 500g of each ore is tested. 2.6kW, 2.45GHz, multimode variable power	The magnetite content in ilmenite and carbonatite gets heated to higher temperature as compared to the other phases present in the ore.[10]
Copper carbonatite ore. Microwave exposure of even a fraction of a second can be very effective in reducing the energy required for comminution	The liberation studies in the + 500 μm size fraction, percentage of locked sulphides of copper decreased (69.2% to 31.8%). Higher power densities with reduced exposure time generates higher thermomechanical stresses and improve the grindability. [11]
Vanadium titano-magnetite (VTM) ore for 140 s	Significant thermal gradient in the material The microcracks formed due to the thermo mechanical stresses in the ore particles increases the grindability and dissociation.[7]

3 Ultrasonic Processing

Sound waves are simple longitudinal pressure waves traversing through a medium. An ultrasound wave has frequency higher than 20 kHz. Ultrasonics with frequency 20–100 kHz are normally used for leaching, chemical reactions, cleaning, etc. Ultrasonic irradiation of ores, such as arsenolite, crocoite, galena, and resulted in 50, 30 and 20 times faster rates of dissolution respectively in comparison to those in the absence of an ultrasonic field. The energy from ultrasonics produces microstreaming in liquids and cavitation.

On the exposure of a liquid to ultrasonics, small bubbles nucleate, grow, move, and collapse as the acoustic pressure increases, as shown in Fig. 2. The bubble burst leads to erosion, fragmentation of ore/ mineral particles in the slurry, thus increasing the reaction rate.

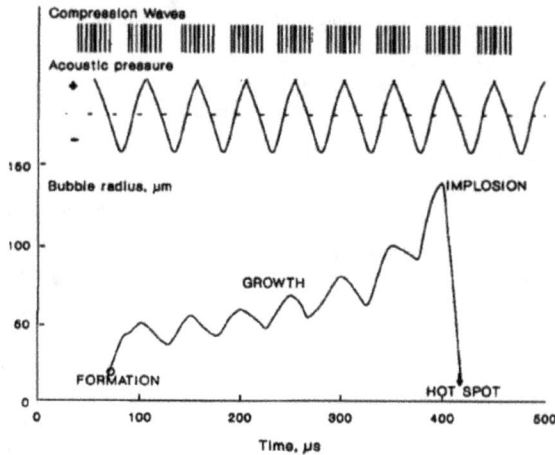

Fig. 2. Bubble growth and implosion during cavitation [12].

The ultrasound decreases effective viscosity, thus increasing mass transfer coefficient, due to decrease in boundary layer thickness. Ultrasound accelerates leaching process through the following effects:

1) product layer passivation
2) enhanced stirring leading to increased mass transfer
3) mineral surface modification/ cracking
4) inter-particle collisions

The sonication system such as choice of transducer, choice of frequency and choice of intensity also influences the ultrasound treated metallurgical processes such as extraction and leaching as shown in Table 2.

3.1 Ultrasonics in Metal Solidification

The ultrasonic treatment of melt refines grain size, and results in better and well distributed equiaxed microstructure. The ultrasonic fields during metal solidification can produce a variety of benefits, such as reduced segregation, better grain refinement, lesser gas and inclusion entrapment, thus improving the quality of the cast product. The mechanical properties of the cast products with ultrasonically treated melt are better than the conventional castings. [19]. Schematic diagram of the apparatus of ultrasonic setup being used for casting is shown in Fig. 3.

Table 2. Results of ultrasonic treatment on metallurgical processes.

System	Frequency (kHz)/ Intensity (W/cm^{-2})	Features
Extraction of Cu mineral	540/ 2kW for 1 h	Ultrasonic vibrations gave ten times better extraction than mechanical stirring.[13]
Leaching of Cu from a) chrysocolla b) chalcocite c) chalcopyrite	18.8/ 2.8	Ultrasound intensified the leaching of Cu 2–3 times in an agitated vessel. For a, b and c, the copper extractions were 99.8% & 50.6%, 71.2% & 46.9% and 3% & 0% respectively with and without sonication.[14]
5% Fe$_2$(SO$_4$)$_3$ + 2% H$_2$SO$_4$ leaching of a) chalcocite b) chalcopyrite	-/-	Dissolution of Cu was better in both the ores using ultrasonic and d.c. fields over ordinary conditions. [15]
Ammonia leaching of Cu	-/-	Leaching rate enhanced 3–4 times in ultrasonic field. [16]
Sulphuric acid leaching of Cu ores	19.18/ 0.5	Using magnetostrictive type ultrasonic transducer, mixing time can be reduced to 5–10 min from 20–60 min with higher extraction of 5–15%. [17]
Leaching of Cu ores from pyrite cake	18, 19, 43 ultrasonic or 0.5V/cm electric field/ -	Decreases leaching time by a factor of 3–12 using ultrasound and 2–3 by d.c. field when compared to mechanically agitated leaching. [18]

The study [21] shows improved grain refining and better melt degassing in 5052 aluminum alloy in direct chill casting under the combined effect of ultrasonics and the feeding of AlTi5%B1% rod. The improvement in microstructure of the ultrasonically treated 5052 alloy is clearly visible in Fig. 4 (c) and Fig. 4 (d) as compared to Fig. 4 (a) and Fig. 4 (b). The ultrasonic treatment reduces the consumption of grain refiner as well as linear sizes of the phases, such as Mg$_2$Si and Al$_3$Fe.

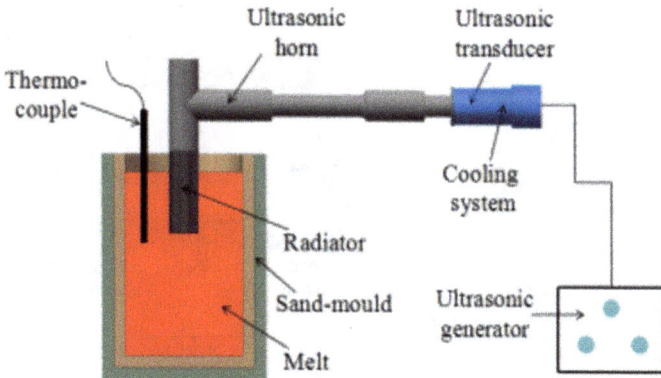

Fig. 3. Schematic diagram of the apparatus of ultrasonic casting [20].

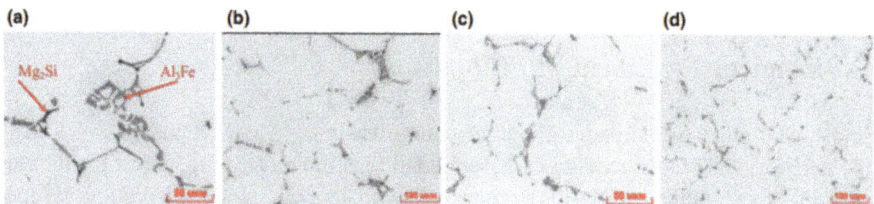

Fig. 4. Typical microstructure of 5052 alloy slab without ultrasonic treatment (a, b) and after ultrasonic treatment (c, d) light microscopy [21].

4 Electrostatic Separation

In the early twentieth century, electrostatic separation became attractive in mineral processing, but there was limitation with regard to capacity, efficiency and complexity of operation and fairly simple froth flotation process took over. With increasing emphasis on environmentally friendly techniques such as limited use of chemicals and with many mineral processing plants (Hindustan Copper, Khetri) facing water crisis, the dry technology such as electrostatic separation could be very attractive.

In electrostatic separation, mineral species are separated based on their electrical conductivity behaviour. The electrostatic field separates into conductor and nonconductor mineral species.

Most sulphides are conductor minerals, that lose the charge to an earthed surface, whereas the insulators retain the charge and gets pinned to the ground surface as shown in Fig. 5 [22].

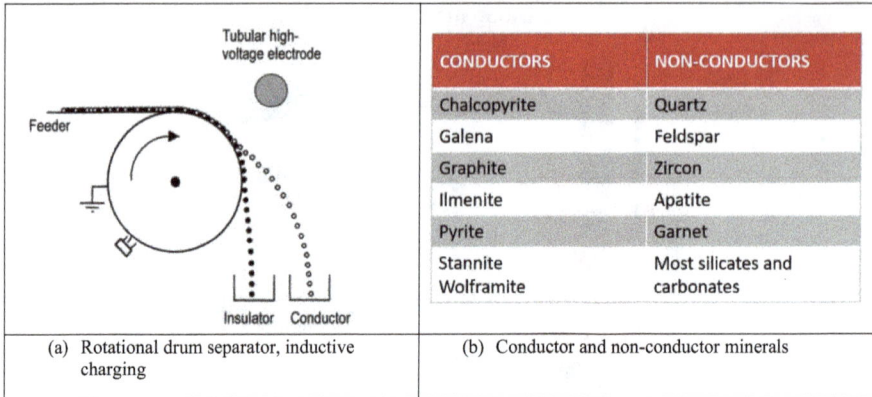

CONDUCTORS	NON-CONDUCTORS
Chalcopyrite	Quartz
Galena	Feldspar
Graphite	Zircon
Ilmenite	Apatite
Pyrite	Garnet
Stannite	Most silicates and
Wolframite	carbonates

(a) Rotational drum separator, inductive charging | (b) Conductor and non-conductor minerals

Fig. 5. Electrostatic separation (a) Mechanism of electrostatic separation (b) Conductor and non-conductor minerals.

5   Biological Treatment of Effluents

Metal containing waste stream includes acid mine drainage (AMD), metal plating, finishing, metallurgical plants, steel making, semiconductor manufacturing, power plants, etc. A variety of techniques such as ion exchange, chemical precipitation, adsorption, reverse osmosis, filtration, electrodialysis, coagulation/flocculation and flotation are used as heavy metals removal technologies. However, these technologies are often unfavourable and uneconomical to remove heavy metals from polluted streams.

Biological mechanisms are oriented to reduce (detoxify, degrade, or transform) concentration of pollutants to harmless state. The success of biological treatment process depends on the environmental factors (nutrient and oxygen concentrations, pH and temperature) and microbial population. Biological treatment includes, algae, fungi, bacteria, plant to detoxify effluent waste stream [23, 24]. Figure 6 shows various biological metal mobilization/ immobilization mechanisms.

Two different approaches are utilized to control toxic pollutants: in-situ, and ex-situ. The organism can carry out the detoxification process through a variety of mechanisms, such as toxic metals as microbes' nutrients, and under different environmental factors such as anaerobic, aerobic, light, pH etc.

Microbes have evolved several types of mechanisms to tolerate the metal-stressed conditions. Biosorption into the cell walls, enveloping in extracellular capsules, precipitation, emanating metal ions outside the cell wall, conversion to a less toxic state accumulation, and forming complexes of metal ions within the cell are some of the mechanisms.

Figure 7. Shows batch reactor to recover metal sulphide precipitates from sulphate containing wastewater with the help of biologically produced sulfide using bioprecipitation. Sulfate reducing bacteria (SRB) can be used for treating contaminated wastewater and process streams. Biologically produced H_2S precipitates metals as metal sulfides, while alkaline bicarbonate can neutralize acidic waters.

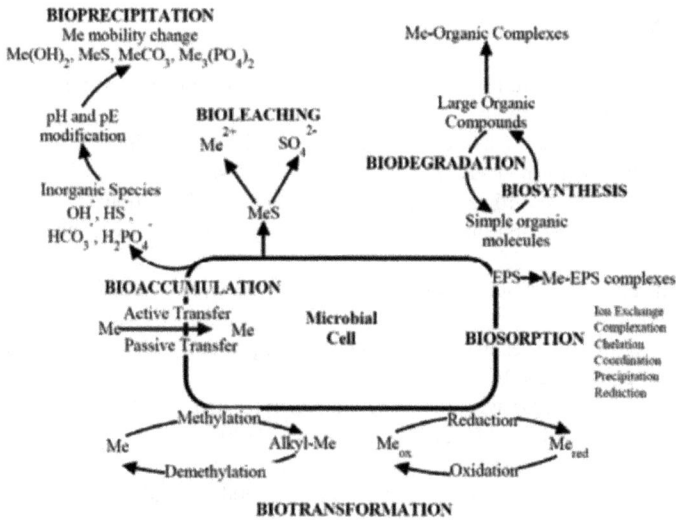

Fig. 6. Biological metal mobilization/immobilization mechanisms [25].

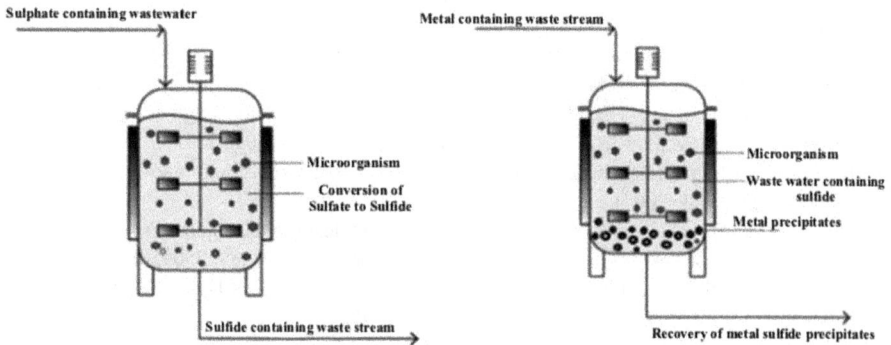

Fig. 7. Batch bioreactor to recover metals [26].

6   Future Outlook

Mineral processors have never faced greater challenges than the present times. Due to various geographical and economic reasons, the operations have shifted focus on producing the products from lower quality reserves or reprocessing waste materials. Environmental factors are equally important, whether that is availability of resources or the effect on the local environment. All these considerations often reduce the degree of freedom and dictate the processing method. The processes elaborated in the paper could form a part of the solution to optimize the resources and make metallurgical industries more sustainable.

References

1. Pickles, C.A.: Microwave in extractive Metallurgy: Part 1-Review of fundamentals. Miner. Eng. **22**, 1102–1111 (2009)
2. Puga, H., etal.: Breakthrough in Ultrasonic assisted industrial continuous casting, Casthouse, Aluminium International Today, September-October 2014
3. Dong, Y., et al.: Study on growth influencing factors and desulfurization performance of sulfate reducing bacteria based on the response surface methodology. ACS Omega **8**, 4046–4059 (2023)
4. Dotterl, M., et al.: Electrostatic Separation, Ullmann's Encyclopedia of Industrial Chemistry, pp. 1–29 (2016)
5. Pickles, C.A.: Microwaves in extractive metallurgy: part 2-a review of applications. Miner. Eng. **22**, 1112–1118 (2009)
6. Singh, S., et al.: Microwave processing of materials and applications in manufacturing industries: a review. Mater. Manuf. Process. **30**, 1–29 (2015)
7. Wang, J.-P., et al.: Influence of microwave treatment on grinding and dissociation characteristics of vanadium titano-magnetite. Int. J. Minerals Metallur. Mater. **26**(2), 160–168 (2019)
8. Amankwah, R.K., Styles, M.T., Nartey, R.S., Al Hassan, S.: The application of direct smelting of gold concentrates as an alternative to mercury amalgamation in small-scale gold mining operations in Ghana. Int. J. Environ. Pollution **41**(3/4), 304 (2010). https://doi.org/10.1504/IJEP.2010.033238
9. Guo, S., et al.: Microwave assisted grinding of ilmenite ore. Trans. Nonferrous Met. Soc. China **21**, 2122–2126 (2011)
10. Kingman, S.W., et al.: The effect of microwave radiation on the magnetic properties of minerals. J. Microw. Power Electromagn. Energy. **35**(3), 144–150 (2000)
11. Kingman, S.W., et al.: Recent developments in microwave-assisted comminution. Int. J. Miner. Process. **74**, 71–83 (2004)
12. Kingman, S.W., et al.: Application of ultrasound in leaching. Mineral Process. Extract. Metallur. Rev. **14**(3–4), 179–192 (1995). https://doi.org/10.1080/08827509508914124
13. Vasil'ev, V.V., et al.: Leningr. Gos. Univ. 50–55, Chem. Abstr. **57**, 14474a (1962)
14. Rusikhina, I.P., et al.: Chem. Abstr. 79, 339194r (1973)
15. Ozolin L., et al.; Zh. Met., Abstr. No. 8G248, Chem. Abst. **70**, 89853w (1968)
16. Khavskii, N.N., et al.: Chem. Abstr. 78, 19129f (1972)
17. Matyskin Yu. D. et al.: Chem. Abstr. **72**, 124125a (1967
18. Rudenko, N.K., et al.: Chem. Anbstr. **77**, 142677q (1972)
19. Yin, P., Xu, C., Pan, Q., Guo, C., Jiang, X.: Effect of ultrasonic field on the microstructure and mechanical properties of sand-casting AlSi7Mg0.3 alloy. Rev. Adv. Mater. Sci. **60**(1), 946–955 (2021). https://doi.org/10.1515/rams-2021-0061
20. Shi, C., Li, F., Mao, D.: Reducing porosity in 35crmo steel sand-cast by ultrasonic processing. IOP Conf. Ser. Mater. Sci. Eng. **452**, 022127 (2018). https://doi.org/10.1088/1757-899X/452/2/022127
21. Kostin, I.V., et al.: The benefits of ultrasonic treatment of molten metal for slabs casting at UC Rusal facilities. In: Martin, O. (ed.) Light Metals 2018, pp. 901–905. Springer International Publishing, Cham (2018). https://doi.org/10.1007/978-3-319-72284-9_117
22. Gill, C.B.: Electrostatic separation. In: Gill, C.B. (ed.) Materials Beneficiation, pp. 141–147. Springer, New York (1991). https://doi.org/10.1007/978-1-4612-3020-5_8
23. Abo-Alkasem, M.I., et al.: Microbial bioremediation as a tool for the removal of heavy metals. Bull. Natl. Res. Centre **47**(1), 131 (2023). https://doi.org/10.1186/s42269-023-01006-z

24. Azubuike, C.C., Chikere, C.B., Okpokwasili, G.C.: Bioremediation techniques–classification based on site of application: principles, advantages, limitations and prospects. World J. Microbiol. Biotechnol. **32**(11), 180 (2016). https://doi.org/10.1007/s11274-016-2137-x
25. Hatzikioseyian, A.: Principles of bioremediation processes, pp. 23–54 (2010)
26. Janyasuthiwong, S.: Bioprecipitation-a promising technique for heavy metal removal and recovery from contaminated wastewater streams. MOJ Civil Eng. **2**(6), 191–193 (2017)

Smelting Reduction Technology – Current Status and Future Outlook

Maneesh C. Srivastava[1](✉) and A. K. Jouhari[2]

[1] Department of Mechanical and Aerospace Engg, ASET, Amity University UP, Lucknow 226028, India
s_maneesh@yahoo.com

[2] A K Jouhari Is Visiting Professor (Aerospace), RRL (Now CSIR IMMT), ASET and Former Scientist G, Bhubaneswar, India

Abstract. The first commercial scale plant based on Smelting Reduction Technology, the Coal reduction process (Corex process) was designed and developed by Germans and promoted by Voest Alpine, Austria. The plant was set up at Pretoria Works, South Africa in 1989. The first Corex process-based plant in India was set up by JSW at Bellary in Karnataka. In this paper, the present state of art of the different smelting reduction (SR) technologies such as Corex, Romelt (Russian), Hismelt (Australia), Finex (Siemens and Posco), along with new low carbon smelting reduction green technologies is discussed.

A brief discussion on fundamentals like foaming of slag during smelting reduction and kinetics of iron oxide reduction in molten slag is also given.

Keywords: Smelting Reduction Technology · slag foaming · kinetics

1 Introduction

The blast furnace continues to dominate the iron making, however in view of high investments, constraints on the availability of metallurgical coke, and the associated environments issues, alternative iron making processes have been explored to supplement the production.

Due to the shortage of good quality coking, various alternative routes of iron making have been tried at various scales and geographies. Alternative iron making processes utilize smelting reduction methods producing the liquid hot metal. These ironmaking processes are being improvised and being made environmentally friendly as more operational experiences are gained. Small scale plants based on alternate routes could contribute to reducing the environmental footprint of iron and steel industry. The fine wastes including mill scale, EAF dust, coke breeze and coal/ coke dust produced in iron and steel industries can be sintered/ pelletized and used as input to these processes.

© The Author(s), under exclusive license to Springer Nature Singapore Pte Ltd. 2024
S. Patra et al. (Eds.): METCENT 2023, *Proceedings of the International Conference on Metallurgical Engineering and Centenary Celebration*, pp. 358–365, 2024.
https://doi.org/10.1007/978-981-99-6863-3_35

2   Smelting Reduction Processes

2.1   Corex Process [1]

Corex technology is a process to produce liquid hot metal of identical properties as that of metal produced in a blast furnace. Lean iron ores and non-coking coal could be used to produce liquid hot metal in a cost-efficient and sustainable manner.

In the Corex process, iron reduction and smelting operations are carried out separately, in reduction shaft and melter-gasifier reactors, respectively, as shown in Fig. 1. Solid DRI is produced in the shaft furnace that is smelted in melter-gasifier. In the bottom reactor, coal is gasified using oxygen, and pre-reduced iron ore/ pellets are melted. Coal directly enters the melter gasifier (950 ~ 1050 °C) and gets converted to char. High pressure oxygen is blown into the melter gasifier, that gasifies the coal generating a reducing gas. The outgoing gas from smelter is taken into upper reactor after cooling (800 to 850 °C) and dedusting, to further reduce iron ore/pellets to sponge iron.

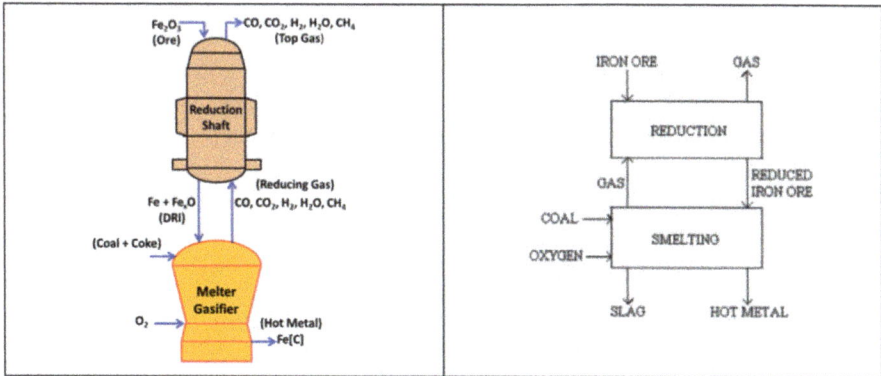

Fig. 1. Two stage Corex process

2.2   Romelt [2]

In Romelt process, iron oxide is reduced to liquid-phase iron as shown in Fig. 2. The iron bearing ore and coal reacts in the smelter, gets agitated by gas and dissolves in slag. Carbon in coal reduces iron oxide from slag. Oxidizing gas bubbled through tuyeres, intensifies heat and mass transfer and enhances the kinetics of the reactions. Gas containing CO and H_2 evolve from the melted slag and combusted at the top. The post combustion heat predominantly provides thermal energy for the reactions in the slag bath. The Romelt process produces large quantities of very high calorific value gas at the exit temperature of 1600 °C, that could be utilized in power plant etc., to make the process economical.

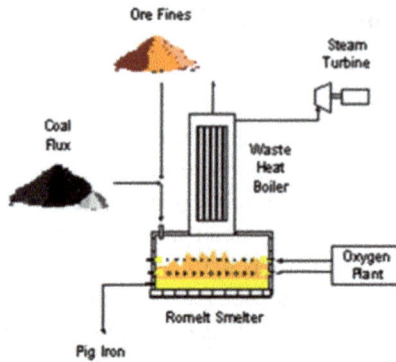

Fig. 2. Romelt process

2.3 HIsmelt [3]

The process HIsmelt is also termed as 'high intensity smelting'. In the process, liquid iron is produced straight from the iron ore. The process (Fig. 3) treats iron ore fines with minimum of pre-treatment, providing a flexibility to quality of iron it can handle. Iron ore fines and non-coking coal with significant impurities could be fed to the process. A wide range of ferrous feeds, high phosphorus ore and steel plant wastes could form as an input to the process.

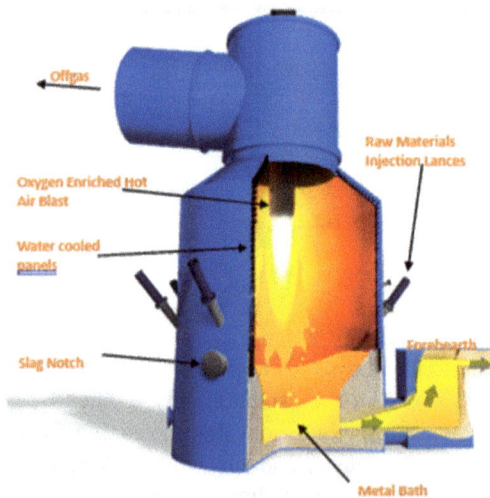

Fig. 3. Smelt Reduction Vessel (SRV) in HIsmelt technology

After more than three decades of research, the HIsmelt technology moved from Australia to China. The HIsmelt plant is located in Shandong Province, Molong Petroleum Machinery Limited.

The Molong HIsmelt plant was fired up in 2016 and is producing 1,600 tonnes per day hot metal at a comparatively lower cost. Molong purchased the HIsmelt intellectual rights from Rio Tinto in 2017.

2.4 Finex [4]

In Finex process, iron is produced in two separate process steps. The fines of iron ore are reduced to iron in successive fluidized-bed reactors, and hot compacted iron (HCI), is then processed in a sequential melter-gasifier. Coal and coal briquettes are gasified in melter-gasifier. Process flow sheet of Finex process and melt gasifier are shown in Fig. 4, and Fig. 5 respectively.

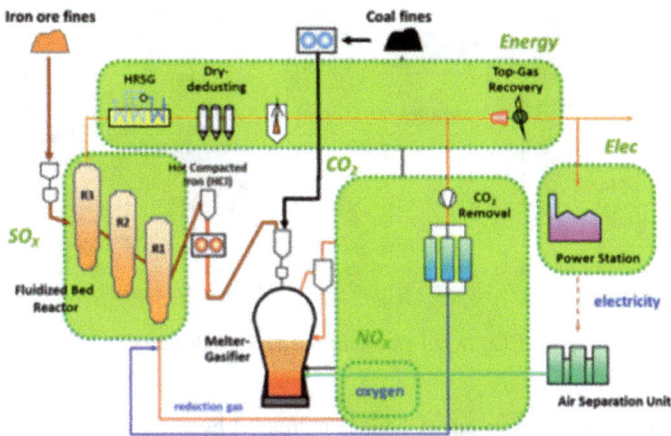

Fig. 4. Process Flowsheet of FINEX

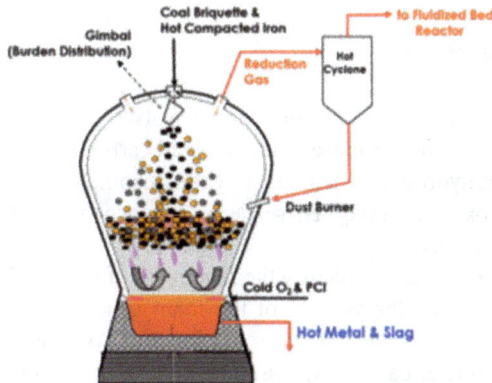

Fig. 5. Finex Melter Gasifier

Direct charging of low-cost non-coking coal, fine ores, generation of clean tail gas, production of quality hot metal, environmentally clean and sustainable process are some of the benefits of the Finex process.

Two commercial FINEX plants are producing hot metal in Pohang Steel Works. Several years of operation and continual improvements have been made in the plant, including in FBR (Fluidized-Bed Reactor), an energy recovery system, that has made the process more energy and resource efficient. Carbon capture and utilization through CO_2 removal system further adds commercial and environmental viability of the process.

3 Commercial capture and utilization of CO_2 [5]

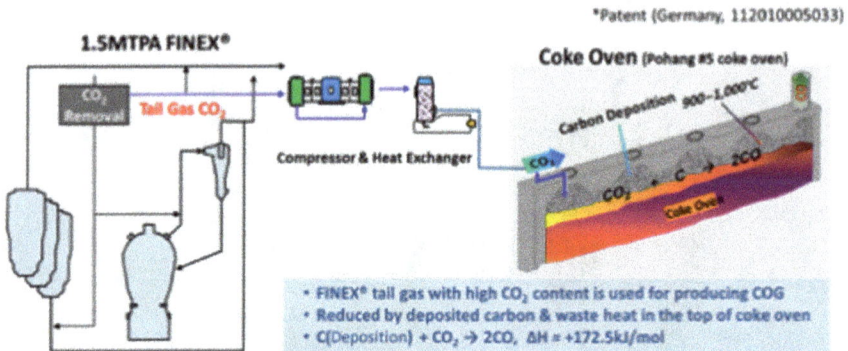

Fig. 6. CO_2 in tail gas being used for producing COG

The amount of coke oven gas (COG) is increased by reacting to high temperature carbon with carbon dioxide by using waste heat generated at a coke oven. Figure 6 shows CO_2 conversion process integrated with Finex process.

4 Foaming of Slag [6]

In the process of smelting reduction, iron oxide dissolved in molten slag is reduced by carbon, that is added as input or present as dissolved carbon in iron. Large amounts of carbon monoxide and hydrogen is generated by the decomposition of coal constituents and reduction of iron oxide-rich slag. These gases influence the slag foaming, and further enhances the production rates.

The foaming index is used to quantify the foaminess of slag. In defining the foaming index Σ, it is assumed that the volume of the foam generated is proportional to the gas flowrate and is termed as constant of proportionality. Assuming the constant cross-sectional area of reactor, Σ can be expressed as ratio of the foaming height h (in cm) and the gas velocity v (in cm s^{-1}).

$\Sigma = h/v$.

The generation of CO gas leads to the slag foaming as per the following reaction in Eq. (1) - Eq. (3):

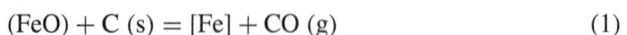

$$(FeO) + C \ (s) = [Fe] + CO \ (g) \tag{1}$$

This reaction takes place in two steps:

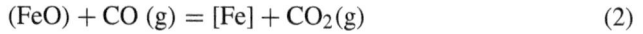

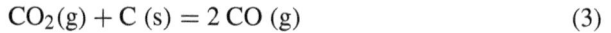

$$(FeO) + CO \text{ (g)} = [Fe] + CO_2(g) \tag{2}$$

$$CO_2(g) + C \text{ (s)} = 2\,CO \text{ (g)} \tag{3}$$

Figure 7 shows the experimental set up of the plasma reactor, in which the set of experiments were conducted, and the results are presented in Figs. 8, 9, 10 and 11.

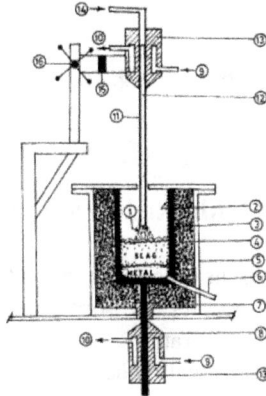

1. Plasma 2. Coating (Zircon) 3. Crucible (Graphite) 4. Alumina 5. Mild Steel Casing 6. Tap hole 7. Bush (Alumina) 8. Electrode (Graphite) 9. Inlet 10. Outlet 11. Electrode (Graphite) 12. Axial Hole 13. Block (Copper) 14. Gas (Plasma Forming) 15. Insulation 16. Mechanical Mechanism

Fig. 7. Plasma reactor

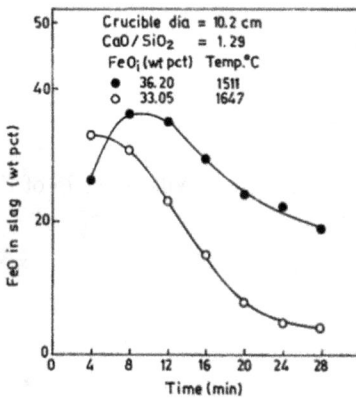

Fig. 8. FeO content in slag with reaction time

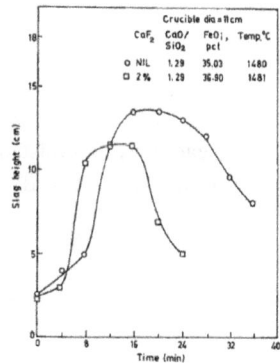

Fig. 9. Slag height with time

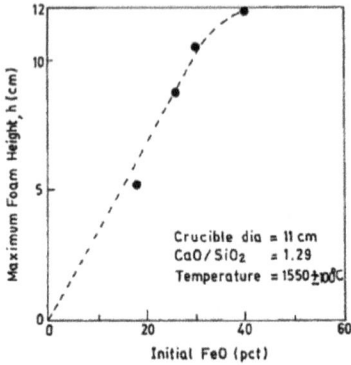

Fig. 10. Maximum foam height h versus initial FeO content in slag

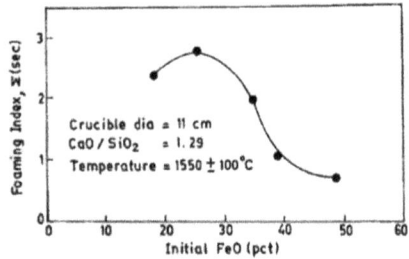

Fig. 11. Foaming index on initial %FeO in slag

5 Kinetic Study of Iron Oxide Reduction in Molten Slag [7]

The kinetics of reaction determines the speed at which a process takes place. The kinetics of iron oxide reduction reaction follows a 1st order reaction which implies that FeO diffusion in the slag phase controls the rate.

Equation (4) represents the fractional reduction of FeO, α, i.e., the fraction of iron oxide reduced

$$\alpha = [(FeO_i - FeO_t)/FeO_i] \tag{4}$$

where, FeO_i = Initial FeO in slag.

FeO_t = FeO in slag at time t.

Equation (5) and Eq. (6) represents the first order kinetic model.

$$d\alpha/dt = k\,(1 - \alpha) \tag{5}$$

$$-\ln\,(1 - \alpha) = kt \tag{6}$$

where, k = reaction rate constant (min^{-1}), t = time (min). With the help of Fig. 12, calculation of rate constant is done.

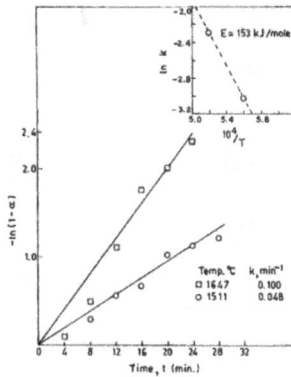

Fig. 12. Estimation of activation energy

6 Conclusion

The paper summarizes the alternative iron making routes, such as Corex, Romelt, Hismelt and Finex process. These processes can handle lean ores, coal, iron/ steel dust and comparatively lower capacity plants (0.5–1.5MT/ yr) could be operated. Heat as well as calorific value of exit gas could be used for heating and power generation. CO_2 capture and utilization could be integrated along with other exhaust gases as discussed in Finex process. FeO reduction in slag is discussed in the light of basic concepts of slag foaming and kinetics.

References

1. Ali, H., et al.: Alternative emerging ironmaking technologies for energy-efficiency and carbon-di-oxide emissions reduction: a technical review. Renew. Sustain. Energy Rev. **33**, 645–658 (2014)
2. Usachev, A.B., Romenets, V.A., Lekherzak, V.E., Balasanov, A.V.: Modern Processes for the coke-less production of Iron. Metallurgist **46**(5/6), 147–151 (2002). https://doi.org/10.1023/A:1020476618338
3. Neil, G.: Hismelt plant ramp-up. J. Iron. Steel Res. Int. **16**, 1–9 (2009)
4. Yi, S.-H., et al.: FINEX® as an environmentally sustainable ironmaking process. Ironmaking Steelmaking. 1–7 (2019)
5. Santos, S.: CO2 capture technologies in Iron making process, pp. 1–40, April 2013
6. Jouhari, A.K., et al.: Foaming during reduction of iron oxide in molten slag. Ironmaking Steelmaking **27**(1), 27–31 (2000)
7. Jouhari, A.K., et al.: Kinetics of iron oxide reduction in molten slag. Scand. J. Metall. **30**, 14–20 (2001)

Green Steel Technology: A Viable Approach for Sustainable World

Arghya Majumder[1] and M. Kalyan Phani[2(✉)]

[1] School of Mines and Metallurgy, Kazi Nazrul University, Asansol, West Bengal, India
director.smm@knu.ac.in
[2] Department of Metallurgical Engineering, OP Jindal University, Raigarh,
Chhattisgarh 496109, India
kalyan.makkuva@opju.ac.in

Abstract. The challenges towards significant climatic changes all over the world lead to the necessity of innovative approaches towards current steel making processes. The approaches are numerous but sustainable solutions are still far away. The other side Indian government has already set its ambitious goals in terms of steel production (i.e., 300 million tons by 2030), Electricity production (i.e., 830 Giga Watt by 2030), 50 per cent of India's energy production to be met by non-fossil fuels, and by 2070 the pledge of 'Net Zero' emissions. In a recent article, it was quoted that by 2050 if we meet the net zero, the country could boost an annual GDP to 7.3% and could also create more than 20 million jobs. In the race of this ambitious goals, steel industries have taken leading measures to adopt sustainable technologies for reduction of carbon foot print. The steel industries are taking new challenging paths in reducing the green-house gases (GHGs) with every ton of steel produced normally emits on average 1.85 tons of carbon dioxide. The main cause for energy inefficiency and environment pollution are due to the outdated steel production technology in use. Green Steel Technology (GST) is a new concept which deals with reducing the harmful greenhouse gases such as carbon dioxide, sulphur dioxide, nitrogen oxides emissions, cuts costs and improves quality of steel. Current steel industry needs are to reduce the carbon foot print by identifying processes which can replace the fossil fuels, utilize energy saving technologies or utilizing renewable energy-based systems, reduce the industrial waste generation and utilizing the waste within the steel industry, encouraging decarbonization technologies and formulation of strict controls for minimizing the industrial pollution. Possible sustainable solutions will pave a way to achieve net zero emissions and increase the possibility of sustainable future.

Keywords: Green Steel Technology · Green House Gases · Sustainable World · Pollution · Steel making

1 Introduction

The steel industry is a vital component of global economic development, providing materials for construction, transportation, and manufacturing [1, 2]. However, traditional steel production processes are associated with significant environmental challenges,

S. Patra et al. (Eds.): METCENT 2023, *Proceedings of the International Conference on Metallurgical Engineering and Centenary Celebration*, pp. 366–374, 2024.
https://doi.org/10.1007/978-981-99-6863-3_36

including high carbon emissions and extensive resource consumption [1, 3]. As the world moves towards a more sustainable future, the need for greener alternatives in the steel industry becomes paramount. Green Steel Technology offers a promising solution by integrating renewable energy, advanced manufacturing techniques, and circular economy principles. This article explores the concept of Green Steel Technology, its potential benefits, and the challenges associated with its implementation [4].

Green Steel Technology, also known as sustainable steel production, refers to the adoption of environmentally friendly practices in the steel manufacturing process. It aims to minimize carbon emissions, reduce energy consumption, and promote the efficient use of resources. Green Steel Technology encompasses several key aspects, including the use of renewable energy sources, innovative production methods, and the application of circular economy principles [5]. The steel industry is a vital component of global economic development, providing materials for construction, transportation, and manufacturing. However, traditional steel production processes are associated with significant environmental challenges, including high carbon emissions and extensive resource consumption. As the world moves towards a more sustainable future, the need for greener alternatives in the steel industry becomes paramount [2]. Green Steel Technology offers a promising solution by integrating renewable energy, advanced manufacturing techniques, and circular economy principles. This article explores the concept of Green Steel Technology, its potential benefits, and the challenges associated with its implementation [6].

2 Understanding Green Steel Technology

Green Steel Technology, also referred to as eco-friendly steel production, pertains to the integration of environmentally conscious practices into the steel manufacturing process [5]. Its primary objective is to minimize carbon emissions, decrease energy consumption, and foster resource efficiency. Green Steel Technology encompasses several crucial elements, including the utilization of sustainable energy sources, innovative production techniques, and the application of principles promoting a circular economy [4].

The notion of Green Steel Technology has garnered significant attention in recent years, propelled by the pressing necessity to combat climate change and mitigate greenhouse gas emissions. It entails reimagining conventional steel production processes and embracing novel technologies and approaches that have a reduced ecological impact [7].

3 Renewable Energy Sources in Steel Production

One of the primary drivers of carbon emissions in traditional steel production is the reliance on fossil fuels, particularly coal, for energy. Green Steel Technology proposes the substitution of fossil fuels with renewable energy sources such as solar, wind, and hydroelectric power. By integrating renewable energy into the steel production process, the carbon footprint can be significantly reduced [3]. Renewable energy sources offer multiple benefits in steel production, including reducing greenhouse gas emissions, enhancing energy efficiency, and providing a sustainable and reliable energy supply. For example, solar power can be harnessed through photovoltaic panels installed in steel plants, while wind energy can be utilized through on-site wind turbines. Hydroelectric power can also be leveraged in steel production processes in areas with suitable water resources. Several steel companies have already begun adopting renewable energy sources to power their operations [8]. For instance, Tata Steel in India has set up wind power projects and has plans to expand its renewable energy capacity. Nippon Steel in Japan has invested in solar power generation systems, reducing its reliance on fossil fuels [9].

4 Advanced Manufacturing Techniques

In addition to renewable energy, Green Steel Technology emphasizes the utilization of advanced manufacturing techniques to improve efficiency and reduce waste. One such technique is direct reduction, which involves using natural gas or hydrogen to remove oxygen from iron ore, resulting in a more energy-efficient and environmentally friendly process. Another approach is electric arc furnaces, which use electricity instead of coal to melt scrap metal, minimizing carbon emissions [4, 10]. These advanced manufacturing techniques not only reduce environmental impacts but also enhance the overall productivity and competitiveness of the steel industry. By adopting such techniques, steel producers can achieve significant energy savings, lower greenhouse gas emissions, and improve the quality of the steel produced [7].

5 Circular Economy Principles in Steel Production

The concept of the circular economy is gaining traction across various industries, including steel production. Green Steel Technology promotes the adoption of circular economy principles, such as recycling and reusing materials, to minimize waste and conserve resources. Steel is an inherently recyclable material, and recycling scrap metal can significantly reduce the need for virgin raw materials and energy-intensive extraction processes [11].

Green Steel Technology encourages the development of efficient recycling infrastructure and the use of recycled steel as a primary input in the manufacturing process. By implementing closed-loop systems, where steel products are collected, recycled, and used to produce new steel, the industry can achieve significant resource savings and reduce its environmental footprint [12]. Circular economy principles play a crucial role in promoting sustainability and resource efficiency in steel production. The steel industry is a significant consumer of resources, and adopting circular economy practices can help minimize waste generation, conserve materials, and reduce environmental impacts. This section explores the application of circular economy principles in steel production and its benefits.

6 Recycling and Reusing Materials

Steel is an inherently recyclable material, and recycling scrap metal is a key component of circular economy principles in the steel industry. Recycling steel reduces the need for virgin raw materials and energy-intensive extraction processes. It also saves resources and helps minimize the environmental impacts associated with mining and ore processing [2, 5].

By establishing efficient recycling infrastructure and encouraging the use of recycled steel as a primary input, the steel industry can close the loop in the material lifecycle, reducing waste and conserving resources [5]. Recycling rates vary across regions and countries, but efforts to improve recycling practices continue to grow.

7 Product Design for Durability and Recyclability

Circular economy principles also emphasize the importance of product design that enables durability and recyclability. Steel products can be designed for longer lifecycles, ensuring they remain in use for extended periods before being recycled. Durability reduces the need for frequent replacements and minimizes waste generation [7].

Additionally, product design should consider ease of disassembly and separation of steel components, facilitating efficient recycling processes. Designing products with standardized connections and using labelling systems for identification can simplify the recycling process and enhance the recovery of steel from end-of-life products [9].

8 Waste Reduction and Resource Optimization

Circular economy principles promote waste reduction and resource optimization throughout the steel production process. By implementing measures to minimize waste generation, such as improving process efficiency, optimizing material use, and reducing energy consumption, the steel industry can reduce its environmental footprint.

Efforts to optimize resources include using by-products or waste materials generated in steel production as inputs in other industries, such as cement manufacturing or road construction. These practices contribute to a more circular approach by ensuring that resources are fully utilized and waste is minimized [13].

9 Collaboration and Value Chain Integration

Circular economy principles require collaboration and integration across the entire steel value chain, including raw material suppliers, steel producers, manufacturers, and consumers. Cooperation among stakeholders is crucial to ensure the efficient flow of materials and the effective implementation of recycling practices [14]. Collaboration can involve establishing partnerships to improve collection and sorting systems for scrap metal, creating closed-loop systems between steel producers and manufacturers, and fostering communication and knowledge sharing to advance circular economy practices throughout the industry. By embracing circular economy principles, the steel industry can achieve several benefits, including reduced waste generation, resource conservation, and environmental sustainability. Implementing circular economy practices requires collective efforts from stakeholders and the development of supportive policies and regulations [15].

10   Potential Benefits of Green Steel Technology

The widespread adoption of Green Steel Technology offers numerous benefits for a sustainable world [8]:

a) Reduced Carbon Emissions: By substituting fossil fuels with renewable energy sources and implementing energy-efficient manufacturing techniques, Green Steel Technology can dramatically decrease carbon emissions. This reduction contributes to global efforts to combat climate change and meet carbon neutrality targets.

b) Resource Conservation: Green Steel Technology promotes the efficient use of resources by maximizing recycling and reducing the extraction of virgin raw materials. This approach helps conserve natural resources and minimizes the ecological impact associated with mining and ore processing.

c) Economic Opportunities: The transition to Green Steel Technology presents significant economic opportunities. Investments in renewable energy projects, advanced manufacturing infrastructure, and recycling facilities can create jobs, stimulate economic growth, and attract sustainable investments.

d) Enhanced Competitiveness: Adopting sustainable practices can enhance the competitiveness of the steel industry in a rapidly evolving global market. Consumers and businesses increasingly demand products with low carbon footprints, and companies that embrace Green Steel Technology will have a competitive edge in meeting these demands.

11   Challenges and Limitations:

Despite its potential, Green Steel Technology faces several challenges and limitations:

a) High Initial Investment: Implementing Green Steel Technology requires substantial investments in renewable energy infrastructure, advanced manufacturing equipment, and recycling facilities. The high initial costs may deter some steel producers, especially those operating in financially constrained environments.

b) Technological Barriers: The widespread adoption of certain advanced manufacturing techniques, such as direct reduction and electric arc furnaces, may require significant technological advancements and industry-wide collaboration. Overcoming these barriers will require research and development efforts, as well as cooperation between steel manufacturers, technology providers, and governments.

c) Scale and Integration: Scaling up Green Steel Technology to meet global steel demand is a complex task. Integration with existing steel production processes and supply chains poses challenges that need to be addressed to ensure a smooth transition.

12   Case Studies

Several steel companies have already made significant progress in adopting Green Steel Technology. One notable example is HYBRIT, a joint venture between Swedish steel producer SSAB, mining company LKAB, and energy supplier Vattenfall. HYBRIT aims to produce fossil-free steel using hydrogen instead of coal, with the goal of achieving commercial-scale production by 2026. Other companies, such as ArcelorMittal and Nippon Steel, have also made commitments to reducing their carbon emissions and exploring sustainable steel production methods. Several case studies showcase the implementation and success of Green Steel Technology, highlighting the feasibility and benefits of sustainable steel production. These examples demonstrate how companies are embracing renewable energy, advanced manufacturing techniques, and circular economy principles to achieve greener steel production. Here are a few notable case studies:

12.1 HYBRIT

HYBRIT (Hydrogen Breakthrough Ironmaking Technology) is a joint venture between Swedish steel producer SSAB, mining company LKAB, and energy supplier Vattenfall. The project aims to produce fossil-free steel by replacing coal with hydrogen as the reducing agent in the ironmaking process. HYBRIT utilizes direct reduction technology powered by renewable energy sources, such as wind and hydroelectric power, to remove oxygen from iron ore and produce high-quality iron pellets with significantly reduced carbon emissions [1, 8, 12].

The project has garnered international attention and support due to its potential for decarbonizing the steel industry. HYBRIT aims to achieve commercial-scale production of fossil-free steel by 2026 and has received funding and support from various stakeholders, including the Swedish government and the European Union.

12.2 ArcelorMittal

ArcelorMittal, one of the world's largest steel producers, has made significant commitments to reducing carbon emissions and exploring sustainable steel production methods. The company has been actively investing in renewable energy projects to power its operations, including wind farms and solar installations [11].

ArcelorMittal has also been adopting advanced manufacturing techniques, such as electric arc furnaces, which use electricity instead of coal to melt scrap metal. This approach reduces carbon emissions and enhances energy efficiency. The company is continuously exploring and investing in research and development to further improve the sustainability of its steel production processes.

12.3 Nippon Steel

Nippon Steel, a leading steel producer in Japan, has been actively pursuing sustainable steel production practices. The company has invested in solar power generation systems to reduce its reliance on fossil fuels and increase the share of renewable energy in its operations.

Furthermore, Nippon Steel has implemented various measures to optimize resource utilization and minimize waste generation. These efforts include the development of technologies to recycle by-products and waste materials generated during steel production, reducing the environmental impact and promoting circular economy principles [7].

12.4 POSCO Steel

To focus on production capabilities and core technologies including technology to produce hydrogen by electrolysis water by 2030. Posco intends to boost production capacity of green hydrogen to 2 million tonnes by 2040 and expand to 5 million tonnes by 2050. To create value chain for hydrogen storage and transportation one such initiative by POSCO is to build a terminal to store hydrogen exclusively. It plans to construct a green hydrogen-based steel works by 2050.

12.5 Tata Steel

Tata steel opts for hydrogen route at its Ijmuiden steel works. Also, in the recent past, 6% of H2 has been injected in the blast furnaces of Tata Steel.

12.6 SAIL

A memorandum of understanding has been signed between SMS group and SAIL (Steel Authority of India Limited), a Maharatna PSU, to collaborate on sustainable steel production and decarbonization initiatives. With a focus on decarbonizing steel production in SAIL's integrated steel plants throughout India, the MoU demonstrates their commitment to driving developments aimed at implementing long-term sustainable steelmaking practises.

This collaboration intends to address the problems associated with cutting carbon emissions and greening the steel sector. In this regard, SMS will offer its technological know-how for design and engineering tasks, equipment supplies, and technical support for projects involving the erection and commissioning at SAIL sites located throughout India.

12.7 JSW

The diversified JSW Group's flagship company, JSW Steel, is on track to increase capacity to 37 million tonnes per year (MTPA) and then to 50 MTPA by the end of this decade. The company has already put 225 MW of solar capacity online and anticipates putting 1,000 MW of renewable energy capacity, including solar and wind, online soon.

For its Vijayanagar steel facility, JSW Steel has teamed up with JSW Energy to secure renewable hydrogen supply. A green hydrogen plant will be built by JSW Energy using 25 MW of renewable energy. In the upcoming 18 to 24 months, the project is anticipated to be operational.

12.8 JSPL

Jindal Steel & Power Ltd. (JSPL) has made plans to produce "green steel" through a number of measures meant to lessen its carbon footprint and support sustainable production.

To accomplish its objective of generating green steel, JSPL has proposed a number of strategies. In the direct reduction method, which turns iron ore into steel, hydrogen is used in place of coal. Carbon capture and storage technologies are also being used to cut greenhouse gas emissions.

Additionally, the business intends to build a 500-megawatt green hydrogen power station in Odisha, India. Hydrogen will be produced at this facility using renewable energy sources and utilised for manufacturing steel afterwards.

These case studies highlight the commitment of steel producers towards transitioning to greener and more sustainable steel production methods [15, 16]. By adopting renewable energy, advanced manufacturing techniques, and circular economy principles, these companies are leading the way in reducing carbon emissions, conserving resources, and creating a more sustainable steel industry. It is important to note that the transition to Green Steel Technology is an ongoing process, and further advancements and case studies will emerge as the industry continues to embrace sustainability as a key driver of its operations.

13   Conclusion

Green Steel Technology holds immense promise in transforming the steel industry into a more sustainable and environmentally friendly sector. By leveraging renewable energy sources, advanced manufacturing techniques, and circular economy principles, Green Steel Technology can significantly reduce carbon emissions, conserve resources, and create economic opportunities. However, overcoming challenges related to initial investments, technological advancements, and integration is essential for the widespread adoption of this approach. In fact, 3 ways to the approach may be divided into.

a. Avoidance of Carbon usage by makeshift, retrofit or transformational technologies
b. Generation and usage of Green Hydrogen
c. Whatever carbon is generated need to be captured and utilized it with a greater purpose.

With concerted efforts from steel producers, governments, and stakeholders, Green Steel Technology can play a pivotal role in building a sustainable world.

Acknowledgement. Dr Arghya Majumder thank the Management of Kazi Nazrul University for the support and encouragement received. MKP thank OPJU for their support and encouragement in performing this review work.

References

1. HYBRIT. https://www.hybritdevelopment.com/
2. ArcelorMittal. https://corporate.arcelormittal.com/sustainability/environment/low-carbon-solutions
3. Nippon Steel. https://www.nipponsteel.com/en/sustainability/environment/ghg_emission.html
4. World Steel Association: Sustainable steel (2021). https://www.worldsteel.org/sustainability/sustainable-steel.html
5. Carbon Disclosure Project: Green Steel (2019). https://www.cdp.net/en/carbon-pricing-works/focus-on/green-steel

6. Tata Steel: Tata Steel accelerates its renewable energy capacity to 350 MW in FY22 (2022). https://www.tatasteel.com/media/newsroom/press-releases/india/2022/tata-steel-acc elerates-its-renewable-energy-capacity-to-350-mw-in-fy22/

7. Nippon Steel: Action against climate change (2022). https://www.nipponsteel.com/en/sustai nability/environment/ghg_emission.html

8. HYBRIT Development AB: About HYBRIT (2022)

9. BSI: Steel Sector Guidance for Implementing Circular Economy Principles (2020). https:// www.bsigroup.com/LocalFiles/en-GB/sectorsandservices/Standards%20and%20Publica tions/Circular-Economy-Steel-Sector-Guidance-2020.pdf

10. Eurofer: Steel, Recycling, and Circular Economy (2021). https://www.eurofer.eu/fileadmin/ user_upload/Eurofer_doc/Policies/Recycling_and_Circular_Economy/20210416_Recycl ing_Brochure_2021.pdf

11. ArcelorMittal: Sustainability (2022). https://corporate.arcelormittal.com/sustainability/env ironment/low-carbon-solutions

12. HYBRIT Development AB: About HYBRIT (2022). https://www.hybritdevelopment.com/ about-hybrit

13. World Steel Association: Regulatory Frameworks for Sustainability (2021). https://www.wor ldsteel.org/sustainability/sustainable-steel/regulatory-frameworks-for-sustainability.html

14. Davis, C.: Steel's Climate Challenge: Is Green Steel Possible? (2020). https://www.spglobal. com/platts/en/market-insights/latest-news/metals/011320-steel-climate-challenge-green- steel

15. European Steel Association (EUROFER): Steel, Recycling, and Circular Economy (2021). https://www.eurofer.eu/fileadmin/user_upload/Eurofer_doc/Policies/Recycling_and_Cir cular_Economy/20210416_Recycling_Brochure_2021.pdf

16. World Steel Association: Circular Economy in Steel (2020). https://www.worldsteel.org/steel- by-topic/sustainability/circular-economy.html

A Review on Use of Biomass as An Alternative to Coal for Sustainable Ironmaking

Amit Kumar Singh, Om Prakash Sinha, and Randhir Singh[✉]

Indian Institute of Technology (BHU), Varanasi 221005, India
randhir.met@itbhu.ac.in

Abstract. Steel is the most important engineering and construction material used globally. However, huge amount of CO_2 emission, $>7\%$ of the total global emissions, from the iron- and steelmaking is posing an eminent threat to the climate goals around the globe. This review article presents a broad picture of historical development, present scenario and possible future for iron and steel making which have to address the binary issues of reduction in CO_2 emission and sustainability. The article shows how iron making, started with charcoal as a reductant, shifted to fossil coal and again it may be switched back to charcoal sustainably. The article mainly focuses on application of biomass as an alternative to coal in different ironmaking routes. An in-depth review of the state of the art on application of biomass as an alternative to coal in ironmaking is presented. Furthermore, it explores possible application of biomass as coal substitute in different components of conventional (i.e., BF-BOF) route and alternative ironmaking route. Biomass being carbon neutral will be potentially useful in short to mid-term to decarbonize the iron and steel industry.

Keywords: Biomass · Charcoal · Green Steel · Sustainable ironmaking · Renewable reductants

1 Introduction

Steel has the widest range of applications ranging from kitchenware to skyscrapers, advanced rockets and spacecrafts. With over 3500 different grades, steel is the most versatile material known to mankind. Use of steel in various sectors are shown in the Fig. 1 [1]. World crude steel production for the year 2021 reached 1950.5 million tonnes which has increased 10-fold in the past 100 years [2]. Steel production by top five countries is shown in Fig. 2. China is the biggest crude steel producer, producing 53% (1032.8 MT) of the global crude steel. India produced 118 million tonnes of steel which was 6.1% of global production.

S. Patra et al. (Eds.): METCENT 2023, *Proceedings of the International Conference on Metallurgical Engineering and Centenary Celebration*, pp. 375–393, 2024.
https://doi.org/10.1007/978-981-99-6863-3_37

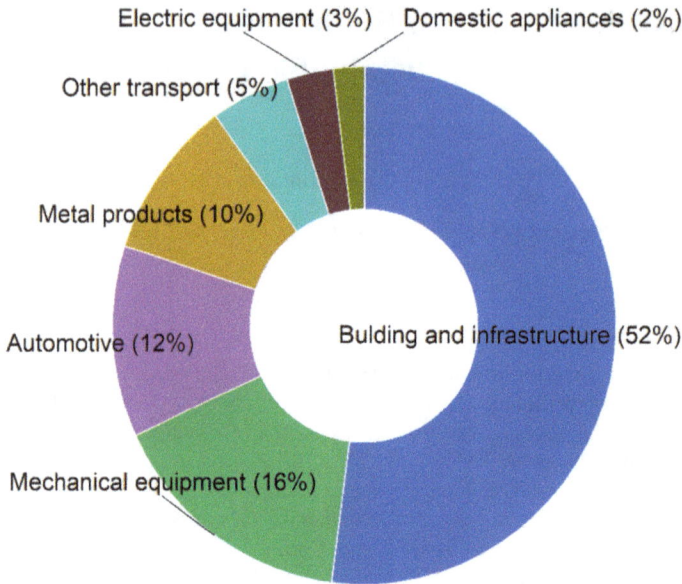

Fig. 1. Sector wise steel consumption in 2019 [1].

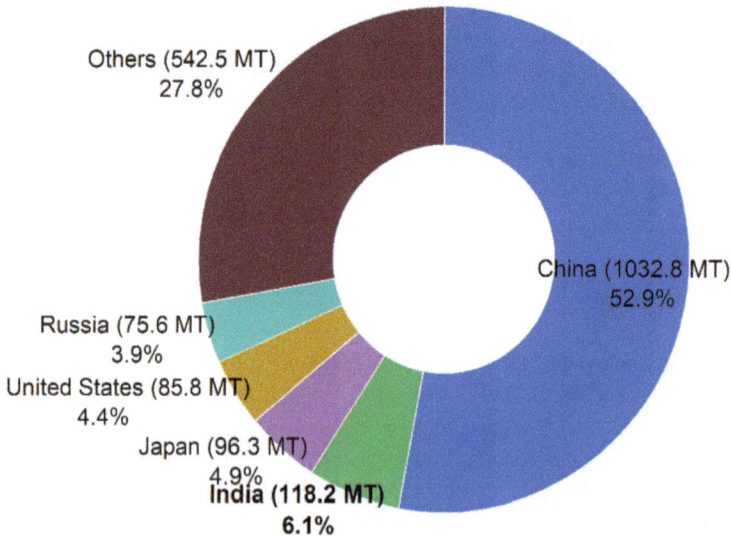

Fig. 2. Crude steel production by top 5 countries [2].

There are three industrial routes for steelmaking practiced around the globe. The first and the most popular is the blast furnace-basic oxygen furnace (BF-BOF) route which constitutes almost 70% of total steel production. Second route is steel scrap melting in electric arc furnace or induction furnace (EAF/IF) and the third one is melting of directly reduced iron in electric arc furnace (DR-EAF). Production of steel requires iron ore/scrap steel, coal and limestone as raw materials. Typically, every tonne of steel produced in a blast furnace consumes 1.6 tonnes of iron ore, 740 kg of coal and 120 kg of limestone and emits 1.85 tonnes of carbon dioxide. Coke rate for the most efficient furnaces in the world is 300–350 kg per ton of hot metal, however, in India the respective value is higher by 50% or more [3]. The reason for that is majorly attributed to higher content of silica and alumina. Typical iron rich hematite ores used in blast furnace contain 64–68% Fe. Metallurgical coal has fixed carbon (FC) content of 86–91%, volatile matter (VM) of 22–32%, moisture content up to 2% and calorific value of 8600–8800 kcal/kg. In general, $CaCO_3$ content in limestone should be greater than 95%, with SiO_2 content less than 5% and MgO & Al_2O_3 each below 3% [4].

Chemistry of the steel produced relies heavily on the quality of raw materials. Specially, in case of blast furnace ironmaking, quality of metallurgical coke has to meet a stringent requirement. Coke produced from metallurgical coal after coking treatment is used in the blast furnace (BF). Coal does not have adequate strength and porosity to make it a suitable fuel for blast furnace application. Coke acts as a reductant, holds the overlying burden and provides thermal energy in the blast furnace. Metallurgical coke should have more than 85% FC, less than 10% ash and less than 2% VM [4].

Exact beginning of the ancient ironmaking process is not clear. However, as per archaeological records, first ironmaking practice dates back to no earlier than 11th century BC [4]. The origin of ironmaking is generally placed at the eastern end of the Mediterranean. Evidences suggest that ancient ironmaking was performed in pit type furnaces in batch operations. In India, delicate surgical instruments evidently were in use before Christ. By 310 AD the knowledge of ironmaking was sufficient to produce non-corrosive iron pillars weighting 6–7 tonnes standing tall presently at Delhi and Dhar [6]. The first blast furnace had appeared in 1340 in Belgium [7]. Wood charcoal was used as reductant and energy source in these blast furnaces and the practice continued till extensive deforestation was apparent in 1676 in the United Kingdom. Dud Dudley in 1619 renewed John Robinson's patent in his own name for 21 years to smelt iron with mineral coal (known as pitcoal back then) [8]. However, due to various reasons pig ironmaking by pitcoal in the form of coke came into general use only after mid of 18th century. Abraham Darby was the first one to use coke successfully in the blast furnace in 1709 [9]. Finney and Mitchell suggested that the industrial revolution began with the transition of wood charcoal to coke as the primary fuel for ironmaking [10]. The blast furnace underwent several modifications and developments to become most efficient reactor for ironmaking. Coke as a fuel for blast furnace continued to be used for over two centuries till middle of 20th century, when researchers realised that reserves of prime coking coal suitable for coke making is limited. Fossil fuels formed over a time span of more than 100 million years is being consumed with very high rate and as per a recent estimate will last only for 100–150 years [11]. Thus, search for an alternate and carbon neutral fuel became essential. Many alternative processes to blast furnace[12], some coke-free, have been developed in the last 50–60 years as a result. These processes are broadly categorised as direct reduction (DR) and smelting reduction (SR) processes. Direct reduction process is the one where removal of oxygen from the iron ore takes place in solid state whereas smelting reduction is the one which yields liquid iron as a product (same as blast furnace). Direct reduction processes can be further classified in two groups based on the type of fuel used i.e., coal and gas based. SL/RN, Finex, Fastmet and Inmetco are some of the coal-based direct reduction processes. Hyl, Midrex and Finmet are some of the gas based processes [13]. COREX, FINEX, ITmk3 and Hismelt are some of the smelting reduction processes. Some of the plants based on these alternative processes are running successfully on commercial scale alongside blast furnace [14]. Fuel consumption under different operating conditions of BF and alternative processes to BF is listed below in Table 1. It is remarkable that the integral total fuel composition for pig-iron smelting in a blast furnace is lower by up to 45% than the alternative SR routes[15]. Thus, SR routes accounts only 0.4% of the total crude steel production [16].

Table 1. Fuel consumption under different operating conditions of BF and alternative processes to BF (calculated results [15]).

Fuel consumption	Blast-furnace smelting				Alternative processes		
	Base	Breqs	$O2 = 60\%$	$pt = 450$ kPa	Corex	Finex	Hismet
coke, kg/ton	332	239	242	328	100	60	–
coal, kg/ton	150	150	300	150	770	870	750
O2, m3/ton	47	47	190	46	455	520	245
*$\sum$F2, kg c.f/ton	608	583	655	603	846	898	693
+$\sum$F3, kg c.f/ton	428	392	422	425	684	680	508

*$\sum$F2, kg c.f/ton: Integral (*the unit fuel costs*) total fuel consumption.
+$\sum$F3, kg c.f/ton: Integral (*the unit fuel costs*) total fuel consumption minus the top gas formed in the process.

2   Fossil Fuel Consumption and CO_2 Emission by Steel Industry

Use of fossil fuel (coke/coal/natural gas) for steelmaking generates CO_2 as one of the by-products. The generation of CO_2 is neither incidental nor accidental, as SO_2 and CO might be considered to be. It is basic and unavoidable during exploitation of fossil fuels. SO_2, NO_x, CO and most of the other industrial pollutants cause problems locally and precautions can be taken to avoid loss of life and property. CO_2 on the other hand cause no local problem and is not considered as a pollutant. However, this CO_2 being a green-house gas causes problems on global scale in the form of global warming by disturbing natural carbon cycle. Among greenhouse gases, CO_2 is special in the sense that it constitutes 70% of the total greenhouse gases and absorbs 15 μm wavelength photons in infrared region which otherwise could easily pass-through atmosphere [17]. In addition, as shown in Figure 3, CO_2 level in the atmosphere is rising rapidly due to heavy CO_2 discharge from anthropogenic activities.

Table 2. Application of biomass derived charcoals for mitigation of CO_2 emission in different ironmaking processes (Mathieson et al., 2011 [41])

Application and replaced carbon source	Typical addition, kg/THM	BM substitution rate, %	BM amount, kg/ THM	Net emission reduction t-CO_2/t-Crude steel	Net reduction in % of CO_2 Emissions
Sintering solid fuel	76.5–102	50–100	38.3–102	0.12–0.32	5–15
Coke making blend	480–560	2–10	9.6–56	0.02–0.11	1–5
BF Tuyere fuel injection (PC)	150–200	0–100	0–200	0.41–0.55	19–25
BF nut coke replacement	45	50–100	22.5–45	0.08–0.16	3–7
BF carbon/ore briquette	10–12	0–100	0–12	0.06–0.12	3–5
Steelmaking recarburiser	0.25	0–100	0–0.25	0.001	0.04
Total	**761.75–919.25**	**0–100**	**70.4–415.25**	**0.69–1.25**	**31–57**

The steel industry accounts for 20% (474 EJ) of industrial fossil fuel consumption [19]. It emits 2.6 Gt CO_2 which is 7% of global anthropogenic greenhouse gas emissions [20]. The recorded level of CO_2 in pre-industrial period was 285.2 ppm (in 1850 [21]) against the current level of 423.68 ppm (in June 2023) [18] in our atmosphere. In response to the rising CO_2 concentration in atmosphere, global average temperature has risen 1° and almost half of the polar ice has already melted [22]. Melting of polar ice has resulted in the rise in sea level in excess of 100 mm. The adverse outcomes of the climate change have not yet unleashed fully. Events of submerging of coastal areas, change in wind & rain pattern, change in vegetation, drought, desertation will follow unless concerted corrective measures are taken globally.

2.1   Global Initiatives at Controlling the Environmental Damage

Recognizing the severity of GHG emission problem, United Nations (UN) has forced nations to sign a treaty to reduce their green-house gas emissions. United nations framework convention on climate change (UNFCCC) was adopted on 9th May 1992. The goal of this body is to impede dangerous anthropogenic interference with climate on Earth. The Paris Agreement, which was enacted in 2015, aims to keep global warming considerably below 2 °C, ideally to 1.5 °C, in comparison to pre-industrial levels. The Paris Agreement is a legally binding international climate change accord. For the first

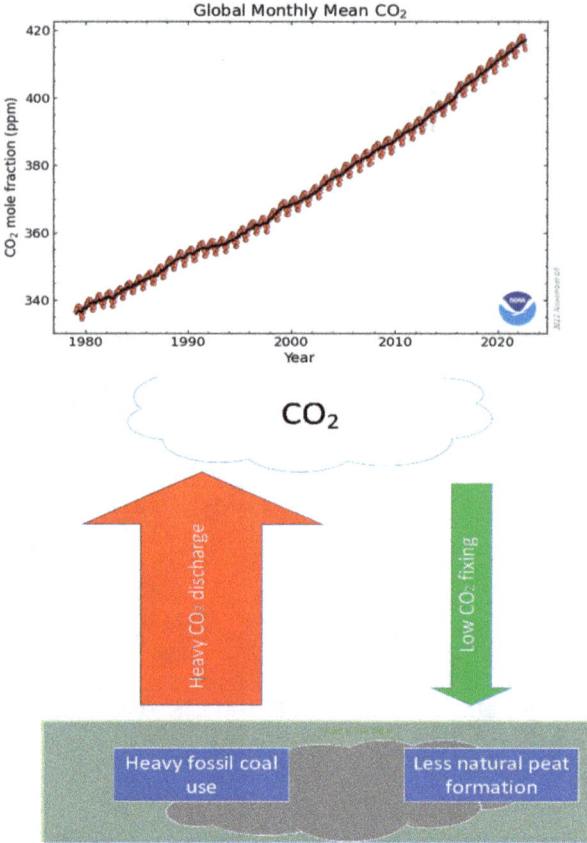

Fig. 3. Rising level of CO_2 in atmosphere [18].

time in Glasgow climate pact 2021, nations were urged to reduce their continued use of coal power and ineffective fossil fuel subsidies.

Governments may enforce rigorous emission standards for the iron and steel industry in the near future, compelling them to employ renewable and environment friendly energy sources. China, for example, already has a "Double Carbon" policy in place [23]. The Double Carbon policy aims to peak carbon emissions by 2030 and attain carbon neutrality by 2060. Governments and industries are actively working towards reducing the carbon and environmental footprint or greening of steelmaking processes [24]. Steel production without the use of fossil fuels is termed as green steel. To reduce 7% of the global CO_2 emission from steelmaking industry we have two alternative renewable non-fossil fuels for green steelmaking i) green hydrogen and ii) renewable biomass.

3   Use of Hydrogen

Hydrogen is an efficient and proven reductant for direct reduction of iron ore at lower temperatures [25–29]. For green hydrogen production water splitting is the only technology for large scale production. Other hydrogen production technologies, like steam-methane reforming, fermentation or photosynthesis are yet to develop for large-scale production [30]. Hydrogen can also be produced from biomass via a variety of techniques[31]. A recent techno-economical assessment of biomass to hydrogen conversion processes [32] found that there are technologies at TRL of 2–8, with a production cost of H_2 in the range of 1.21–4.51 \$/kg. Currently, water electrolysis consumes around 5 kWh of electricity to make one m^3 of H_2 [33]. According to Kainersdorfer, it takes 810 Nm^3 of H_2 to make one ton of steel which translates to 4.1 MWh of electricity per ton of steel production.

Lab and pilot plant scale trials of iron ore reduction using hydrogen has begun in past few years [34–38]. Use of hydrogen plasma for smelting reduction in HPSR– The Sustainable Steel (SuSteel) project has been carried out successfully at a laboratory scale [34]. During smelting reduction of hematite fine ore with hydrogen thermal plasma in lab settings, the degree of hydrogen utilization rose up to 60% [35]. ArcelorMittal has successfully tested the partial replacement of natural gas with green hydrogen for the production of DRI at its steel plant in Contrecoeur, Quebec [37]. A joint venture of SSAB, LKAB and Vattenfall, sponsored by European union under the name HYBRIT (Hydrogen Breakthrough Ironmaking Technology) is conducting pilot plant trials on direct reduction of iron ore pellets with hydrogen in Luleå, Sweden [38]. Apart from production difficulties of hydrogen, some problems related to easy transportation and storage have to be solved. We may have to wait for a while until generation of hydrogen without using fossil fuel (directly or indirectly) becomes commercially feasible.

4   Use of Biomass

As discussed earlier, ironmaking process started off with the use of wood charcoal as a fuel and it soon became universal fuel for the smelting of iron ore. The same practice continued till about middle of the 18th century. The possibility of using wood charcoal for ironmaking could be re-examined specially, due to its renewability and non-polluting nature. Biomass can be made renewable by short rotation forestry or energy plantation. An energy plantation is one where some of the fast-growing trees are grown for their fuel value. It consists of raising larger than normal number of plants per unit area and harvesting them in shorter than normal (say 5 to 10 years) period of time [39]. In tropical countries like India with huge landmass and around 300 sunny days per year energy plantation becomes more favorable [40]. The biomass is renewable, carbon-neutral, highly reactive, material having low sulfur, low ash, high specific surface area and stable pore structure. A comparison of formation and consumption time cycles for coal and biomass are shown in Fig. 4.

Firstly, biomass is carbon neutral in the sense that it releases the exact amount of CO_2 on pyrolysis which it utilized during photosynthesis while growing. Consumption of fossil fuels contributes to the atmospheric CO_2. Secondly, being lower in sulfur content than fossil fuels, biomass is not likely to pollute the atmosphere with SO_2. Also, steel made

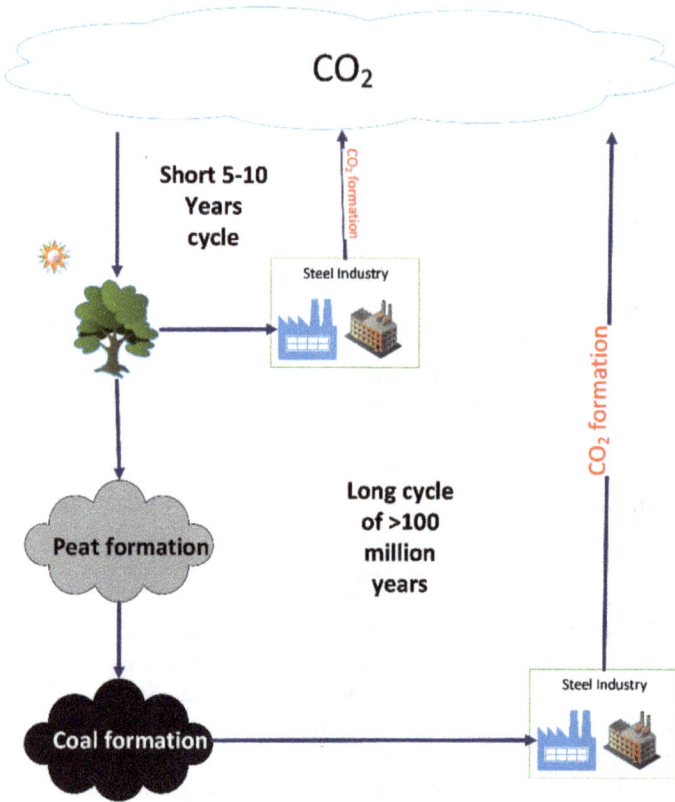

Fig. 4. Formation and consumption time cycles for coal and biomass

using biomass will have lower sulfur level and might not need external desulphuriza-tion. Biomass cultivation and utilization would be a positive step in reducing atmospheric pollution and could substitute for fossil fuel used in ironmaking. The following section discusses recent research conducted across the globe on biomass application as a reduc-tant and energy source in different ironmaking processes for possible replacement of fossil fuel and its potential in reducing the fossil CO_2 emission.

4.1 Blast Furnace – Basic Oxygen Furnace (BF-BOF) Route

Various researchers have tried partially substituting fossil fuel by biomass in blast furnace process. It was pointed out that important area of application for biomass in blast furnace are coke making, sinter making, pulverized injection through tuyeres and nut coke. Table 1 lists typical biomass addition, possible substitution rate, amount of charcoal needed and net CO_2 emission reduction in the above-mentioned areas.

Coke cannot be fully replaced due to its physical, chemical, and thermal roles in the process. Lab and pilot scale studies show that biomass can substitute 5–15% of coal in the coal blend used to make bio-coke [42–45]. Addition of biomass reduces coke strength and increases reactivity. However, biomass after pretreatment like pyrolysis can be added in higher quantities without deteriorating the coke quality. Particle size, density and VM should be considered before adding to the coal blend.

Coke breeze is primary fuel used for sintering of iron ore fines. Due to low calorific value and high moisture content, substitution of coke breeze with biomass is not possible. However, pyrolyzed biomass i.e., charcoal can partially replace it and decrease SO_x and NO_x generation during sintering [46, 47]. Addition of charcoal has some limitations in terms of decreased sinter strength, yield and granulation [48]. Coke breeze up to 25 wt% can be replaced by charcoal and up to 40–60 wt% can be replaced by coke-charcoal composite [49, 50]. The amount of charcoal can be increased by increasing particle size, density and decreasing its reactivity [50].

Biomass with mill scale can be used as cold bonded briquettes for top charging in the blast furnace. Mechanical strength of these briquettes decreases with increase in the amount of biomass [51]. This issue has been tackled by increasing the binder (cement) content. Biomass along with iron ore in the form of composite pellet is also an alternative which can be used for top charging in the blast furnace. Reduction process is rapid due to close contact of carbon and iron oxide particles in these composites which acts as highly active microreactors [52]. The iron oxide was completely reduced above 1000 °C. Rate of the reduction increased with rise in reduction temperature due to improved decomposition of hydrocarbons into reducing gases in conjunction with more heat available for endothermic reduction reaction [53]. A complete reduction of iron ore to metallic iron required 30 wt% of sawdust [54]. Low crushing strength of these composite pellets limits its practical applications in the blast furnace. Top charging of these briquettes and composite pellets decreases CO_2 emission and energy consumption in the blast furnace.

The easiest option for biomass application in the blast furnace is injection of biomass/charcoal through tuyeres as mechanical strength is not a requirement. Because of properties such as strong reactivity, low ash, low sulfur and high hydrogen content, biomass tuyere injection is advantageous in terms of low slag volume, high injection rates, improved process efficiency, and therefore higher production rate. Wood pellets can replace 20% while torrefied biomass 22.8% of pulverized coal [55, 56]. There are some limitations as well like low calorific value, grindability and cost [57]. Opportunities for using wood charcoal in steelmaking are [58]:

  i. Coal-blend component
 ii. Sintering solid fuel
iii. Nut coke replacement
 iv. C-ore composites
  v. Tuyere injectant, and
 vi. Liquid steel recarburiser

Safarian [102] analyzed the extent to which biochar could replace coal and coke in steel industries. The three of the areas where biochar could partially replace coal were coke-making, sintering and blast furnace. The extent of biochar use and emission reduction potential in these areas are summarized in Table 3.

Table 3. Processes where biochar can be used, the extent of use and CO_2-emmision reduction potential [102].

Processes where biochar can be used	Extent of use (%)	CO_2-emmision reduction (%)
Coke-making	2–10	1–5
Sintering	40–60 (of the coke breeze)	---
Blast furnace	60–75 (of the total pulverized coal injection)	25

4.2 Alternative Routes

Apart from the above-mentioned studies, work is also being carried out to evaluate charcoal as a reductant in alternative ironmaking processes. Owing to higher carbon content, lower phosphorous and sulfur, low ash and no constraint on mechanical strength enables charcoal as a replacement of coal in alternative processes such as direct reduction. The European Union has developed ULCOS, or ultra-low CO_2 steelmaking [30]. The World Steel Association made considerable efforts through its CO2 breakthrough initiative, including the Australian program co-led by Bluescope Steel, OneSteel, and CSIRO. Iron ore-biomass composite might be used effectively in lower height blast furnaces as well in addition to reduction techniques like rotary hearth furnaces and rotary kiln furnaces [59–62]. The available options for direct reduction of iron ore based on raw materials, energy sources and reactor technologies are shown in Fig. 5. Biomass / charcoal can directly be applied, at least in part, or can be gasified and used as a heat source and as a reductant.

Researchers have studied direct reduction of iron ore with hardwood biomass such as sawdust and charcoal powder in composite pellets or packed bed type setup. Decomposition of biomass produces CO and H_2 along with CO_2, CH_4 and other lower hydrocarbons, which act as reducing agent to iron ore [64, 65]. Hardwood biomass sawdust mixed with iron ore starts reduction reaction at 670 °C and completes at 1200 °C [54, 66]. Additionally, carbon deposition on iron ore can induce reduction to occur at temperatures above 500 °C [67]. Generation of the CO and H_2 gases increases with rise in reduction temperature [53]. Increasing the heating rate promotes reduction rate efficiently up to 20 °C/min [68]. Reduction of iron ore by biomass is a two-stage process: reduction by product of volatile cracking followed by reduction by non-volatile carbon. Singh et al. and Bagatini et al. [69, 70] have utilized the biomass volatiles to reduce iron ore and found that the carbon from biomass to be more effective for iron ore reduction than conventional carbon sources such as coke and coal. Singh et al. [69] have used bed type reduction setup where iron ore pellets were placed on top of biomass chips. They found that up to 93% of reduction of the iron ore was possible at a reduction temperature of 1050 °C. Bagatini et al. [70] reported that Boudouard reaction or carbon gasification reaction ($CO_2 + C \rightarrow 2CO$) controls the overall reduction reaction with activation energies close to 400 kJ/mol. Diffusion of reducing gas controls the first stage and carbon gasification controls the second stage according to Wei et al. [71]. Other authors reported that diffusion controls the reduction reaction at lower temperature i.e., 800–1000 °C, chemical

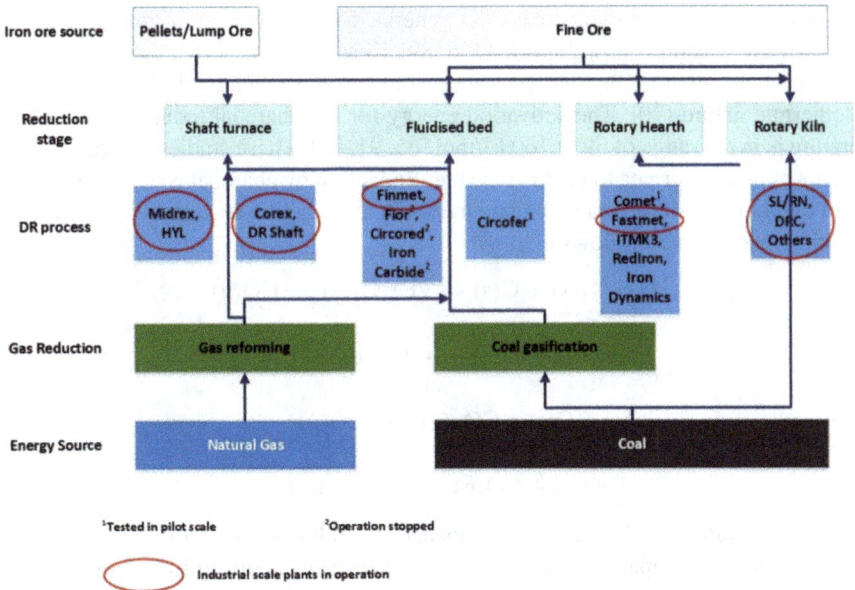

Fig. 5. Classification of direct reduction processes (after Schenk, 2019 [63]).

reaction controlling the reduction at higher temperatures [72, 73]. The activation energy for the reduction is reported to be in the range of 84 to 185 kJ/mol [71, 73].

Biomass derived charcoals have higher reactivity towards CO_2 than fossil coal enabling over 100 °C lower reduction start temperature (about 800 °C) [74, 75]. Higher reactivity is due to amorphous nature of wood charcoal [76, 77]. The maximum reaction rate in case of charcoal is approx. 1.5 times the rate obtained with coal and concomitant with much higher reduction extent [74, 75]. Ye et al. reduced iron oxide (mill scale) with charcoal and reported a total iron content of 98.56%, with the metallization ratio of 99.25% [78]. A reduction temperature of 1200 °C and reduction duration of 20 min was reported to be sufficient for complete reduction of composite pellets by Liu et al. [79]. Iron ore composite pellets reduced at 1000 °C with charcoal derived from hardwood gives reduction over 90% [80, 81]. Increasing the amount of charcoal in composite pellets increases the rate and extent of reduction [75]. Increasing the heating rate promotes reduction rate efficiently up to 20 °C/min [82]. Reduction sequence is the same as in reduction of iron ore and coal composite pellet i.e., Fe_2O_3 -> Fe_3O_4 -> FeO -> Fe [82–85]. Iron ore and charcoal composite pellets first expand and then shrink same as coal [86]. Preheating at lower temperature before actual reduction process increases extent of reduction of iron ore composite pellets by creating pores and increasing the contact area between iron oxide and reducing gas for the subsequent reduction process [87]. Increasing the composite pellet size from 8 to 16 mm resulted in decrease of reduction rate [76, 87]. It is agreed that initially solid carbon reacts with hematite and forms magnetite and CO gas (Eq. 1). CO gas then reacts with magnetite and forms wustite and CO_2 (Eq. 2). CO_2 generated from gas-solid reaction reacts with solid carbon to produce CO which is known as Boudouard or carbon gasification reaction (Eq. 3). Gasification and iron oxide reactions take place simultaneously. CO generated from gasification reaction reduces wustite to iron metal as shown in Eq. 4 [88, 89]. Researchers reported that the reduction reaction is controlled by Boudouard reaction or carbon gasification [88] with its highly endothermic nature [90]. The activation energy for the charcoal reduction reported in literature is in the range of 59 to 160 kJ/mol [65, 91, 92]. These studies indicate that use of biomass as an alternate to coal is feasible. Reduction by charcoal is better in terms of extent and kinetics of the reaction and environment. Reaction sequence for the iron ore reduction reaction is as follows.

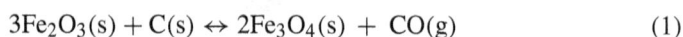

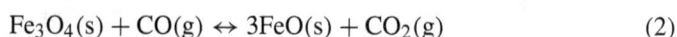

$$3Fe_2O_3(s) + C(s) \leftrightarrow 2Fe_3O_4(s) + CO(g) \tag{1}$$

$$Fe_3O_4(s) + CO(g) \leftrightarrow 3FeO(s) + CO_2(g) \tag{2}$$

$$C(s) + CO_2(g) \leftrightarrow 2CO(g) \tag{3}$$

$$FeO(s) + CO(g) \leftrightarrow Fe(s) + CO_2(g) \tag{4}$$

Different studies performed on various hardwood biomasses for their potential use in ironmaking are compared in Table 4. Researchers have studied different aspects such as biomass characterization, carbonization of wood charcoal, characterization of wood charcoals for its use in ironmaking and the actual use of wood charcoal for iron ore reduction. It is noticeable that Kumar and Gupta from 1989 to 1999 [39, 93–97] have systematically studied and the former wrote a thesis on the use of *Acacia nilotica* and *Eucalyptus* in ironmaking [98]. They have characterized the biomass, studied the carbonization process of the biomasses and product characteristics. This Indian group, however, did not perform the reduction of iron ore using the produced charcoals.

In a study using biomass volatiles and biochar from ground nut shell and corn-cob as a reducing agent for iron ore pellets, Das et al. [99] could achieve a maximum of 92% reduction (1000 °C; 75 min; charcoal from corn-cob). Very recently, Anand et al. [100] proposed bio-coke (refers to a product obtained after carbonizing inferior grade coal with biochar and molasses as a binder) as a sustainable solution to the unavailability of metallurgical coke in India. Jaiswal et al. [101] proposed a novel approach of utilizing the waste biomass in the magnetizing roasting for recovery of iron from goethitic iron ore containing 54.36% Fe. Post roasting and magnetic separation, iron ore grade reached 72.93% with Fe-recovery of 77.34%.

The main bottleneck in the large-scale promotion of biomass in ironmaking and metallurgical industries in general, is the development of conversion technology and key equipment to yield a product (say charcoal, torrefaction liquid) with specific thermal and chemical properties optimized for a particular process/application [23]. Conversely, a conventional process in ironmaking may need to be adapted to utilize a specific biomass/charcoal.

5 Conclusions

There is an urgent need for CO_2 emission reduction from iron and steel industry. Biomass is a renewable energy source and reductant with potential to decarburize ironmaking industry. There are two possible applications for biomass: i) in blast furnace and ii) in

Table 4. Prior research on use of different biomasses in ironmaking.

Biomass	Biomass Characterization	Carbonization	Charcoal Characterization	Reduction
Acacia nilotica	Kumar and Gupta 1992; Kumar et al., 1998	Kumar and Gupta 1992	Kumar and Gupta 1989; Kumar et al., 1992; Kumar et al., 1993; Kumar et al., 1994	-
Eucalyptus	Kumar and Gupta 1992; Kumar et al., 1998	Kumar and Gupta 1992, S Kumar 2014; Noumi et al., 2014	Kumar and Gupta 1989; Kumar et al., 1992; Kumar et al., 1993; Kumar et al., 1994, S Kumar 2014;	-
Pine	-	Paris et al., 2005; Matsumura et al., 2008, Park et al., 2012;	-	-
Bamboo		Du et al., 2014	Du et al., 2014	Yuan et al., 2021
Spruce	-	Paris et al., 2005; Demirbas et al., 2016, Surup et al., 2019, A Phounglamcheik, 2020	Surup et al., 2019, A Phounglamcheik, 2020	-
Beech	-	Demirbas et al., 2016		
Ailantus	-	Demirbas et al., 2016	-	-
Oak	-	Surup et al., 2019	Surup et al., 2019	
Cypress	-	Konishi et al., 2010	-	Konishi et al., 2010; Usui et al., 2018
Cedar				Usui et al., 2018
Sawdust		Strezov et al., 2007, Xiong et al., 2014, M Wilk, 2016		Zuo et al., 2013; Xiaoming et al., 2017

alternative routes of ironmaking. Scope of this review is limited to use of biomasses in alternative route of ironmaking i.e., direct reduction of iron ore. Researchers have utilized various biomasses and wastes for reduction of iron ore. Any biomass species out of forest or from local sources cannot sustain huge scale of ironmaking. The practice will lead to deforestation as was the case in the past that led to shift from wood charcoal to fossil coal. To use biomass for sustainable ironmaking, it needs to come from short rotation forestry or man-made forests. After reviewing the state of the art for biomass application in ironmaking, it is apparent that systematic studies on various hardwood biomasses are required. A breakthrough in sustainable biomass production via short-rotation forestry may enable short to medium-term decarbonization of steelmaking industry.

The invigorated research activities around using biomass to generate hydrogen, charcoal, power and power/biomass to X-approaches, motivated primarily by the climate concerns, bodes well for the greening of economies around the globe. Metallurgical industries, iron- and steel-making in particular, need to expedite the development and scale up of promising *biomass*-[23]/*hydrogen*-metallurgy [36] processes in order, at least to remove the "most-polluting" tag from them.

References

1. Department SR. Distribution of steel end-usage worldwide in 2019, by sector. In: 2019 Gsubs, ed.: statista.com (2022)
2. Worlsteel.org. 2022 World Steel in Figures. World Steel Association. **2022**, 7–23 (2022)
3. Rai, S.: This Coke Ain't No Cool (2023). https://www.niti.gov.in/coke-aint-no-cool#:~: text=While%20the%20Coke%20rate%20(Specific,in%20the%20Indian%20iron%20ore. Accessed 08 Aug 2023
4. Tupkary, R., Tupkary, V.: An introduction to Modern Iron Making. Khanna Publishers, New Delhi (2016)
5. Dragna, E.C., Ioana, A., Constantin, N.: Aspects regarding the capitalization of the powdery wastes in steelmaking. IOP Conf. Ser. Mater. Sci. Eng. **294**, 012016 (2018). https://doi.org/ 10.1088/1757-899X/294/1/012016
6. Anantharaman, T.: The Iron Pillar at Delhi. Vigyan Prasar, New Delhi (1997)
7. den Ouden, A.: The introduction and early spread of the blast furnace in Europe. Wealden Iron Res. Group Bull. **5**, 21–35 (1985)
8. King, P.: Dud Dudley's contribution to metallurgy. Hist. Metall. **36**(1), 43–53 (2002)
9. Flinn, M.W.: Abraham Darby and the coke-smelting process. Economica **26**(101), 54–59 (1959)
10. Finney, C., Mitchell, J.: History of the coking industry in the United States. JOM **13**(4), 285–291 (1961)
11. Company TBP: Statistical review of world energy 2021 (2022). https://www.bp.com/en/glo bal/corporate/energy-economics/statistical-review-of-world-energy/coal.html.html#coal-reserves
12. Zang, G., et al.: Cost and life cycle analysis for deep CO2 missions reduction for steel making: direct reduced iron technologies. Steel Res. Int. **94**(6), 2200297 (2023). https://doi. org/10.1002/srin.202200297
13. Sridhar, S., Sohn, H.: Descriptions of high-temperature metallurgical processes. In: Seetharaman, S. (ed.) Fundamentals of Metallurgy. CRC Press (2005). https://doi.org/10.1201/978 1439823613.ch1
14. Hasanbeigi, A., Arens, M., Price, L.: Alternative emerging ironmaking technologies for energy-efficiency and carbon dioxide emissions reduction: a technical review. Renew. Sustain. Energy Rev. **33**, 645–658 (2014)

15. Kurunov, I.: The blast-furnace process–is there any alternative? Metallurgist **56**(3–4), 241–246 (2012)
16. Safarian, S.: To what extent could biochar replace coal and coke in steel industries? Fuel **339**, 127401 (2023)
17. Nda, M., Adnan, M.S., Ahmad, K.A., Usman, N., Razi, M.A.M., Daud, Z.: A review on the causes, effects and mitigation of climate changes on the environmental aspects. Int. J. Integr. Eng. **10**(4), 1–7 (2018). https://doi.org/10.30880/ijie.2018.10.04.027
18. Global monitoring laboratory Noaa. Trends in Atmospheric Carbon Dioxide (2022). https://gml.noaa.gov/ccgg/trends/graph.html
19. Ye, L., Peng, Z., Wang, L., Anzulevich, A., Bychkov, I., Kalganov, D., et al.: Use of biochar for sustainable ferrous metallurgy. JOM **71**(11), 3931–3940 (2019)
20. Kim, J., et al.: Decarbonizing the iron and steel industry: a systematic review of sociotechnical systems, technological innovations, and policy options. Energy Res. Soc. Sci. **89**, 102565 (2022)
21. Etheridge, D.M., Steele, L., Langenfelds, R.L., Francey, R.J., Barnola, J.M., Morgan, V.: Natural and anthropogenic changes in atmospheric CO2 over the last 1000 years from air in Antarctic ice and firn. J. Geophys. Res. Atmos. **101**(D2), 4115–4128 (1996)
22. NASA C. Global climate change : vital signs of the planet 2022. https://climate.nasa.gov/. Accessed 30 Dec 2022
23. Zhang, J., Fu, H., Liu, Y., Dang, H., Ye, L., Conejio, A.N., et al.: Review on biomass metallurgy: pretreatment technology, metallurgical mechanism and process design. Int. J. Miner. Metall. Mater. **29**(6), 1133–1149 (2022)
24. Mallett, A., Pal, P.: Green transformation in the iron and steel industry in India: Rethinking patterns of innovation. Energ. Strat. Rev.. Strat. Rev. **44**, 100968 (2022)
25. Scharm, C., Küster, F., Laabs, M., Huang, Q., Volkova, O., Reinmöller, M., et al.: Direct reduction of iron ore pellets by H2 and CO: in-situ investigation of the structural transformation and reduction progression caused by atmosphere and temperature. Miner. Eng. **180**, 107459 (2022)
26. Yi, L., Hu, Z.H., Peng, T.J.: Action rules of H2 and CO in gas-based direct reduction of iron ore pellets. J. Central South Univ. **19**(8), 2291–2296 (2012). https://doi.org/10.1007/s11771-012-1274-0
27. Zhang, T., Lei, C., Zhu, Q.: Reduction of fine iron ore via a two-step fluidized bed direct reduction process. Powder Technol. **254**, 1–11 (2014)
28. Man, Y., Feng, J.: Effect of gas composition on reduction behavior in red mud and iron ore pellets. Powder Technol. **301**, 674–678 (2016)
29. Oh, J., Noh, D.: The reduction kinetics of hematite particles in H2 and CO atmospheres. Fuel **196**, 144–153 (2017)
30. Babich, A., Senk, D.: Low carbon ironmaking technologies: an European approach. In: Iron Ore, pp. 777–816. Elsevier (2022). https://doi.org/10.1016/B978-0-12-820226-5.00011-2
31. Balat, H., Kırtay, E.: Hydrogen from biomass–present scenario and future prospects. Int. J. Hydrogen Energy **35**(14), 7416–7426 (2010)
32. Lepage, T., Kammoun, M., Schmetz, Q., Richel, A.: Biomass-to-hydrogen: a review of main routes production, processes evaluation and techno-economical assessment. Biomass Bioenerg.Bioenerg. **144**, 105920 (2021)
33. Birat, J.-P., Borlée, J., Korthas, B., Van der Stel, J., Meijer, K., Günther, C., et al.: ULCOS program: a progress report in the Spring of 2008. In: 3rd International Conference on Process Development in Iron and Steelmaking (2008). www. ulcos. org/en/docs/Ref09
34. Sormann, A., Seftejani, M., Schenk, J., Spreitzer, D.: Hydrogen: the way to a carbon free steelmaking. AdMet (2018)
35. NaseriSeftejani, M., Schenk, J., Zarl, M.A.: Reduction of haematite using hydrogen thermal plasma. Materials **12**(10), 1608 (2019)

36. Tang, J., Chu, M., Li, F., Feng, C., Liu, Z., Zhou, Y.-S.: Development and progress on hydrogen metallurgy. Int. J. Minerals Metallur. Materials **27**(6), 713–723 (2020). https://doi.org/10.1007/s12613-020-2021-4

37. ArcelorMittal. ArcelorMittal successfully tests partial replacement of natural gas with green hydrogen to produce DRI. https://corporate.arcelormittal.com/media/news-articles/arcelormittal-successfully-tests-partial-replacement-of-natural-gas-with-green-hydrogen-to-produce-dri2022

38. HYBRIT. A fossil-free development. https://www.hybritdevelopment.se/en/a-fossil-free-development/. Accessed 03 Jan 2023

39. Kumar, M.: Studies on wood carbonisation and gasification of resultant wood chars for application in ironmaking. Metallurgical Engineering, p. 294. Ph.D. Institute of Technology, Banaras Hindu University, Varanasi (1991)

40. Solar Energy (2021). www.ireda.in/solar-energy. Accessed 15 Sep 2021

41. Mathieson, J.G., Rogers, H., Somerville, M., Ridgeway, P., Jahanshahi, S.: Use of biomass in the iron and steel industry–an Australian perspective. EECR-METEC InSteelCon, p.1 (2011)

42. Montiano, M., Díaz-Faes, E., Barriocanal, C., Alvarez, R.: Influence of biomass on metallurgical coke quality. Fuel **116**, 175–182 (2014)

43. Suopajärvi, H., Dahl, E., Kemppainen, A., Gornostayev, S., Koskela, A., Fabritius, T.: Effect of charcoal and Kraft-lignin addition on coke compression strength and reactivity. Energies **10**(11), 1850 (2017)

44. Ng, K.W, Giroux, L., MacPhee, T., Todoschuk, T.: Incorporation of charcoal in coking coal blend-a study of the effects on carbonization conditions and coke quality. In: Proceedings of the AISTech 2012, pp. 225–236 (2012)

45. Qin, L., Han, J., Ye, W., Zhang, S., Yan, Q., Yu, F.: Characteristics of coal and pine sawdust co-carbonization. Energy Fuels **28**(2), 848–857 (2014)

46. Mousa, E., Ahmed, H.: Utilization of biomass as an alternative fuel in iron and steel making. In: Iron Ore, pp. 665–690. Elsevier (2022). https://doi.org/10.1016/B978-0-12-820226-5.00020-3

47. Lu, L., Adam, M., Kilburn, M., Hapugoda, S., Somerville, M., Jahanshahi, S., et al.: Substitution of charcoal for coke breeze in iron ore sintering. ISIJ Int. **53**(9), 1607–1616 (2013)

48. Mathieson, J., Norgate, T., Jahanshahi, S., Somerville, M., Haque, N., Deev, A., et al.: The potential for charcoal to reduce net greenhouse gas emissions from the Australian steel industry (2012)

49. Mousa, E., Babich, A., Senk, D.: Iron ore sintering process with biomass utilization. In: Proceedings of the METEC and 2nd European Steel Technology and Application Days Conference (METEC and 2nd ESTAD), pp. 1–13 (2015)

50. Fan, X., Ji, Z., Gan, M., Chen, X., Yin, L., Jiang, T.: Characteristics of prepared coke–biochar composite and its influence on reduction of NOx emission in iron ore sintering. ISIJ Int. **55**(3), 521–527 (2015)

51. Mousa, E., Lundgren, M., Sundqvist Ökvist, L., From, L.-E., Robles, A., Hällsten, S., et al.: Reduced carbon consumption and CO2 emission at the blast furnace by use of briquettes containing torrefied sawdust. J. Sustain. Metallur. **5**(3), 391–401 (2019)

52. Kasai, A., Matsui, Y.: Lowering of thermal reserve zone temperature in blast furnace by adjoining carbonaceous material and iron ore. ISIJ Int. **44**(12), 2073–2078 (2004)

53. Ueki, Y., et al.: Reaction behavior during heating biomass materials and iron oxide composites. Fuel **104**, 58–61 (2013). https://doi.org/10.1016/j.fuel.2010.09.019

54. Strezov, V.: Iron ore reduction using sawdust: experimental analysis and kinetic modelling. Renew. Energy **31**(12), 1892–1905 (2006)

55. Wang, C., Mellin, P., Lövgren, J., Nilsson, L., Yang, W., Salman, H., et al.: Biomass as blast furnace injectant–Considering availability, pretreatment and deployment in the Swedish steel industry. Energy Convers. Manage. **102**, 217–226 (2015)

56. Wang, C., Larsson, M., Lövgren, J., Nilsson, L., Mellin, P., Yang, W., et al.: Injection of solid biomass products into the blast furnace and its potential effects on an integrated steel plant. Energy Proc. **61**, 2184–2187 (2014)

57. Feliciano-Bruzual, C., Mathews, J.: BIO-PCI, charcoal injection in blast furnaces: state of the art and economic perspectives. Rev. Metal. **49**(6), 458–468 (2013)

58. Jahanshahi, S., Somerville, M., Deev, A., Mathieson, J.: Biomass: providing a low capital route to low net CO2. In: IEAGHG/IETS Iron and Steel Industry CCUS and Process Integration Workshop. Tokyo, pp. 5–7 (2013)

59. Ueda, S., Yanagiya, K., Watanabe, K., Murakami, T., Inoue, R., Ariyama, T.: Reaction model and reduction behavior of carbon iron ore composite in blast furnace. ISIJ Int. **49**(6), 827–836 (2009)

60. Kowitwarangkul, P., Babich, A., Senk, D.: Reduction behavior of self-reducing pellet (SRP) for low height blast furnace. Steel Res. Int. **85**(11), 1501–1509 (2014). https://doi.org/10.1002/srin.201300399

61. Kowitwarangkul, P., Babich, A., Senk, D.: Reduction kinetics of self-reducing pellets of iron ore. AISTech Proc **1**, 611–622 (2014)

62. Ahmed, H.M., Viswanathan, N., Bjorkman, B.: Composite pellets - a potential raw material for iron-making. Steel Res. Int. **85**(3), 293–306 (2014). https://doi.org/10.1002/srin.201300072

63. Schenk, J.: Direct reduction and smelting reduction. In: 5th International Seminar on Ironmaking, Duisburg, Germany (2019)

64. Nunes, L.J., Matias, J.C.D.O., Catalao, J.P.D.S.: Torrefaction of biomass for energy applications: from fundamentals to industrial scale. Academic Press (2017)

65. Mayyas, M., Nekouei, R.K., Sahajwalla, V.: Valorization of lignin biomass as a carbon feedstock in steel industry: iron oxide reduction, steel carburizing and slag foaming. J. Clean. Prod. **219**, 971–980 (2019)

66. Wang, R., Zhao, Y., Babich, A., Senk, D., Fan, X.: Comprehensive study on the reduction of biomass embedded self-reducing pellets (SRP) under H2 involved conditions by TG-DTA. Powder Technol. **407**, 117654 (2022)

67. Zhao, H., Li, Y., Song, Q., Liu, S., Ma, L., Shu, X.: Catalytic reforming of volatiles from co-pyrolysis of lignite blended with corn straw over three iron ores: effect of iron ore types on the product distribution, carbon-deposited iron ore reactivity and its mechanism. Fuel **286**, 119398 (2021)

68. Zulkania, A., Rochmadi, R., Hidayat, M., Cahyono, R.B.: Reduction reactivity of low grade iron ore-biomass pellets for a sustainable ironmaking process. Energies **15**(1), 137 (2021)

69. Singh, A.K., Mishra, B., Kumar, S., Sinha, O.P., Singh, R.: Reduction behaviour of iron ore pellets using hardwood biomasses as a reductant for sustainable ironmaking. Biomass Convers. Bioref. (2022). https://doi.org/10.1007/s13399-022-03407-y

70. Bagatini, M.C., Kan, T., Evans, T.J., Strezov, V.: Iron ore reduction by biomass volatiles. J. Sustain. Metallur. **7**, 215–226 (2021)

71. Wei, R., Cang, D., Bai, Y., Huang, D., Liu, X.: Reduction characteristics and kinetics of iron oxide by carbon in biomass. Ironmaking Steelmaking **43**(2), 144–152 (2016)

72. Guo, D., Hu, M., Pu, C., Xiao, B., Hu, Z., Liu, S., et al.: Kinetics and mechanisms of direct reduction of iron ore-biomass composite pellets with hydrogen gas. Int. J. Hydrogen Energy **40**(14), 4733–4740 (2015)

73. Das, D., Anand, A., Gautam, S.: Effect of rice husk volatiles in iron ore reduction and its kinetic study. Energy Sourc. Part A Recov. Utiliz. Environ. Effects **44**(3), 6321–6333 (2022)

74. Bagatini, M.C., Zymla, V., Osório, E., Vilela, A.C.F.: Carbon gasification in self-reducing mixtures. ISIJ Int. **54**(12), 2687–2696 (2014)
75. Zuo, H., Hu, Z., Zhang, J., Li, J., Liu, Z.: Direct reduction of iron ore by biomass char. Int. J. Minerals Metallur. Mater. **20**(6), 514–521 (2013). https://doi.org/10.1007/s12613-013-0759-7
76. Usui, T., et al.: Evaluation of carbonisation gas from coal and woody biomass and reduction rate of carbon composite pellets. Adv. Mater. Sci. Eng. **2018**, 1–14 (2018). https://doi.org/10.1155/2018/3807609
77. Singh, A.K., Singh, R., Sinha, O.P.: Characterization of charcoals produced from Acacia, Albizia and Leucaena for application in ironmaking. Fuel **320**, 123991 (2022)
78. Ye, Q., Zhu, H., Zhang, L., Ma, J., Zhou, L., Liu, P., et al.: Preparation of reduced iron powder using combined distribution of wood-charcoal by microwave heating. J. Alloys Compd. **613**, 102–106 (2014)
79. Liu, Z., Bi, X., Gao, Z., Liu, W.: Carbothermal reduction of iron ore in its concentrate-agricultural waste pellets. Adv. Mater. Sci. Eng. **2018**, 1–6 (2018)
80. Konishi, H., Ichikawa, K., Usui, T.: Effect of residual volatile matter on reduction of iron oxide in semi-charcoal composite pellets. ISIJ Int. **50**(3), 386–389 (2010)
81. Singh, A.K., Sinha, O.P., Singh, R.: Reduction behavior and kinetics of iron ore-charcoal composite pellets for sustainable ironmaking. Metall. Mater. Trans. B **54**(2), 823–832 (2023)
82. Wang, G., Zhang, J., Zhang, G., Wang, H., Zhao, D.: Experiments and kinetic modeling for reduction of ferric oxide-biochar composite pellets. ISIJ Int. **57**(8), 1374–1383 (2017)
83. Ubando, A.T., Chen, W.-H., Ong, H.C.: Iron oxide reduction by graphite and torrefied biomass analyzed by TG-FTIR for mitigating CO2 emissions. Energy **180**, 968–977 (2019)
84. El-Tawil, A., Ahmed, H.M., El-Geassy, A., Bjorkman, B.: Effect of volatile matter on reduction of iron oxide-containing carbon composite. In: The 54th Annual Conference of Metallurgists (COM 2015) was held at the Fairmont Royal York in Toronto, Ontario, Canada, on August 23–26th, pp. 1–14 (2015)
85. El-Geassy, A., Halim, K.A., Bahgat, M., Mousa, E., El-Shereafy, E., El-Tawil, A.: Carbothermic reduction of Fe2O3/C compacts: comparative approach to kinetics and mechanism. Ironmaking Steelmaking **40**(7), 534–544 (2013)
86. Han, H., Duan, D., Yuan, P., Li, D.: Biomass reducing agent utilisation in rotary hearth furnace process for DRI production. Ironmaking Steelmaking **42**(8), 579–584 (2015)
87. Luo, S., Zhou, Y., Yi, C.: Two-step direct reduction of iron ore pellets by utilization of biomass: effects of preheating temperature, pellet size and composition. J. Renew. Sustain. Energy **5**(6), 063114 (2013)
88. Kumar, U., Maroufi, S., Rajarao, R., Mayyas, M., Mansuri, I., Joshi, R.K., et al.: Cleaner production of iron by using waste macadamia biomass as a carbon resource. J. Clean. Prod. **158**, 218–224 (2017)
89. Yuan, P., Shen, B., Duan, D., Adwek, G., Mei, X., Lu, F.: Study on the formation of direct reduced iron by using biomass as reductants of carbon containing pellets in RHF process. Energy **141**, 472–482 (2017)
90. Khaerudini, D.S., Chanif, I., Insiyanda, D.R., Destyorini, F., Alva, S., Pramono, A.: Preparation and characterization of mill scale industrial waste reduced by biomass-based carbon. J. Sustain. Metallur. **5**(4), 510–518 (2019)
91. Suman, S., Yadav, A.M.: Biomass derived carbon for the reduction of iron ore pellets. In: Pal, S., Roy, D., Sinha, S.K. (eds.) Processing and Characterization of Materials. SPM, vol. 13, pp. 85–94. Springer, Singapore (2021). https://doi.org/10.1007/978-981-16-3937-1_9
92. Yuan, X., Luo, F., Liu, S., Zhang, M., Zhou, D.: Comparative study on the kinetics of the isothermal reduction of iron ore composite pellets using coke, charcoal, and biomass as reducing agents. Metals **11**(2), 340 (2021)

93. Kumar, M., Gupta, R.C.: Properties of acacia and eucalyptus woods. J. Mater. Sci. Lett. **11**, 1439–1440 (1992)

94. Kumar, M., Gupta, R., Sharma, T.: Effects of carbonisation conditions on the yield and chemical composition of Acacia and Eucalyptus wood chars. Biomass Bioenergy **3**(6), 411–417 (1992)

95. Kumar, M., Gupta, R.C.: Influence of carbonization conditions on the gasification of acacia and eucalyptus wood chars by carbon dioxide. Fuel **73**(12), 1922–1925 (1994)

96. Kumar, M., Verma, B.B., Gupta, R.C.: Mechanical properties of Acacia and eucalyptus wood chars. Energy Sourc. **21**(8), 675–685 (1999). https://doi.org/10.1080/00908319950014425

97. Kumar, M., Gupta, R.C.: Correlation of reactivity and properties of wood chars. Fuel **73**(11), 1805–1806 (1994). https://doi.org/10.1016/0016-2361(94)90173-2

98. Kumar, M.: Studies on wood carbonisation and gasification of resultant wood chars for application in ironmaking, p. 294. Department of Metallurgical Engineering. PhD. Institute of Technology Banaras Hindu University, Varanasi (1991)

99. Das, D., Anand, A., Gautam, S., Rajak, V.K.: Assessment of utilization potential of biomass volatiles and biochar as a reducing agent for iron ore pellets. Environ. Technol. **2022**, 1–12 (2022). https://doi.org/10.1080/09593330.2022.2102936

100. Anand, A., Gautam, S., Kundu, K., Ram, L.C.: Bio-coke: a sustainable solution to Indian metallurgical coal crisis. J. Anal. Appl. Pyrol.Pyrol. **171**, 105977 (2023)

101. Jaiswal, R.K., Soren, S., Jha, G.: A novel approach of utilizing the waste biomass in the magnetizing roasting for recovery of iron from goethitic iron ore. Environ. Technol. Innov.Innov. **31**, 103184 (2023)

102. Safarian, S.: To what extent could biochar replace coal and coke in steel industries? Fuel **339**, 127401 (2023)

Basics of Iron Ore Sintering

A. K. Jouhari[✉]

An Alumnus of Metallurgy, IIT BHU and Former Scientist G, CSIR-Institute of Minerals and
Materials Technology, Bhubaneswar 751013, India
ak_jouhari@yahoo.com

Abstract. In iron making process, sintering of iron ore fines is an integral step
to utilize not only the ore fines but also to introduce a part of flux (limestone and
dolomite) along with the sinter in blast furnace burden. In this way, the productivity
of the furnace considerably improves. In this paper, some fundamental aspects
such as the vertical sintering speed, influence of mixing and granulation, etc.
are discussed. The author has worked for almost three decades in this area and
has extended the sintering technology to manganese ore fines and chromite fines.
In case of chromite, the temperature required for sintering is about 200^0C more
than that required for iron ore sintering. The process conditions for manganese ore
sintering are almost similar to that for iron ore. The author has actively contributed
to projects for setting up of batch type (capacity 15000 tpa) sinter plants at industry
site. The first plant was setup for iron ore fines, the second unit was setup for
utilization of manganese ore fines and the third plant was setup for utilization of
manganese ore bag house dust. The manganese ore sinters thus produced were
used as part of the burden for production of ferromanganese.

Keywords: sintering · agglomeration · pan-sintering

1 Introduction

The iron ore fines and concentrate cannot be directly utilized in blast furnace, and
therefore, its agglomeration i.e. enlargement of particle size is necessary before it is
used as feed for iron making. Sintering is one of the methods of agglomeration. All
iron oxide bearing wastes and coke breeze is used for sintering purpose. The use of
limestone in blast furnace is also avoided by using fluxed sinters of suitable basicity.
This results in increasing the productivity of the furnace for iron making. Almost all the
iron bearing wastes generated at plant site can be recycled as sinter charge mix. However,
one must examine the presence of undesirable elements such as sodium, potassium, lead,
zinc, etc. in the waste. Blue dust, a pure form of iron oxide with very little gangue and
rejects of sponge iron plant have also been utilized in sinter making. The iron oxide fines
containing small amount of pellet fines have also been used to produce quality sinters.

© The Author(s), under exclusive license to Springer Nature Singapore Pte Ltd. 2024
S. Patra et al. (Eds.): METCENT 2023, *Proceedings of the International Conference
on Metallurgical Engineering and Centenary Celebration*, pp. 394–399, 2024.
https://doi.org/10.1007/978-981-99-6863-3_38

2   The Sintering Process

The main sintering process parameters are:

a. Sintering bed temperature, and is controlled by coke breeze content in charge mix.
b. Bed permeability and the vertical sintering speed, which is controlled by the moisture content in charge mix.
c. Heat economy and sinter quality, are controlled by the mode and time given for charge bed ignition.

The two-phenomenon governing the sintering operation are:

a. Heat transfer in bed, which is related to vertical sintering speed.
b. Reactivity of coke breeze.

The coke content in charge mix, suction below the grate bars, moisture content, vertical sintering speed, granulation and bed permeability are some of the process parameters controlling the productivity & quality of sinters.

The mineral formation reactions, other than reduction or oxidation, are the following:

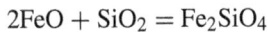

$2FeO + SiO_2 = Fe_2SiO_4$
(formation of fayalite)
$2CaO + SiO2 = Ca2SiO4.$
(formation of calcium ortho-silicate)
$(2-x)FeO + xCaO + SiO_2 = Ca_xFe_{2-x}SiO_4$
(formation of iron calcium olivines where x varies from 0 to 2)
$Fe_2O_3 + CaO = CaFe_2O_4$
(formation of mono-calcium ferrite)
$Fe_2O_3 + 2CaO = Ca_2Fe_2O_5$
(formation of di-calcium ferrite)

It is observed, from thermodynamic analysis, that the mono-calcium formation reaction is feasible, whereas the di-calcium formation reaction is not. With presence of CaO in the sinter, formation of iron-calcium olivines is more feasible than fayalite formation. Kinetics of each reaction also plays an important role. The propagation of heat transfer and combustion zones through the bed during sintering, gas flow, heat & mass transfer needs modeling studies and detailed experimental investigations. While the requirement of coke breeze for manganese ore sintering is almost similar to that of iron ore, the requirement of coke breeze for chromite fines is more because the sintering temperature is higher than that required for iron ore. A good sinter should have higher softening temperature and less difference between the softening temperature and melting temperature. The temperature at which the first drop of hot metal forms is also important.

At the Institute (CSIR-IMMT}, a sintering pot grate furnace facility of 400 × 400 × 730 mm size, with a suction blower having specifications, air flow rate, 0.6 m^3/s and suction pressure, 13.6 kPa, driven by a 37 kW motor is used for carrying out laboratory experiments. On an average 175 kg of iron ore sinters were produced in each experiment, the productivity varies with change in bed height and the sinters thus produced were subjected to stabilization test, shatter test, tumbler test, relative reducibility test and reduction disintegration test.

3 Sinter Quality Evaluation

The sinters produced in pot grate furnace were dropped three times from a height of 2 m onto a steel plate and + 5.6 mm fraction is taken as usable sinter i.e. yield and the balance as return sinter. The material for hearth layer is drawn from the usable sinter.

3.1 Shatter Test

20 kg of 10–40 mm size sinter is taken for shatter test. The material is allowed to fall four times from a height of 2 m on to a steel plate, +10 mm fraction survived after the test is reported as Shatter Index (SI). The test equipment is $560 \times 420 \times 200$ mm with droppable bottom.

3.2 Tumbler Test

15 kg of 10–40 mm size sinter is taken for the test. The material is subjected to tumbling in a drum having 1000 mm diameter 500 mm width. The drum is provided with two lifters of 50 mm height on opposite ends. The drum is rotated at 25 rpm for 8 min. The +6.3 mm fraction is reported as Tumbler Index (TI) and −0.5 mm fraction is reported as Abrasion Index (AI).

3.3 Relative Reducibility

The relative reducibility is carried out as per ISO 7215. In this test 500 g sized sample (10–12.5 mm) is taken. The sample is kept in 75 mm diameter retort and the reduction is carried out at 900 °C for a period of 3 h, passing reducing gas having 30% CO and 70% N_2 at a rate of 15 lit./min. The results are reported as relative reducibility (R_f) which may be defined as:

R_f = (Loss in wt.in g $\times 10^{-4}$) / Wt. of sample in g $\times$ (0.43 $\times$ wt. % total Fe-0.112 $\times$ wt.% FeO)

3.4 Reduction Disintegration Index

This test is carried out as per ISO 4696 method where the sample is subjected to cold tumbling after static reduction. The 500 g sample is reduced at 500 °C for a period of one hour in 75 mm diameter retort. The reduced sample was subjected to tumbling in a drum of 130 mm diameter and 200 mm width with lifters at 30 rpm for 10 min. The material was then sieved and −3 mm fraction is reported as Reduction Disintegration Index (RDI).

Sintering experiments were carried out with the charge containing ore fines, fluxes, ferruginous wastes, coke breeze, etc. Ignition of the bed was done by spreading a layer of about 650 g charcoal on top of the charge bed and igniting it at a lower suction of 500 mm WG for duration of 1.5 to 2.0 min. The suction was then raised, to the desired level, such that the vertical sintering speed of 22–23 mm/min is maintained and the sintering operation is completed.

4   Results and Discussion

Using the pot grate furnace facility available at the Institute, the iron ore fines procured from various sources, ferruginous waste material, given fluxes and coke breeze, large number of experiments were carried out. The basicity of sinter was fixed on the basis of percentage of sinter charged in blast furnace. The level of silica in sinter was decided depending upon the end use of hot metal produced for steelmaking or for foundry purpose. The MgO content in sinter is based on the blast furnace slag composition. All limestone and dolomite is added in sinter to get higher productivity in blast furnace. Though the quality of ore fines and other raw materials used, process parameters and experimental conditions have been very widely varied, there is a correlation observed in the quality and quantity of sinters produced. Some results are given in the Table 1 below.

Table 1. Process parameters, coke rate, quality and productivity of sinters.

S.No	Basicity, CaO/SiO2	Bed height, mm	Suction, mm WG	Coke rate kg/t sinter	Productivity t/m^2.day
1	1.80	450	950	68	29.5
2	2.43	500	1200	65	32.4
3	2.80	650	1400	46	34.0

With increase in bed height and suction below the grate bars, the coke rate decreases and the productivity increases.

5   Design of a Batch Type Pan Sintering Plant

Keeping in view the quantity of sinter requirement, for low shaft and mini blast furnace, 50 to 100 tonne per day, design of a batch type pan sinter plant was developed and the units were installed at industry site. The same is also applicable to Ferro-chrome and Ferro-manganese industries. A 4 m^2 area circular (2.25 m diameter) pan with two stage gas cleaning system, material mixer cum granulator, suction blower, ducts and material handling equipment were installed. A cyclone separator followed by a wet scrubber was set up for cleaning the exit gases from pan before entry to the suction blower. In cyclone separator, while large diameter reduces pressure drop, the small diameter has a higher collection efficiency of dust particles for the same entrance conditions and pressure drop. In wet scrubber, a water film is formed on the inner surface which collects the finer left over dust particles. With two stage gas cleaning, the dust level in exit gas was brought down to a level of below 80 mg/m^3 gas. Figures 1 and 2 give the views of sintering pan and the shed showing the gas cleaning system. For more details the reader may refer the two references [1, 2] given below.

Fig. 1. The Sintering Pan

Fig. 2. Front View of the Sinter Plant Shed

6 Conclusions

On the basis of extensive experimental research work carried out for various industries at the Institute, the following conclusions may be drawn.

i) Most of the wastes generated at the industry site can be utilized for the production of quality sinters.
ii) Higher suction and higher bed height give higher productivity and lower coke rate.
iii) Sinter productivity of 1.4 t/m^2 – hr, assuming 90% availability of sintering facility, is estimated.
iv) The pan sintering units are ideally suited, where the sinter requirement is low.

References

1. Jouhari, A.K., Datta, P., Ray, H.S.: Use of iron and steel industry wastes in sinter making. In: Gupta, R.C. (ed.) EMMI 2000, BHU Varanasi, pp. 169–172. Allied Publishers, New Delhi (2000)
2. Jouhari, A.K.: Sintering of ore fines-principles and a few case studies. In: Misra, V.N., Reddy, P.S.R., Mohapatra, B.K. (eds.) Mineral Characterisation and Processing, pp. 237–243. Allied Publishers, New Delhi (2004)

Author Index